Computational and Statistical Methods for Chemical Engineering

In recent decades, emerging new molecular measurement techniques and online databases have allowed easier retrieval of the associated data by the chemical analyst. Before the data revolution, most books focused either on mathematical modeling of chemical processes or exploratory chemometrics. ***Computational and Statistical Methods for Chemical Engineering*** aims to combine these two approaches and provide aspiring chemical engineers a single, comprehensive account of computational and statistical methods.

The book consists of four parts:

- **Part I** discusses the necessary calculus, linear algebra, and probability background that the student may or may not have encountered before.
- **Part II** provides an overview on standard computational methods and approximation techniques useful for chemical engineering systems.
- **Part III** covers the most important statistical models, starting from simple measurement models, via linear models all the way to multivariate, non-linear stochiometric models.
- **Part IV** focuses on the importance of designed experiments and robust analyses.

Each chapter is accompanied by an extensive selection of theoretical and practical exercises. The book can be used in combination with any modern computational environment, such as R, Python and MATLAB. Given its easy and free availability, the book includes a bonus chapter giving a simple introduction to R programming.

This book is particularly suited for undergraduate students in Chemical Engineering who require a semester course in computational and statistical methods. The background chapters on calculus, linear algebra and probability make the book entirely self-contained. The book takes its examples from the field of chemistry and chemical engineering. In this way, it motivates the student to engage actively with the material and to master the techniques that have become crucial for the modern chemical engineer.

Computational and Statistical Methods for Chemical Engineering

Wim P. Krijnen
University of Groningen, The Netherlands

Ernst C. Wit
Universita della Svizzera italiana (USI), Switzerland

CRC Press
Taylor & Francis Group
Boca Raton London New York

CRC Press is an imprint of the
Taylor & Francis Group, an **informa** business
A CHAPMAN & HALL BOOK

First edition published 2023
by CRC Press
6000 Broken Sound Parkway NW, Suite 300, Boca Raton, FL 33487-2742

and by CRC Press
4 Park Square, Milton Park, Abingdon, Oxon, OX14 4RN

Library of Congress Cataloging-in-Publication Data

Names: Krijnen, Wim P., author. | Wit, Ernst, author.
Title: Computational and statistical methods for chemical engineering / Wim P. Krijnen, University of Groningen, The Netherlands, Ernst C. Wit, Universita della Svizzera italiana (USI), Switzerland.
Description: First edition. | Boca Raton : C&H CRC Press, 2023. | Includes bibliographical references and index.
Identifiers: LCCN 2022034270 (print) | LCCN 2022034271 (ebook) | ISBN 9781032013244 (paperback) | ISBN 9781032013268 (hardback) | ISBN 9781003178194 (ebook)
Subjects: LCSH: Chemical engineering--Data processing.
Classification: LCC TP184 .K75 2023 (print) | LCC TP184 (ebook) | DDC 660/.28--dc23/eng/20220901
LC record available at https://lccn.loc.gov/2022034270
LC ebook record available at https://lccn.loc.gov/2022034271

ISBN: 978-1-032-01324-4 (hbk)
ISBN: 978-1-032-01326-8 (pbk)
ISBN: 978-1-003-17819-4 (ebk)

DOI: 10.1201/ 9781003178194

Typeset in CMR10
by KnowledgeWorks Global Ltd.

Publisher's note: This book has been prepared from camera-ready copy provided by the authors.

To Marloes, Steven, and Hidde

Hopefully you agree, Isabella, Carolina and Cristina,
it is nice to see it in print.

Contents

V Appendix 351

Foreword

When a few years back, the director of Chemical Engineering at the University of Groningen, Francesco Picchioni, asked us to teach an undergraduate course on computational and statistical methods for chemical engineers, we responded "yes" before realizing a serious dilemma. Teaching a combination of computational algorithms and statistics in chemistry is, in principle, perfectly sensible: many modern chemical production processes are accompanied by detailed data acquisition that requires computational and statistical mastery to monitor yield, quality, or other material aspects. However, comprehensive textbooks describing these topics are rare. In fact, most existing textbooks either focus on computational methods or on statistical approaches for chemical engineers. However, our brief was clear: one course combining both computational and statistical methods. When no single book could be found, we decided to write the notes ourselves. Over the course of a number of years the notes grew out to a rough framework for a book and when the editor, Rob Calver, approached us for book ideas we suggested that this may be an unexplored opportunity.

The aim of this book is to introduce students and researchers from chemical engineering into subjects of numerically solving differential equations, optimization, error propagation, statistical estimation, testing, and (non-)linear model fitting. The idea is not to be comprehensive but rather to develop a semester course around a broad and important number of topics. The book starts with introducing mathematics and probability theory as these play a role throughout. Basic ideas are alternated with more advanced topics with the aim of bringing the students to a level in which they are able to understand and execute high-level computational and statistical procedures. Examples of these are Taylor approximations, characteristic functions, stiff algebraic differential equations, diagnostics in regressions analysis, model selection based on information criteria, inverse regression, additive modeling, mixed effect modeling, bootstrapping, chemodynamics, multivariate analysis, cross-validation, analysis of experimental design, and robust estimation.

Models come with assumptions. The models described in this book are no exception to that rule. For this reason, in this book we stress the importance of checking model validity, for example by means of model diagnostics. These diagnostic tools will be explained when the models are introduced. Care was taken to select examples in Chapters 5 through 11, for which the model assumptions generally do hold. Data from chemical experiments with violations of model assumptions can be analyzed by more advanced methods.

In Chapter 12, we will introduce robust methods that make fewer model assumptions in order to deal with data outliers.

Throughout the book we use vectors and matrices in order to make notation more compact and comprehensible, as well as to give insight in how computational algorithms work for analyzing chemical data. It is crucial for the fledgling industrial engineer to follow such reasoning.

Frequently we illustrate a new method by a relatively simple example so that different perspectives on the problem easily come to mind. Care was taken to incorporate examples from various chemical fields such as chromatography, spectrography, production technology, as well as related fields such as agriculture and medicine. These examples mainly involve data from publicly available resources that can be accessed via the R project. Software implementations and references for further study are briefly mentioned at the end of each chapter.

All computations in this book are performed with the freeware language R, available in the `UsingRforCSMCE.R` file associated with this book. For most of the applications predefined functions were used making the computations and reproducing the computations fairly straightforward. However, with a few additional programming skills the scope of application of computational and statistical methods to chemical engineering problems can be greatly enhanced. Although many computations in this book can be performed using computational languages such as MATLAB® or Python, mastering R is a useful investment of time when becoming a chemical researcher. R contains many pre-defined functions in combination with standard programming capabilities, which allow the user to implement more bespoke algorithms. The book is mainly based on standard methods that come together with the core R distribution. However, there exists a plethora of other methods that can readily be installed as free add-on packages.

The Preliminaries (Part I) start in Chapter 2 with an introduction to the mathematics necessary for understanding the analysis of chemical data. The basic idea of a Taylor representation or approximation plays an important role throughout the book. Other mathematical concepts such as derivatives, integrals, matrix inverses, and Newton algorithm are defined and illustrated and will appear frequently in the later chapters.

Chapter 3 gives an introduction to probability theory. In this chapter, the concepts of a probability space, random variables, probability densities and moments are introduced. These concepts are the basis for stochastic processes and statistical inference procedures that are introduced in the subsequent chapters.

Numerics and Error Propagation (Part II) explains and applies algorithms for finding the maximum of a function, e.g., expressing the yield of a chemical process, for approximating the solution of an ordinary differential equation, or for the propagation of measurement error.

In Chapter 4 the method of fixed point iteration is used for finding the roots of a function. Applied to the first-order derivative of a function, this procedure may give the maximum of a function, e.g., expressing a yield or some other objective function of interest to the chemical engineer. The fixed point method nicely generalizes to Newton's iterative method, by which a maximum can be found often in just a few iterations. Chemical reactions can be represented by means of a system of differential equations that often do not have an analytic paper-and-pen solution. An intuitive method developed by Euler can quite easily be generalized to that of Runge-Kutta which is more precise and stable. By even more advanced methods, we can solve differential algebraic equations for approximate solutions of chemical reactions. The software to implement these methods is easy to use and generally available.

Various Types of Models and their Estimation are the subject of Part III. Under this heading fall general measurement models, linear and non-linear regression models, chemodynamics and multivariate statistics.

In Chapter 6 the measurement model is introduced making it possible to investigate the precision and possible bias of chemical measurements in an experiment. In almost all of science an important role is played by the Law of Large Numbers and the Central Limit Theorem as these reveal the degree of approximation of an empirical average, and its distribution, respectively, to the unknown target of interest. These ideas are generalized to a general inference paradigm, which allows for summarizing data into confidence intervals, for testing arbitrary chemical hypotheses and for comparing different models for the data.

In Chapter 7 you learn how to define, to estimate, and to evaluate a linear model for the yield of a chemical process. After an extensive inspection of the model diagnostics, it is possible to draw valid scientific conclusions. Basic model specifications such as simple regression can be used for practical chemical engineering tasks, such as calibration of measurement instruments. Non-linear curves, for example using polynomial regression, can fit more complex relationships between the response and the variables of interest. Modern penalization methods, such as the lasso, can be used to select important variables. These ideas have become well-established, both mathematically and empirically. We also describe the possibility of introducing random effects into a regression model. These effects can be used to describe, e.g., the difference between labs or individual operators within a lab, as well as temporal dependence between observations. The concept of random effects is essential for modern experimental industrial engineers.

The solution of a qualitative ODE yields a class of functions from which we aim to find the curve that best fits the data from a corresponding chemical experiment. In Chapter 8 you learn how the parameters of such models can estimated and evaluated. Calibration problems such as estimating the concentration of a chemical substance from the measurements of the absorption of light, can be tackled by inverse regression. If the class of functions which produced the yield is unknown, and a predictive model is needed, then a flexible spline type of approach can be employed. In a setting where the outcome of an experiment consists of counts, such as the number of insects killed by various doses of an insecticide, a different regression approach is called for. Generalized linear models are regression-type methods able to model response variables that are not normally distributed, such as counts or success/failure binary responses.

Although ODEs are excellent tools to describe non-linear processes where the uncertainty is limited to measurement noise, there are classes of processes where additional process uncertainty is present. An archetypal case is that of chemical reactions involving small numbers of atoms. In this case the reaction uncertainty results in possibly dramatically different dynamic trajectories, where ODE modeling would fail. Modeling such dynamic processes is explained and illustrated in Chapter 9.

In Chapter 10 techniques are given by which the dependencies between several measurements can be investigated. Often, low-dimensional directions are present in multivariate measurements. Identifying these structures can be interesting in their own right, but they can also be used as more stable predictors for some outcome of interest.

The Analysis of Designed Experiments in Part IV explains the basic techniques to analyze data from experiments designed to test the causal effect of certain interventions, ideas to optimize such designs as well as modern estimation techniques when there are outliers in the data.

In Chapter 11 variables affecting the response, known as factors, are investigated by designing, conducting, and analyzing data from experiments. The idea of optimally designed experiments is to maximize the information learned from a certain number of available measurements. This is particularly useful in the exploratory part of the experiment. In a more confirmatory phase the goal is to design and analyze an experiment providing optimal settings for the factors, for example to be able to find the maximum of the response surface of some yield of interest. The ideas of linear modeling can be generalized to non-linear modeling with mixed random as well as fixed effects providing great flexibility and precision to the engineer.

In Chapter 12 we deal with the problem of outlying observations from an experiment. In any experimental setting it may happen that the model diagnostics indicate that a small proportion of the data has an unreasonably large influence on the parameter estimates. We discuss techniques to deal with outliers for almost any model relevant for the chemical engineer.

We would like to thank a number of people that have been helping us along the way. We were lucky to be able to be inspired by some prior notes on error propagation from Franjo Weissing. The contributions of the students Bettina Soos, Alexandra Meerovici, Mohammad Siam Shahkhan, and Stefan Kuipers to improving the exercises are gratefully acknowledged. Never downtrodden by our lack of progress, Rob Calver and assistant editors from CRC always encouraged us to complete the book. A previous version was reviewed by two anonymous referees, whose constructive criticism helped us to expand the book, basically doubling its content, making it more independently readable. Moreover, our respective academic institutions uphold the long-standing tradition to let us work on these stimulating projects for the benefit of learning worldwide. A warm thank you to our families for their general support and understanding should never be forgotten.

Wim P. Krijnen and Ernst C. Wit
Scheemda (NL) and Muzzano (CH),
11 May 2022

Symbols

Symbol Description

Symbol	Description
$\|x\|$	absolute value of scalar x
f	function f
$\\|x\\|$	Euclidian norm of vector x
$\exists$	there exists
f^{-1}	inverse function of f
$\mathbf{A}^{-1}$	inverse of square matrix $\mathbf{A}$
$\frac{df}{dx}(x)$	first-order derivative of f with respect to x
$y'(x)$	first-order derivative of y with respect to x
$\frac{\partial f}{\partial x}(x,y)$	first-order derivative of f with respect to x
H_f	Hessian of function f
J_f	Jacobian of function f
∇	gradient operator ∇
$\lim_{x \to x_0}$	limit as x goes to x_0
$\mathbf{x}^T$	Transpose to vector $\mathbf{x}$
$\forall$	for all
$\mathbb{R}^p$	real p-dimensional space
$n!$	n factorial
$\zeta(p)$	Riemann zeta function
$\int_a^b f$	integral of f over the interval $[a,b]$
$\int f$	indefinite integral
$A \cup B$	union of sets A and B
$A \cap B$	intersection of sets A and B
A^c	complement of set A
$\|A\|$	cardinality of set A, i.e., number of elements in A.
$\emptyset$	empty set
$\bigcup_{n=1}^{\infty} A_n$	union of set A_1 to set A_∞
$\sigma(A)$	sigma field generated by the set A
$P(A)$	probability of event represented by set A
$I_A(\omega)$	function that indicates if ω is a member of set A
$P(B\|A)$	probability of event B given that A happened
$E[X]$	expected value of random variable X
$F_X(x)$	Distribution function of random variable X
$f_X(x)$	Probability density function of X, sometimes written as $f_\theta(x)$ to stress the dependence on a parameter θ
$\mathrm{Var}(X)$	variance (matrix) of X
σ^2	(error) variance
$\mathrm{SD}(X)$	standard deviation of X
$M_X(t)$	moment generating function of X evaluated at t
$\binom{n}{y}$	binomial coefficient
$f_X(x)$	density function of random variable X evaluated at x
$\sum_{i=1}^{n} x_i$	Summation of x_1 to x_n, i.e., $x_1 + x_2 \cdots + x_n$
$\Gamma(x)$	gamma function
$N(\boldsymbol{\mu}, \boldsymbol{\Sigma})$	multivariate normal distribution with mean $\boldsymbol{\mu}$ and variance matrix $\boldsymbol{\Sigma}$
$N(\mu, \sigma)$	univariate normal distribution with mean μ and variance σ
Φ	Cumulative distribution function of a standard normal distribution $N(0,1)$

z_α	$1-\alpha$ quantile of a $N(0,1)$ distribution, i.e., $\Phi(z_\alpha) = 1-\alpha$.
$t(n)$	t-distribution with n degrees of freedom
$F(m,n)$	F-distribution with m, n degrees of freedom
$\chi^2(m)$	chi-squared distribution with m degrees of freedom
$\varphi_Y(t)$	characteristic function of random variable Y evaluated at t
s_{xy}	sample covariance between $\mathbf{x}$ and $\mathbf{y}$
r_{xy}	sample correlation between $\mathbf{x}$ and $\mathbf{y}$
s_x^2	sample variance of $\mathbf{x}$
s_x	sample standard deviation of $\mathbf{x}$
$\overline{\mathbf{x}}$	sample average of $\mathbf{x}$
$\xrightarrow{D}$	convergence in distribution
$\xrightarrow{P}$	convergence in probability
$\sim$	left-hand side is distributed as right-hand side
IID	independent and identically distributed
$\widehat{\boldsymbol{\beta}}$	estimator of $\boldsymbol{\beta}$
$I_A(x)$	indicator function of set A
RSS	residual sums of squares
H_0	null hypothesis
H_A	alternative hypothesis
sign(x)	sign function of x
$(x)_+$	positive part of x
$\ell(\theta)$	log-likelihood of θ
$\dot{\ell}(\theta)$	derivative of log-likelihood w.r.s. θ
$\ddot{\ell}(\theta)$	second derivative of log-likelihood w.r.s. θ
$I(\theta)$	Fisher information of θ

Author Bios

Wim P. Krijnen is a lecturer at the Faculty of Science and Engineering at the University of Groningen. He has been teaching a course called "Computational and Statistical Methods" for several years to undergraduate students in Chemical Engineering. In addition, he has taught courses on linear algebra, probability theory, mathematical statistics, statistical modeling, statistical consulting to bachelor's and master's students in various fields. He is a professor of Applied Statistical Research Methods at the Hanze University of Applied Sciences in Groningen.

Ernst C. Wit is the Fondazione Leonardo Professor of Data Science at the Universita della Svizzera italiana in Switzerland. He has 30 years of experience in teaching statistics, applied mathematics and data science courses to students from many fields, including chemical engineering. His teaching philopsophy is about combining theoretical insights with practical skills, as he believes that the former without the latter is pointless whereas the latter without the former aimless.

Part I

Preliminaries

1

What to Expect in This Book?

Paraphrasing Galileo Galilei, also the laws of chemistry are written in the language of mathematics. Considering the concentration of a chemical substance that changes in time, we may want to describe its value by a function $y(t)$ in time t. Often, we wish to know about the properties of the function to understand whether it is increasing, decreasing, or converging to a fixed value in time. Alternatively, we store several sequences of chemical measurements in a matrix, or for approximating a non-linear curve, we need a complete matrix with derivatives. Using an integral, we can express not only the cumulative concentration function but also the probability of an event we are interested in. For developing algorithms, e.g., for optimization problems, we frequently need the Taylor approximation of a function in its first or second order. Many chemical reaction equations or processes may be summarized well by differential equations. Such mathematical objects and their properties are defined and elaborated in Chapter 2.

Chemical measurements are subject to uncertainty around their intended target. This uncertainty can be described by means of probability theory, of which random variables are its work horse. Random variables can be described by their probability density function. The more precise the measurement, the higher the probability density of its associated random variable in an interval around the target. The rest of the book will show that both theoretical concepts of probability and the ability to simulate probabilistic experiments is very useful in order for the chemical engineer to learn from data obtained from real chemical experiments. The essentials of probability and random variables are given in Chapter 3.

Suppose the concentration $y(t)$ of a chemical substance starts at zero, then increases to a maximum, and next decreases to zero with time t. This can be expressed in the form of an ordinary differential equation (ODE) as $y' = (b/t - K)y$ with time $t > 0$, and positive parameters b and K. At the start, of time $\boldsymbol{t}$, the value of b/t is larger than K so that y' is positive and y increases, whereas for larger t, the value of b/t is smaller than K so that y' is negative and y decreases. The solution to such an ODE consists of a class of concentration curves in time t, given by

$$y(t) = at^b e^{-Kt} \text{ for any value } a > 0.$$

Generally speaking, we may try to find the exact solution of an ODE by an analytic method or by a symbolic algebra system such as WolframAlpha. In case

DOI: 10.1201/9781003178194-1

this is unsuccessful or impossible, we need numerical methods to approximate the solution to an ODE with sufficient precision, as introduced in Chapter 4.

We may also be interested in further properties of the obtained concentration function such as the time point at which the maximum occurs and its value. By setting the first-order derivative of a function to zero and checking that the second-order derivative is negative, we can find a local maximum. Such a calculus type of approach may be successful for relatively simple functions, but for more difficult problems, we need numerical procedures to approximate the function value at which the maximum occurs.

Another relevant industrial engineering question is the total amount of some chemical reaction product produced in a time interval, say between zero and 100 seconds. This amount can be represented as an integral. At high school, you learned to compute an integral such as $\int_0^{100} e^{-t} dt$. During a calculus course at university, you probably learned the technique of integration by parts to compute more complex integrals, such as $\int_0^{100} t^2 e^{-t} dt$. However, in practical situations we could have real numbers for the parameters such as $a = 0.2214$, $b = 1.2295$, and $K = 0.0959$ in the previous example, so that the integral of the concentration curve becomes

$$\int_0^{100} 0.2214 \cdot t^{1.2295} \cdot e^{-0.0959 \cdot t} dt.$$

Such an integral cannot be computed by straightforward methods from calculus, so that numerical methods become necessary to solve the problem with sufficient precision.

Suppose we wish to determine a chemical quantity μ in a liquid; however, the precision of its measurements x is not perfect. We may want to know how many measurements do we need for the precision of the average $\bar{x}$ to be sufficient. If instead, we are interested not directly in x itself, but in some functional transformation of x, say $g(x)$, then how does the uncertainty in x propagates to $g(x)$. Such types of questions will be answered in Chapter 5 and 6. Light is absorbed in water in varying amounts depending of the concentration of lead in the water. This presents an opportunity to determine the lead concentration of water by determining the functional relationship between lead concentration and light absorption. Giri et al. (1983) performed an experiment measuring the absorption of light for various lead concentrations in aqueous solutions through electrochemical atomic absorption spectrometry with graphite-probe atomization. This yielded the measurements reported in Table 1.1. From these data, we observe a clear increase of absorption due to an increase in concentration. The values of the concentrations are plotted against the corresponding absorbance values together with the best fitting line through the points in Figure 1.1. This line is drawn by the values of the parameters of a simple linear model $y = \hat{\beta}_0 + \hat{\beta}_1 \cdot x$, with concentration x, absorbance y, intercept $\hat{\beta}_0 = 0.0703$, and slope $\hat{\beta}_1 = 0.0046$. The line just found from the data is called the calibration curve. Some questions to be asked are as follows: how well does the calibration curve fit to the data? How precise

TABLE 1.1
Lead concentration and absorption.

Measurement	**1**	**2**	**3**	**4**	**5**	**6**
Lead concentration, ng ml^{-1}	10	25	50	100	200	300
Absorbance	0.05	0.17	0.32	0.60	1.07	1.40

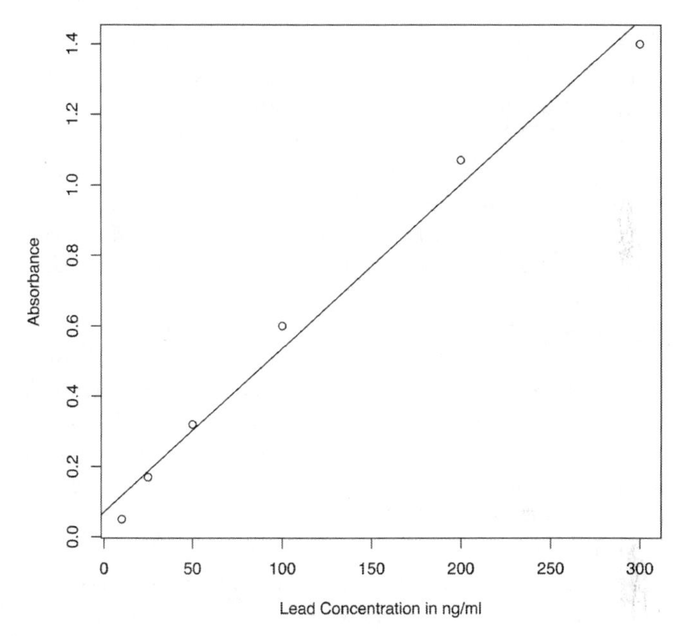

FIGURE 1.1
Lead concentration by light absorbance measurements with the best fitting line.

are the estimates? Such questions are answered in Chapter 7. Suppose that for a future determination of lead concentration, a spectrometry measurement yields an absorbance of 0.50. The question now becomes, how precisely can we predict the value of the concentration? Such "inverse" regression types of questions are answered in Chapter 8, in addition to finding non-linear curves that best fit the measurements.

Often, in an experiment the outcome of a chemical process potentially depends upon several other variables, but we do not know exactly which are really important. In the Belle Ayr Liquefaction experiment (Cronauer et al., 1978), for example, the dependency of the outcome CO_2 is investigated for seven explanatory variables. Since each of these can be either part of a model or not, we would have to consider $2^7 = 128$ different models. How to evaluate these? In Chapter 7, we will see that for such type of problems, a model selection criterion based on an information criterion or on minimizing the prediction error may be of great help.

Chemistry is the science of reactions, whereby reactants are transformed into products. The quantitative relationship between reactants and products is called stoichiometry. In Part III of the book, we study the relationships between the quantities of various substrates by means of a number of models. We start off with the simplest linear model and move on to more realistic non-linear models. However, there is one important component missing from our models so far, which is the stochasticity of the process itself. When two containers contain the same amount of reactants, the randomness of the movement of the particles will mean that eventually the two systems will diverge in their relative behavior. This is particularly important when the quantities under consideration are small. In Chapter 9, therefore, we consider an extension of the ordinary differential equation model in order to study the sets of chemical reactions. The resulting *stochastic differential equations* (SDEs) are a powerful tool to study apparently divergent trajectories within the same quantitative framework.

If the dependency between a relatively large number of chemical variables is to be investigated, the idea of a direction of maximal variation in the data may be of interest. Investigating the axes of maximal variation in the data brings us to the general idea of eigenvalues, which are of great importance in various branches of chemical analysis. Moreover, when many predictive variables are available for some quantity of interest, the aim would be to find a low-dimensional subspace, which is highly predictive for the outcome. In Chapter 10, such type of topics are treated and applied to chemical data.

In an exploratory stage of designing an industrial process, often several factors are hypothesized which have an effect on the outcome. A general problem is how to design an experiment by which main effects, as well as interactions between these, are to be investigated. When the set of potentially influential factors is identified and their effects are estimated, often the question about outcome optimization becomes urgent. For example, does the stationary point indicate for a global optimum or a saddle point? In Chapter 11, these types of questions are answered.

It frequently happens in research that there are a few outlying data points which are relatively far away from the bulk of the observations. Leaving the data inside standard inference procedures may negatively affect the conclusions. It is therefore important to be able to diagnose these outliers and to devise methods that are able to deal with them. Methods robust against outliers are nowadays available for almost all statistical techniques for analyzing chemical data. Robust estimation methods are the topic of Chapter 12.

2

Calculus and Linear Algebra Essentials

Calculus is the mathematical study of functions and limits, whereas linear algebra deals with vectors and matrices. Together they are the cornerstone of any computational or statistical analysis, and therefore it is expected that the student of this course is able to manipulate them. In order to refresh the reader of this book, we include in this chapter the main calculus concepts that will be used tacitly throughout the book.

In this chapter you will learn the following:

- The concepts of scalars, vectors, and matrices, and how they can be added to and multiplied with each other in Section 2.1.
- How certain finite as well as infinite sums can be calculated in Section 2.2.
- How functions connect one space, called the domain, with another space, called the image, via a graph in Section 2.3, and also, how for the special subclass of multivariate linear functions, inverse transformations correspond to matrix inverse multiplications.
- How the derivative of a function describes the instantaneous increase or decrease of the graph in Section 2.4.
- How derivatives can be used to find extremes of a function in Section 2.5.
- How the integral of a function can be used to calculate the area under a graph in Section 2.6.
- How integration can be used to solve a special class of equations, called differential equations, that appear frequently in engineering in Section 2.7.

2.1 Scalars, Vectors, and Matrices

Numbers come in different shapes and sizes. You may be familiar with the difference between natural numbers $\mathbb{N}$ (e.g., $0, 1, 2, \ldots$), rational numbers $\mathbb{Q}$ (e.g., 0.5, $\frac{-23}{17}$, $-0.5947, \ldots$), real numbers $\mathbb{R}$ (all the rational numbers plus everthing in between, e.g., $\pi, e, \sqrt{2}, \ldots$), and complex numbers $\mathbb{C}$ (all the real

DOI: 10.1201/9781003178194-2

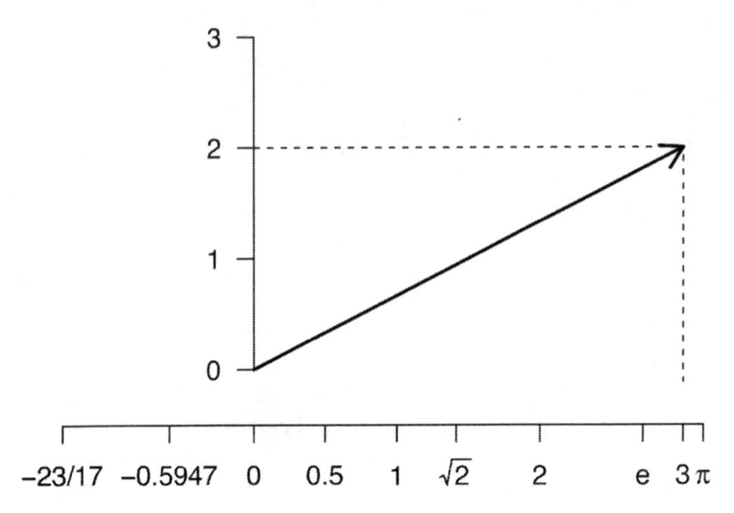

FIGURE 2.1
The real numbers $\mathbb{R}$ are the main set of scalars we will deal with in this book. It can be represented by an infinite line. Vectors are multidimensional objects, which in the case of $\mathbb{R}^2$ and $\mathbb{R}^3$ can be represented by arrows.

numbers plus, e.g., $2+3i, e+\pi i, \ldots$). For the moment, we will be mainly interested in the set of real numbers, which we will be indicating by means of the symbol $\mathbb{R}$. We can represent each real number by a point on an infinite line as shown in Figure 2.1. The real numbers are closed under addition, subtraction, multiplication, and division. This means that if you add, subtract, multiply, or divide two real numbers, you obviously obtain another real number. We will use real numbers as our fundamental set of one-dimensional **scalars** to construct more complex features.

A **vector** is a quantity that is defined by multiple scalars. For example, each point in the two-dimensional Cartesian plane can be represented by a vector (x, y), where $x, y \in \mathbb{R}$. A vector field is not only a collection of vectors but also comes with a number of operations. For example, two vectors can be added or subtracted in the following component-wise way:

$$\begin{bmatrix} x_1 \\ x_2 \end{bmatrix} + \begin{bmatrix} y_1 \\ y_2 \end{bmatrix} = \begin{bmatrix} x_1 + y_1 \\ x_2 + y_2 \end{bmatrix} \quad \begin{bmatrix} x_1 \\ x_2 \end{bmatrix} - \begin{bmatrix} y_1 \\ y_2 \end{bmatrix} = \begin{bmatrix} x_1 - y_1 \\ x_2 - y_2 \end{bmatrix}.$$

Furthermore, various types of multiplications exist in vector fields. The most common one is **scalar multiplication**, which involves shortening or lengthening a vector by a scalar. This is again done in an element-wise way, i.e., in two dimensions

$$k \in \mathbb{R}: \quad k \begin{bmatrix} x_1 \\ x_2 \end{bmatrix} = \begin{bmatrix} kx_1 \\ kx_2 \end{bmatrix}.$$

Furthermore, if the space is an inner product space, such as Euclidean space, including the Cartesian 2D plane, then there exists an **inner product** between two vectors in order to introduce important geometric features, such as norms and angles. The inner product of two vectors is a scalar, defined as $< x, y >= \sum_i x_i y_i$, which for a two-dimensional vector is given as

$$\left\langle \begin{bmatrix} x_1 \\ x_2 \end{bmatrix}, \begin{bmatrix} y_1 \\ y_2 \end{bmatrix} \right\rangle = x_1 y_1 + x_2 y_2.$$

The **norm** of a vector is, intuitively speaking, its length and is defined by means of its associated inner product. In general, the $\|x\| = \sqrt{< x, x >}$ and, in particular, for a two-dimensional vector,

$$\left\| \begin{bmatrix} x_1 \\ x_2 \end{bmatrix} \right\| = \sqrt{x_1^2 + x_2^2}.$$

The **distance** between two vectors is the norm between the difference vector, $d(x, y) = \|x - y\|$, whereas the **angle** θ between two vectors is defined as $\cos\theta = \frac{<x,y>}{\|x\|\|y\|}$. There are a number of useful inequalities that are often used in practice. The **Cauchy-Schwarz inequality** states that the size of the inner product is always less than or equal to the product of the norms, i.e., $| < x, y > | \leq \|x\|\|y\|$. The **triangle inequality** states that any side of a triangle is less than or equal to the sum of the other two sides, i.e., $\|x + y\| \leq \|x\| + \|y\|$.

We finish this section with an introduction to matrices. A **matrix** is a rectangular table of scalars. We typically will use capital roman letters to indicate a matrix. For example,

$$A = \begin{bmatrix} a_{11} & \dots & a_{1p} \\ \vdots & & \vdots \\ a_{n1} & \dots & a_{np} \end{bmatrix}$$

with $\{a_{ij} \mid a_{ij} \in \mathbb{R}\}$ is a $n \times p$ matrix that belongs to the space of matrices with n rows and p columns, indicated as $\mathbb{R}^{n \times p}$. Matrices of the same size can be added or subtracted, which is defined in an element-wise way, e.g.,

$$\begin{bmatrix} a_{11} & \dots & a_{1p} \\ \vdots & & \vdots \\ a_{n1} & \dots & a_{np} \end{bmatrix} + \begin{bmatrix} b_{11} & \dots & b_{1p} \\ \vdots & & \vdots \\ b_{n1} & \dots & b_{np} \end{bmatrix} = \begin{bmatrix} a_{11} + b_{11} & \dots & a_{1p} + b_{1p} \\ \vdots & & \vdots \\ a_{n1} + b_{n1} & \dots & a_{np} + b_{np} \end{bmatrix}.$$

Similar as for vectors, matrices can be **multiplied by a scalar** in the usual way, i.e., $k \in \mathbb{R}$

$$k \begin{bmatrix} a_{11} & \dots & a_{1p} \\ \vdots & & \vdots \\ a_{n1} & \dots & a_{np} \end{bmatrix} = \begin{bmatrix} ka_{11} & \dots & ka_{1p} \\ \vdots & & \vdots \\ ka_{n1} & \dots & ka_{np} \end{bmatrix}.$$

An important operation in many computational algorithms is to turn the matrix on its side, the so-called **transpose** of A of size $p \times n$,

$$A^t = \begin{bmatrix} a_{11} & \dots & a_{n1} \\ \vdots & & \vdots \\ a_{1p} & \dots & a_{np} \end{bmatrix},$$

where the superscript t stands for *transpose*. Clearly, turning a matrix on its side two times in a row gets us back to the original matrix, $A^{tt} = A$. Matrices can be multiplied with each other. If A is a matrix with p *columns*, say an $n \times p$ matrix, and B is a matrix with p *rows*, say a $p \times m$ matrix, then the **product** AB exists and is a matrix of size $n \times m$, defined as

$$\begin{bmatrix} a_{11} & \dots & a_{1p} \\ \vdots & & \vdots \\ a_{n1} & \dots & a_{np} \end{bmatrix} \begin{bmatrix} b_{11} & \dots & b_{1m} \\ \vdots & & \vdots \\ b_{p1} & \dots & b_{pm} \end{bmatrix} = \begin{bmatrix} \sum_i a_{1i}b_{i1} & \dots & \sum_i a_{1i}b_{im} \\ \vdots & & \vdots \\ \sum_i a_{ni}b_{i1} & \dots & \sum_i a_{ni}b_{im} \end{bmatrix}.$$

Note that unless $n = m$, the product BA does not exist. Moreover, even if $m = n$, the product does not commute in general, i.e., $AB \neq BA$.

2.2 Sequences and Series

A sequence is an indexed set of numbers. We are particularly interested in *infinite* sequences. A series is a sum of an infinite sequence.

2.2.1 Sequences

In mathematics, a **sequence** is an indexed or ordered set of numbers. Whereas in a set the order of the numbers does not matter, in a sequence it does. So, whereas $(1, 2, 3, 4, 5, 6)$ and $(1, 3, 5, 2, 4, 6)$ constitute the same set, they constitute two different sequences. Sequences can be finite, such as in the previous example, but we will focus our attention on infinite sequences. Two examples are the sequence of *odd numbers*

$$1, 3, 5, 7, 9, \dots$$

or the sequence of *prime numbers*

$$1, 2, 3, 5, 7, 11, 13, 17, \dots$$

When we indicate each element of the sequence by means of a variable that is indexed by the set of natural numbers, say x_n with $n \in \mathbb{N}$, it is possible to keep track of individual elements of the sequence. For example, for the prime numbers, we have

$$x_1 - 1,\ x_2 = 2,\ x_3 = 3,\ x_4 = 5,\ x_5 = 7,\ x_6 = 11,\ x_7 = 13,\ x_8 = 17, \ldots$$

and the sequence of prime numbers would be written as $(x_n)_{n \in \mathbb{N}}$, where we implicitly assume that the reader is familiar with the fact that there are a countably infinite number of prime numbers. Another famous sequence is the sequence of *Fibonacci numbers*, given as

$$x_1 = 1, x_2 = 1, x_3 = 2, x_4 = 3, x_5 = 5, x_6 = 8, x_7 = 13, x_8 = 21, \ldots$$

Fibonacci can be defined conveniently via a *recursive* formula,

$$x_1 = 1,\ x_2 = 1, \forall i \geq 3 : x_i = x_{i-2} + x_{i-1}.$$

We return to this sequence when we talk about *series.*

Having introduced the sequence, it is now possible to define a number of properties that are important in practice. In particular, a sequence can be **increasing**, **decreasing**, or neither of the two. The sequence of prime numbers as defined above is clearly an increasing sequence. In fact, given that each following number is strictly larger than the previous one, it is called a **strictly increasing** sequence. Increasing sequences where subsequent numbers are sometimes also equal to the previous one are called **monotonically increasing** sequences. Formally, the Fibonacci sequence is only monotonically increasing, because $x_1 = x_2 = 1$.

Secondly, a sequence can be **bounded** if a sequence stays within a certain region. Formally, a sequence is **bounded from above** by some U, if for all $n \in \mathbb{N}$ the values of the sequence are below that value, $x_n \leq U$. Similarly, a sequence is **bounded from below** by some L, if for all $n \in \mathbb{N}$ the values of sequence are above that value, $x_n \geq L$. Finally, a sequence is bounded, if it is bounded from both below and above. It is clear that the sequence of prime numbers is bounded from below by, e.g., $L = 1$. However, it is not bounded from above: there is no largest prime number. Therefore, the sequence of prime numbers is not bounded.

Finally, the concept of boundedness is related to the concept of **convergence**, i.e., the sequence of its long-run behavior. Clearly, if a sequence is unbounded, then it does not converge. On the other hand, even if a sequence is bounded, then the sequence may fluctuate and still not settle down to a particular value.

Definition 2.1 *A sequence of real numbers* (x_n) **converges** *to a value* x, *i.e.,*

$$\lim_{n \to \infty} x_n = x,$$

if for all $\epsilon > 0$, *there exists a value* n_ϵ *such that for all* $n \geq n_\epsilon$

$$|x_n - x| < \epsilon.$$

A sequence is **divergent** if it is not convergent, or, more formally, for all $x \in \mathbb{R}$ there exists some $\epsilon > 0$, such that for all $n \in \mathbb{N}$ there exists some $n_\epsilon > n$ such that $|x_{n_\epsilon} - x| > \epsilon$. Practically this either means that a divergent sequence either keeps on fluctuating or that it wanders of to negative or positive infinity.

A sequence is **unbounded** if there does not exist a finite number B such that $|x_n| < B$, for all n. An example is $x_n = \sqrt{n}$.

There are a number of standard sequences that are useful in practice:

1. **Constant.** Any constant sequence is convergent and converges to that particular constant C.
   - $\lim_{n\to\infty} C = C$,
     for example,
     - $\lim_{n\to\infty} 2 = 2$
     - $\lim_{n\to\infty} e = e$
2. **Infinite denominator.** A ratio sequence where the numerator a_n is bounded and the denominator b_n is unbounded converges to zero.
   - $\lim_{n\to\infty} \frac{a_n}{b_n} = 0$, where (a_n) is bounded and (b_n) is divergent.
     For example,
     - $\lim_{n\to\infty} \frac{1}{n} = 0$
     - $\lim_{n\to\infty} \frac{\sin n}{\sqrt{n}} = 0$
3. **Infinite numerator and denominator.** A ratio sequence where both the numerator a_n and the denominator b_n are unbounded converges to zero, if the numerator is a "slower" sequence than the numerator sequence.

   $$\lim_{n\to\infty} \frac{a_n}{b_n} = 0,$$

   where (a_n) and (b_n) are unbounded, but (a_n) is slower than (b_n).

   The "speed" of a sequence is defined by its place on the following list, from fastest to slowest:

   (a) n^n
   (b) $n!$
   (c) C^n with $|C| > 1$
   (d) n^C
   (e) $\log(n)$

   The sums of the elements on the list take the ranking of their highest placed constituent. Functions multiplied by a scalar remain on the same place on the list. Some examples are as follows:

   - $\lim_{n\to\infty} \frac{n!}{n^n} = 0$
   - $\lim_{n\to\infty} \frac{2e^n + 3n^{41}}{\log(n) + 0.01n!} = 0$

4. **Some famous sequences.**

   •**Exponential:** A famous sequence is the one that generates the irrational number $e = 2.718282\ldots$.

$$e = \lim \left(1 + \frac{1}{n}\right)^n$$

   •$\sqrt{\mathbf{2}}$: Repeated fractions are a good way to generate interesting sequences. Consider, for example,

$$x_1 = 1, \; x_2 = 1 + \frac{1}{2}, \; x_3 = 1 + \frac{1}{2 + \frac{1}{2}}, \; x_4 = 1 + \frac{1}{2 + \frac{1}{2+\frac{1}{2}}}, \ldots$$

   Then, we have that

$$\lim_{n \to \infty} x_n = \sqrt{2}.$$

Furthermore, if the sequences (a_n) and (b_n) are convergent, then we have the additional rules,

1. **Sums.**

$$\lim_{n \to \infty} (a_n + b_n) = \lim_{n \to \infty} a_n + \lim_{n \to \infty} b_n$$

2. **Products.**

$$\lim_{n \to \infty} (a_n b_n) = \lim_{n \to \infty} a_n \times \lim_{n \to \infty} b_n,$$

   in particular, for any scalar $c \in \mathbb{R}$,

$$\lim_{n \to \infty} c a_n = c \lim_{n \to \infty} a_n.$$

3. **Fractions.** Provided that $\lim_{n \to \infty} b_n \neq 0$,

$$\lim_{n \to \infty} \frac{a_n}{b_n} = \frac{\lim_{n \to \infty} a_n}{\lim_{n \to \infty} b_n}.$$

4. **Powers.** If $p > 0$ and $a_n > 0$, then

$$\lim_{n \to \infty} a_n^p = \left(\lim_{n \to \infty} a_n\right)^p.$$

2.2.2 Series

A mathematical **series** is defined as the sum of an infinite sequence. Despite perplexing some of the ancient philosophers (as exemplified by Zeno's paradox) and modern audiences alike (see modern-day movies about strange Ramanujan series on the internet), the development of modern calculus since the 17th century has put this concept on a solid footing. Given a sequence (a_n), a series is written as the infinite sum of its constituents,

$$\sum_{k=1}^{\infty} a_k = a_1 + a_2 + a_3 + \dots.$$

As the right-hand side value may depend on the order in which the elements are summed, we need a formal definition for the value of the left-hand side.

Definition 2.2 *Consider an* **infinite series** $\sum_{k=1}^{\infty} a_k$. *Given the associated sequence of partial sums* (s_n), *where*

$$s_n = \sum_{k=1}^{n} a_k,$$

the **value of a series** *is defined as the limit of the partial sum sequence* (s_n), *if it exists, i.e.,*

$$\sum_{k=1}^{\infty} a_k = \lim_{n \to \infty} s_n.$$

A series is said to be **convergent** *if the limit of the partial sum exists. It is* **divergent** *if it does not exist. A series is said to* **converge absolutely** *if the series of absolute values* $\sum_{k=1}^{\infty} |a_k|$ *converges.*

The importance of absolute convergence lies in the fact that no reordering of an absolute converging sequence would alter its value.

There are many tests that can be used to check if a series converges. Here we discuss only the *ratio test*, which is simple and powerful: if there exists a constant $p < 1$, such that there exists a value n_0 such that for all $n \geq n_0$

$$\left| \frac{a_{n+1}}{a_n} \right| < p < 1,$$

then the series $\sum_{k=1}^{\infty} a_i$ converges. This rule is closely connected to the famous *geometric series*, i.e., for any $|p| \leq 1$,

$$\sum_{k=0}^{\infty} p^k = \frac{1}{1-p}.$$

There are other standard series or other partial sums that often appear in analysis:

- **Exponential.** For any value $x \in \mathbb{R}$, the exponential function can be written as a series,

$$\sum_{k=0}^{\infty} \frac{x}{k!} = e^x.$$

This is closely connected with the Taylor expansion of the exponential function. There are a number of related series. These two are useful for calculating the moments of a Poisson random variable,

$$\sum_{k=0}^{\infty} k\frac{x}{k!} = xe^{x}$$

$$\sum_{k=0}^{\infty} k^2\frac{x}{k!} = (x+x^2)e^{x}.$$

- **Leibniz formula for π.** Although first discovered by the Indian mathematician Madhava of Sangamagrama in the 14th century, the formula is named after the German mathematician Gottfried Leibniz who lived 3 centuries later:

  $$\sum_{k=1}^{\infty} \frac{(-1)^{k+1}}{2k-1} = \frac{\pi}{4}.$$

  Other sequences involving π also exist, such as

  $$\sum_{k=1}^{\infty} \frac{1}{k^2} = \frac{\pi^2}{6}.$$

- **p-series.** The p-series is a converging series for all $p > 1$. Its value is the Riemann zeta function, introduced in 1859, which still holds many unsolved mysteries,

  $$\sum_{k=1}^{\infty} \frac{1}{k^p} = \zeta(p).$$

  For values of $p \leq 1$, the series is divergent. In particular, the harmonic series $\sum_{k=1}^{\infty} \frac{1}{k}$ is divergent.

- **Natural logarithm.** The natural logarithm is associated with an alternating series,

  $$\sum_{k=1}^{\infty} \frac{(-1)^{k+1}}{k} = \log 2.$$

- **Trigonometric functions.**

$$\sum_{k=0}^{\infty} \frac{(-1)^k x^{2k+1}}{(2k+1)!} = \sin x \tag{2.1}$$

$$\sum_{k=0}^{\infty} \frac{(-1)^k x^{2k}}{(2k)!} = \cos x. \tag{2.2}$$

- **Partial sums.** Although not formally series, these partial sums are often useful in practice:

$$\sum_{k=1}^{n} k = \frac{n(n-1)}{2}$$

$$\sum_{k=1}^{n} (2k-1) = n^2$$

$$\sum_{k=1}^{n} k^2 = \frac{n(n+1)(2n+1)}{6}$$

$$\sum_{k=1}^{n} k^2 = \frac{n^2(n-1)^2}{4}$$

$$\sum_{k=m}^{n} p^k = \frac{p^m - p^{n+1}}{1-p}, \quad \text{for } |p| < 1$$

$$\sum_{k=0}^{n} \binom{n}{k} = 2^n$$

$$\sum_{k=0}^{n} (-1)^k \binom{n}{k} = 0$$

$$\sum_{k=0}^{n} \binom{k}{m} = \binom{n+1}{m+1}$$

2.3 Functions

Functions are operators that map a **domain** into a certain **range** or **image** by assigning for each value in the domain a corresponding value in the range. Typically we use small roman letters, such as f, g, and h, to indicate a particular function, and to express that it maps a domain D into a range Im, we use the $\longrightarrow$ symbol, as in

$$f : \text{Domain} \longrightarrow \text{Range}$$

For example, the function that takes a value and squares it can be denoted as

$$f : \mathbb{R} \longrightarrow [0, \infty)$$
$$f(x) = x^2.$$

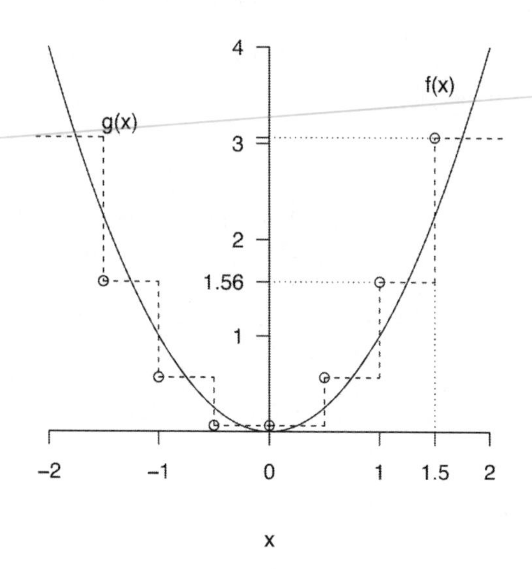

FIGURE 2.2
The graph associated with the convex and continuous function $f(x) = x^2$ and with the discontinuous and neither convex nor concave stepfunction g that approximates f.

The set of pairs $\{(x, f(x)) \mid x \in \mathbb{R}\}$ is called a **graph** and can be visualized in a two-dimensional figure, describing the function f as the well-known parabola shape in geometry, as shown in Figure 2.2.

Convexity and concavity

A function is called *convex* if a straight line connecting any two points on the graph lies above the graph between the two points. It is cup shaped $\cup$, like the parabola in Figure 2.2.

Definition 2.3 *Formally, a function* $f : \mathbb{R}^n \longrightarrow \mathbb{R}$ *is* **convex** *if for all* $x_1, x_2 \in D_f \subset \mathbb{R}^n$ *and for all* $0 \leq t \leq 1$,

$$f(tx_1 + (1-t)x_2) \leq tf(x_1) + (1-t)f(x_2).$$

Convex functions are an important class of functions in many applications, particularly related to optimization. As can be seen from the parabola, convex functions on a closed domain (such as $\mathbb{R}$ or some finite closed interval $[a, b]$) have a unique minimum. Moreover, the inequality inherent in the definition of a convex function gives rise to many very useful other inequalities, such as Jensen's.

The flipside of convexity is called *concavity* if it has a $\cap$-shaped graph. Or in more formal terms, a function f is concave if for all $x_1, x_2 \in D_f$ and for all $0 \leq t \leq 1$, we have $f(tx_1 + (1-t)x_2) \geq tf(x_1) + (1-t)f(x_2)$. There are also functions that are neither concave nor convex. The step function g in Figure 2.2 is an example of such function.

2.3.1 Continuity

There is another property of a function that is immediately intuitive from its association graph. The function $f(x) = x^2$, shown in Figure 2.2, is continuous on its entire domain. Practically this means that there are no jumps in the graph. Formally, we say that a function is *continuous* if for all elements x_0 in the domain the function approaches from both sides the value $f(x_0)$. Before we can introduce the concept of continuity formally, we need to extend the concept of a *limit* that we first encountered in Definition 2.1 in the previous section.

Definition 2.4 *The* **limit** *of a function f at x_0 is y, i.e.,*

$$\lim_{x \to x_0} f(x) = y,$$

if

$$\forall \epsilon > 0, \exists \delta > 0, \forall x : ||x - x_0|| < \delta \implies ||f(x) - y|| < \epsilon.$$

Here we have introduced the concept of a limit. It captures the intuitive meaning of "approaching" a certain value. Practically speaking, the limit at x_0 is that value that you will approach, the closer you get to x_0. For example, $\lim_{x \to 2} x^2 = 4$ or $\lim_{x \to \infty} \frac{1}{x} = 0$.

The concept of a limit now allows us to introduce the concept of continuity formally.

Definition 2.5 *A function is* **continuous** *on set $X \subset D_f$ if*

$$\forall x_0 \in X : \lim_{x \to x_0} f(x) = f(x_0).$$

Practically speaking, it means that a function is continuous in x_0 if approaching x_0 from below or above the function f gets closer and closer to $f(x_0)$.

Functions do not have to be continuous. A very useful class of functions that are not continuous is the class of piece-wise constant functions. As shown in Figure 2.2, the step function g approximates the continuous function f. It is clear that at the places where the function g jumps, we have that the limit from below is not equal to the limit from above for certain values. For example,

$$\lim_{x \uparrow 1.5} g(x) = 1.56 \neq 3.06 = \lim_{x \downarrow 1.5} g(x).$$

However, notice that also the piece-wise linear function is continuous in most of its domain: only at the jump points is the function discontinuous.

2.3.2 Composition of functions

Under certain circumstances, it is possible to combine functions. If the image of g is contained in the domain of f, i.e., $Im_g \subset D_f$, then it is possible to put the output of g as input into the function f. We call this *composition*. Formally, $f \circ g : D_g \longrightarrow Im_f$ is the composition of g and f, defined as

$$f \circ g(x) = f(g(x)).$$

For example, if $g(x) = e^x$ and $f(x) = x^2$, then

$$f \circ g(x) = (e^x)^2 = e^{2x}.$$

It is useful to know that a composite function of two continuous functions is continuous and that a composite function of two convex functions is convex.

2.3.3 Inverse functions and solving equations

A function is a way to connect an input with an output. If there exists a function that maps the output back to the input, then this is called the **inverse** or inverse function. Formally, the *inverse function* is $f : D \longrightarrow I$, which is often indicated as $f^{-1} : I \longrightarrow D$, such that for all $x \in D$,

$$f^{-1} \circ f(x) = x.$$

Example 2.1 *Consider the function $f(x) = ax + b$. How can you obtain the inverse function? One general way involves writing the output of the function as y,*

$$y = ax + b,$$

and then rework the equation in order to express x as a function of y:

$$ax = y - b \tag{2.3}$$

$$x = \frac{y - b}{a}. \tag{2.4}$$

Therefore, the inverse function f^{-1} of f is

$$f^{-1}(y) = \frac{y - b}{a}.$$

In other examples, the logarithmic function is the inverse of the exponential function,

$$\log e^x = x.$$

Or the square is the inverse of the square root function,

$$(\sqrt{x})^2 = x.$$

As one can see from the examples, it is important to stress that in general $f^{-1} \neq 1/f$, with only one exception: multiplication by a scalar,

$$\frac{1}{c}(cx) = x,$$

as long as $c \neq 0$. The condition for the existence of an inverse is that each value in the image of f corresponds to only one value in its domain. In other words, f needs to be an *injection* or *one-to-one*. For example, $f(x) = x^2$ is not injective, as both -2 and $+2$ map to 4, and therefore there is no inverse for f on its entire domain. However, for the restricted function $f : (-\infty, 0] \longrightarrow [0, \infty)$ with $f(x) = x^2$, it can be seen that the inverse is given as $f^{-1}(y) = -\sqrt{y}$.

Inverse functions are particularly useful in practice to solve equations. Consider the following equation,

$$f(x) = c.$$

In order to find for which value of x this is true, we can apply the inverse function to both the left and the right in order to obtain,

$$x = f^{-1} \circ f(x) = f^{-1}(c).$$

Example 2.2 *If the aim is to know for which x the following holds,*

$$e^{\sqrt{x}} = 2,$$

then we can use the fact that the inverse is given as $f^{-1}(y) = (\log y)^2$, and therefore,

$$x = (\log 2)^2 \approx 0.48.$$

The inverse operation for a number of special functions has been studied for centuries. Although the Babylonians and Euclid developed geometric methods of "completing a square," it were Hindu mathematicians such as Brahmagupta (598–665 AD) that developed a method for solving quadratic equations. For the quadratic equation

$$ax^2 + bx + c = 0,$$

the inverse operation

- either does not exists, namely, if $b^2 - 4ac < 0$;
- or is a unique inverse, if $b^2 - 4ac = 0$, and given by $x = -b/2a$;
- or has two solutions, namely, if $b^2 + 4ac > 0$, given by

$$x = \frac{-b \pm \sqrt{b^2 - 4ac}}{2a}. \tag{2.5}$$

In the first half of the 16th century, Tartaglia and Cardano solved the general form of a cubic equation, which are implemented in modern mathematical software.

2.3.4 Multivariate functions

In engineering it is rare for a single input to completely determine a certain output, and therefore, it is sensible to consider more general functions, where the response depends on a number of variables or where the output itself is multivariate. Examples of such functions are

$$f(x, y) = xy^2, \quad g(\theta, r) = \begin{bmatrix} r \sin \theta \\ r \cos \theta \end{bmatrix},$$

where $f : \mathbb{R}^2 \longrightarrow \mathbb{R}$ and $g : \mathbb{R}^2 \longrightarrow \mathbb{R}^2$. In general, we will consider functions from, say, $\mathbb{R}^p$ to $\mathbb{R}^n$, $f : \mathbb{R}^p \longrightarrow \mathbb{R}^n$

$$f(x_1, \ldots, x_p) = \begin{bmatrix} f_1(x_1, \ldots, x_p) \\ \vdots \\ f_n(x_1, \ldots, x_p) \end{bmatrix}.$$

So a multivariate function $f : \mathbb{R}^p \longrightarrow \mathbb{R}^n$ basically stacks n functions $f_1, \ldots, f_n$ on top of each other, where $f_i : \mathbb{R}^p \longrightarrow \mathbb{R}$. Often it is possible to analyze each of these subfunctions separately.

When it comes to multivariate functions, an important special case is the linear function,

$$\begin{aligned} f(x_1, \ldots, x_p) &= \begin{bmatrix} a_{11}x_1 + \ldots + a_{1p}x_p \\ \vdots \\ a_{n1}x_1 + \ldots + a_{np}x_p \end{bmatrix} \\ &= \begin{bmatrix} a_{11} & \ldots & a_{1p} \\ \vdots & & \vdots \\ a_{n1} & \ldots & a_{np} \end{bmatrix} \begin{bmatrix} x_1 \\ \vdots \\ x_p \end{bmatrix} \\ &= Ax, \end{aligned}$$

where A is the $n \times p$ matrix consisting of the scalars $\{a_{ij}\}_{ij}$ and $x = (x_1, \ldots, x_p)^t$ a $p \times 1$ dimensional vector.

The definition of an **inverse** is fully general and can be applied to the multivariate case. For example, consider the function

$$f(x_1, x_2) = \begin{bmatrix} \sqrt{x_1^2 + x_2^2} \\ \arccos(x_1/\sqrt{x_1^2 + x_2^2}) \end{bmatrix}$$

that maps the Euclidean coordinates $(x_1, x_2) \neq (0, 0)$ into the polar coordinates (r, θ), representing the length of the vector (x_1, x_2) and its angle with the $x_1 - axis$ as shown in Figure 2.3. The inverse is given by the function,

$$f^{-1}(r, \theta) = \begin{bmatrix} r \sin \theta \\ r \cos \theta \end{bmatrix}.$$

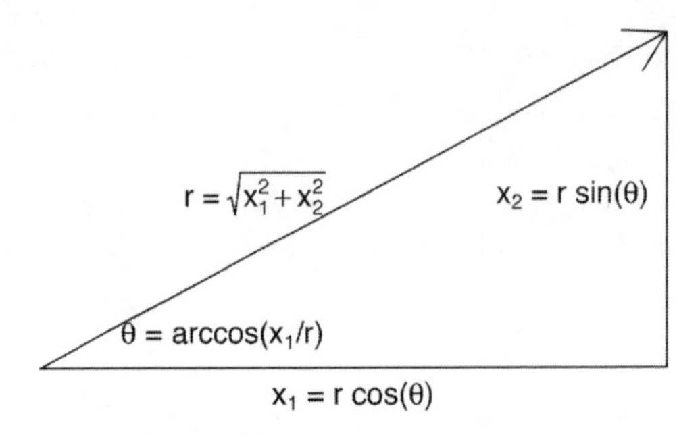

FIGURE 2.3
The Cartesian coordinates (x_1, x_2) can be mapped into the polar coordinates (r, θ), and by means of its inverse function, the latter can be mapped back into the former.

2.3.5 Linear transformations, matrix inverses, and matrix decompositions

An important special case of a multivariate function is the linear transformation $f : \mathbb{R}^p \longrightarrow \mathbb{R}^p$, defined via the $p \times p$ square matrix A,

$$f(x) = Ax.$$

What is the inverse function in this case? We saw above that if the function is univariate, i.e., $p = 1$, then the inverse $f^{-1}(y) = A^{-1}y$, as long as $A \neq 0$. For the multivariate case $p > 1$, we will define a $p \times p$ **inverse matrix** A^{-1}, such that

$$A^{-1}A = AA^{-1} = I,$$

where I is the identity matrix. There is no easy way to tell what the inverse of a matrix is, or even if it is invertible at all. There are some special cases that are useful in practice, such as the inverse of a 2×2 matrix,

$$\begin{bmatrix} a & b \\ c & d \end{bmatrix}^{-1} = \frac{1}{ad - bc} \begin{bmatrix} d & -b \\ -c & a \end{bmatrix}.$$

So, the inverse only exists if $ad - bc$, known as the **determinant** of a 2-by-2 matrix is not equal to zero. The existence of an inverse A^{-1} allows one to solve the following system of p equations $Ax = b$ in a *unique* way,

$$x = A^{-1}b.$$

If the matrix A is not invertible, then there are two possible sets of solutions to the problem $Ax = b$: either (1) there are multiple solutions (consider the example that $A = 0$ and $b = 0$) or (2) there are no solutions (consider e.g., $A = 0$ and $b = 1$).

Gauss-Jordan elimination

Not for every matrix does an inverse exist. But if it does exist, it can be found by means of Gauss-Jordan elimination. In this section we describe the method in detail and give a few examples.

We define the following three **elementary row operations** that can be applied to the rows of a matrix,

1. Swap two rows;
2. Multiply a row by a non-zero constant;
3. Add a multiple of a row to another row.

Consider a square $p \times p$ matrix A and the identity matrix I of the same size. Write the extended matrix

$$\begin{bmatrix} A & | & I \end{bmatrix}.$$

If by applying any number of elementary row operations to this extended matrix, it is possible to obtain the following form

$$\begin{bmatrix} I & | & B \end{bmatrix},$$

then B is the inverse of A. We begin by reordering the rows using operations of type 1 such that the first non-zero element in each row, also called the *pivot* moves to the right when moving down the rows. This is the so-called *row-echelon* form. Then by iterative operations of types 2 and 3, one aims to obtain a upper diagonal matrix on the left working one's way down the rows. Repeating this, but then moving back up the rows, one aims to eliminate the left upper diagonal entries to reach the identity matrix.

Let's consider the following example:

$$A = \begin{bmatrix} 0 & -0.5 & 0.5 \\ -0.5 & 1 & -0.5 \\ 0.5 & 0.5 & 1 \end{bmatrix}$$

Then the first stage involves bringing the matrix in the row-echelon form, such that the zeroes in the matrix are concentrated in the lower left corner:

$$[A|I] = \begin{bmatrix} 0 & -0.5 & 0.5 & 1 & 0 & 0 \\ -0.5 & 1 & -0.5 & 0 & 1 & 0 \\ 0.5 & 0.5 & 1 & 0 & 0 & 1 \end{bmatrix} \Longrightarrow \begin{bmatrix} -0.5 & 1 & -0.5 & 0 & 1 & 0 \\ 0.5 & 0.5 & 1 & 0 & 0 & 1 \\ 0 & -0.5 & 0.5 & 1 & 0 & 0 \end{bmatrix}.$$

After this we eliminate iteratively the lower-diagonal matrix on the left.

$$\Longrightarrow \begin{bmatrix} -0.5 & 1 & -0.5 & 0 & 1 & 0 \\ 0 & 1.5 & 0.5 & 0 & 1 & 1 \\ 0 & -0.5 & 0.5 & 1 & 0 & 0 \end{bmatrix} \Longrightarrow \begin{bmatrix} -0.5 & 1 & -0.5 & 0 & 1 & 0 \\ 0 & -0.5 & 0.5 & 1 & 0 & 0 \\ 0 & 0 & 2 & 3 & 1 & 1 \end{bmatrix}.$$

We then multiply the rows so that the left diagonal contains ones:

$$\Longrightarrow \begin{bmatrix} 1 & -2 & 1 & 0 & -2 & 0 \\ 0 & 1 & -1 & -2 & 0 & 0 \\ 0 & 0 & 1 & 1.5 & 0.5 & 0.5 \end{bmatrix}.$$

Then moving our way back up, we eliminate iteratively the upper-diagonal matrix on the left:

$$\Longrightarrow \begin{bmatrix} 1 & -2 & 0 & -1.5 & -2.5 & -0.5 \\ 0 & 1 & 0 & -0.5 & 0.5 & 0.5 \\ 0 & 0 & 1 & 1.5 & 0.5 & 0.5 \end{bmatrix} \Longrightarrow \begin{bmatrix} 1 & 0 & 0 & -2.5 & -1.5 & 0.5 \\ 0 & 1 & 0 & -0.5 & 0.5 & 0.5 \\ 0 & 0 & 1 & 1.5 & 0.5 & 0.5 \end{bmatrix}$$
$$= [I|A^{-1}].$$

So, from this we can read off that the inverse matrix of A is given as

$$A^{-1} = \begin{bmatrix} -2.5 & -1.5 & 0.5 \\ -0.5 & 0.5 & 0.5 \\ 1.5 & 0.5 & 0.5 \end{bmatrix}.$$

Sometimes, however, the Gauss-Jordan elimination cannot be completed: in the elimination of the entries of lower-diagonal matrix on the left, a complete row becomes zero. This means that the matrix A has linearly dependent rows and, therefore, it is not invertible.

Eigendecomposition

The calculation of inverses is an important task in many practical problems. For this reason, a number of useful techniques have been developed to aid this process. In this paragraph we describe the eigendecomposition of a matrix that can be used to diagnose and calculate matrix inverses in practice.

A (non-zero) $p \times 1$ vector u is an **eigenvector** of the $p \times p$ square matrix A if for some non-zero $\lambda \in \mathbb{R}$ the following holds,

$$Au = \lambda u,$$

where λ is called the **eigenvalue**. A $p \times p$ matrix A can have up to p linearly independent eigenvectors. If it does have exactly p linearly independent eigenvectors, $u_1, \ldots, u_p$, then we can collect these vectors in a $p \times p$ matrix $U = [u_1; \ldots; u_p]$ for which holds,

$$AU = U\Lambda,$$

where Λ is a diagonal matrix with the eigenvalues $\lambda_1, \ldots, \lambda_p$ on the diagonal. By multiplying both sides with the inverse of U, we obtain

$$A = U\Lambda U^{-1}.$$

This is called the **eigendecomposition** of A. This eigendecomposition is particularly useful in case A is a *symmetric* matrix, i.e., when $A^t = A$, and the columns of U are rescaled to have length 1. In that case

$$U^{-1} = U^t$$

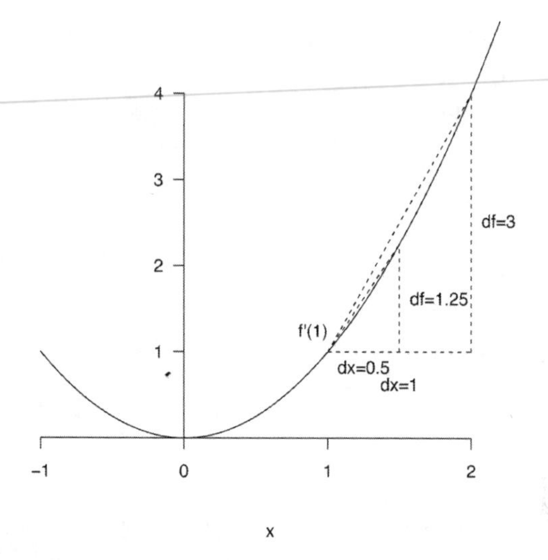

FIGURE 2.4
The slope graph associated with the continuous function $f(x) = d^2$ and with the discontinuous step function g that approximates f.

and no calculation of the inverse of U is needed. When A is symmetric, the eigenvalues Λ and eigenvectors U allow us to calculate the inverse of A easily,

$$A^{-1} = U\Lambda^{-1}U^t.$$

The matrix Λ^{-1} is simply a diagonal matrix with the values $1/\lambda_1, \ldots, 1/\lambda_p$ on the diagonal.

2.4 Differentiation

We can ask ourselves the question how quickly a function rises or falls on any value of its domain. The *derivative* of a function expresses the rate of increase of the function for each value in the domain. In Figure 2.4 we can see that the slope of the quadratic function $f(x) = x^2$ in $x = 2$ can be approximated by a series of line segments connecting $(2, f(2))$ and $(2 + dx, f(2 + dx))$ on the parabola for ever smaller values of dx. In other words, just as continuity before, the slope or **derivative** of a function is a concept defined as a limit.

Definition 2.6 *A function $f : D_f \longrightarrow Im_f$ has the derivative $f'(x_0)$ at $x_0 \in D_f$ if the following limit exists*

$$f'(x_0) = \lim_{x \to x_0} \frac{f(x) - f(x_0)}{x - x_0}.$$

Another common way of writing the derivative is $\frac{df}{dx}(x_0) = f'(x_0)$.

Clearly, if a function is discontinuous in x_0, then the limit may explode and the derivative does not exist. Even if a function is continuous in x_0, there is no guarantee that the function is differentiable. The function $f(x) = |x|$, for example, is continuous, but not differentiable in $x = 0$.

We return to the function $f(x) = x^2$ and try to find its derivative. Note,

$$\frac{df}{dx} = \frac{(x+dx)^2 - x^2}{dx} = \frac{2xdx + dx^2}{dx} \tag{2.6}$$

$$= 2x + dx \tag{2.7}$$

$$\longrightarrow 2x \qquad \text{as } dx \to 0 \tag{2.8}$$

Therefore, the limit exists on the entire domain $\mathbb{R}$, and the derivative of $f(x) = x^2$ is given as

$$f'(x) = 2x.$$

There are a large amount of standard derivatives that an engineer ought to know. We have collated the most important ones in Table 2.1. Furthermore, there are a number of standard rules that allow you to calculate derivatives from combinations of functions. For example, it is easy to see from the definition of the derivative that the derivative of a sum of functions is the sum of the derivatives. Similarly, there are rules for products, division and composition. The rules are shown in the same table.

To show how these rules can be applied, we work out an example for a moderately complicated function that uses these rules, as indicated by the row indices of Table 2.1 above the equality signs:

$$\begin{aligned}
f(x) &= e^{x^2 \sin x + \frac{x}{\arccos x}} \\
f'(x) &\overset{d,a}{=} \left(\frac{d}{dx} x^2 \sin x + \frac{d}{dx} \frac{x}{\arccos x} \right) e^{x^2 \sin x + \frac{x}{\arccos x}} \\
&\overset{e,b,c}{=} \left(\sin x \frac{dx^2}{dx} + x^2 \frac{d \sin x}{dx} + \frac{\arccos x \frac{dx}{dx} - x \frac{d \arccos x}{dx}}{\arccos^2 x} \right) e^{x^2 \sin x + \frac{x}{\arccos x}} \\
&\overset{2,7,14}{=} \left(2x \sin x + x^2 \cos x + \frac{\arccos x + \frac{x}{\sqrt{1-x^2}}}{\arccos^2 x} \right) e^{x^2 \sin x + \frac{x}{\arccos x}}
\end{aligned}$$

In the first equality, we used the composition rules and the fact that the derivative of a sum is the sum of the derivatives. In the second equality, we used the rules for the derivative of a multiplication and a ratio of two functions. In the final equality, we used the specific rules for the various known functions.

Higher-order derivatives.

It is possible to repeatedly differentiate the same function. Where the derivative indicates the change of the function, the derivative of the derivative

TABLE 2.1
Table of standard derivatives and integrals.

	$f(x)$	$f'(x) = \frac{df}{dx}$	$\int f(x)\,dx$
1)	c (constant)	0	cx
2)	$x^k (k \neq -1)$	kx^{k-1}	$\frac{1}{k+1}x^{k+1}$
3)	x^{-1}	$-x^{-2}$	$\log\|x\|$
4)	e^x	e^x	e^x
5)	c^x	$\log c c^x$	$\frac{c^x}{\log c}$
6)	$\log x$	$\frac{1}{x}$	$x \log x - x$
7)	$\sin x$	$\cos x$	$-\cos x$
8)	$\cos x$	$-\sin x$	$\sin x$
9)	$\tan x$	$\sec^2 x = \frac{1}{\cos^2 x}$	$-\log\|\cos x\|$
10)	$\sec x$	$\sec x \tan x$	$\log\|\sec x + \tan x\|$
11)	$\cot x$	$-\csc^2 x = \frac{-1}{\sin^2 x}$	$\log\|\sin x\|$
12)	$\csc x$	$-\csc x \cot x$	$-\log\|\csc x + \cot x\|$
13)	$\arcsin x$	$\frac{1}{\sqrt{1-x^2}}$	$x \arcsin x + \sqrt{1-x^2}$
14)	$\arccos x$	$\frac{-1}{\sqrt{1-x^2}}$	$x \arccos x - \sqrt{1-x^2}$
15)	$\arctan x$	$\frac{1}{1+x^2}$	$x \arctan x - \frac{1}{2}\log(1+x^2)$
16)	$\sinh x$	$\cosh x$	$\cosh x$
17)	$\cosh x$	$\sinh x$	$\sinh x$
19)	$\tanh x$	$\operatorname{sech}^2 x = \frac{1}{\cosh^2 x}$	$\log(\cosh x)$
a)	$f+g$	$f'+g'$	$\int f + \int g$
b)	fg	$fg' + f'g$	$f\int g - \int(f'\int g)$
c)	ff'	$ff'' + f'^2$	$\frac{1}{2}f^2$
d)	$\frac{f}{g}$	$\frac{fg'-f'g}{g^2}$	
e)	$f \circ g$	$(f' \circ g)g'$	

indicates how the derivative changes,

$$\frac{d^2 f}{dx^2}(x) = \frac{d}{dx} f'(x).$$

For example, the second derivative of the quadratic function $f(x) = x^2$ is given as

$$\begin{aligned}\frac{d^2 f}{dx^2}(x) &= \frac{d}{dx}(2x) \\ &= 2.\end{aligned}$$

So for every increase in x by 1 the change in function f will increase by 2. The fact that the second derivative is strictly positive, means that the quadratic function is **convex**. If the second derivative of a function is strictly negative, then that function is **concave**.

Higher-order derivatives can be defined recursively, i.e., for $k = 2, 3, 4, \ldots$,

$$\frac{d^k f}{dx^k}(x) = \frac{d}{dx} f^{(k-1)}(x),$$

where $f^{(1)}(x) = f'(x)$ and $f^{(k)}(x)$ is the kth derivative of f.

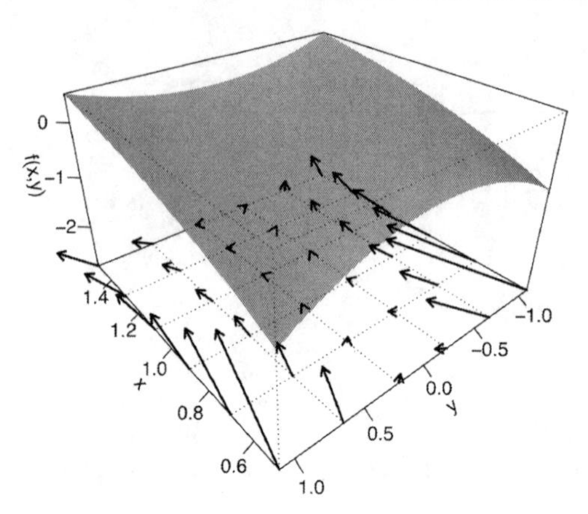

FIGURE 2.5
The gradient vectors for the function $f(x, y) = y^2 \log x$ "live" in the (x, y) plane and they point in the direction of the steepest increase of the function f at that location; furthermore, the larger the gradient vector, the steeper the increase.

2.4.1 Multivariate derivatives

As we saw above, any function $f : \mathbb{R}^p \longrightarrow \mathbb{R}^n$ can be seen as stacking n functions $f_i : \mathbb{R}^p \longrightarrow \mathbb{R}$ with $i = 1, \ldots, n$. When it comes to differentiation, it is helpful to focus on each of these individual subfunctions. Consider the function $f : \mathbb{R}^2 \longrightarrow \mathbb{R}$,

$$f(x, y) = y^2 \log x.$$

It is possible to ask how this function changes when moving in the x-direction or how it changes when moving in the y-direction. In particular,

$$\frac{\partial f}{\partial x}(x, y) = \frac{y^2}{x}$$
$$\frac{\partial f}{\partial y}(x, y) = 2y \log x$$

Although f maps into the one-dimensional space $\mathbb{R}$, in order to describe how f changes in each point (x, y), we require a mapping $\nabla f : \mathbb{R}^2 \longrightarrow \mathbb{R}^2$, called the **gradient**, where

$$\nabla f(x, y) = \begin{bmatrix} \frac{y^2}{x} \\ 2y \log x \end{bmatrix}.$$

The gradient vector can be interpreted as the direction and rate of fastest increase in the function f for each point (x, y). Figure 2.5 shows the gradient vectors for the function $f(x, y) = y^2 \log x$ in the (x, y) plane. The longer the vector, the steeper the slope at that (x, y) location. Furthermore, the direction of the gradient vector points towards the steepest increase in the function.

The gradient operator ∇ is the generalization of the derivative and straightforward generalizations of the **sum, product, and chain rules** exist,

$$\begin{aligned}\nabla(af+bg)(x) &= a\nabla f(x)+b\nabla g(x)\\ \nabla(fg)(x) &= g(x)\nabla f(x)+f(x)\nabla g(x)\\ \nabla(f\circ g)(x) &= \nabla f(g(x))\cdot g'(x).\end{aligned}$$

The derivative of a general multivariate function $f:\mathbb{R}^p\longrightarrow\mathbb{R}^n$ can be obtained by stacking the transposed gradients into an $n\times p$ matrix. This is called the **Jacobian**, i.e.,

$$J_f=\begin{bmatrix}\nabla f_1^t\\ \vdots\\ \nabla f_n^t\end{bmatrix}=\begin{bmatrix}\frac{\partial f_1}{\partial x_1} & \cdots & \frac{\partial f_1}{\partial x_p}\\ \vdots & & \vdots\\ \frac{\partial f_n}{\partial x_1} & \cdots & \frac{\partial f_n}{\partial x_p}\end{bmatrix}=\begin{bmatrix}\frac{\partial f_1}{\partial x^t}\\ \vdots\\ \frac{\partial f_n}{\partial x^t}\end{bmatrix}.$$

Higher-order derivatives for multivariate functions $f:\mathbb{R}^p\longrightarrow\mathbb{R}$ can be derived in an iterative way. In particular, $\frac{\partial^{\sum_i k_i} f}{\partial x_1^{k_1}\ldots\partial x_p^{k_p}}=\frac{\partial^{\sum_i k_i-1} f}{\partial x_1^{k_1-1}\ldots\partial x_p^{k_p}}$. In practice, we will often rely on the second derivative of a multivariate function, called the **Hessian**. The Hessian is a $p\times p$ matrix valued function

$$H_f=\begin{bmatrix}\frac{\partial^2 f}{\partial x_1^2} & \frac{\partial^2 f}{\partial x_1 x_2} & \cdots & \frac{\partial^2 f}{\partial x_1 x_p}\\ \frac{\partial^2 f}{\partial x_1 x_2} & \frac{\partial^2 f}{\partial x_2^2} & \cdots & \frac{\partial^2 f}{\partial x_2 x_p}\\ \vdots & \vdots & \ddots & \vdots\\ \frac{\partial^2 f}{\partial x_1 x_p} & \frac{\partial^2 f}{\partial x_2 x_p} & \cdots & \frac{\partial^2 f}{\partial x_p^2}\end{bmatrix}=\frac{\partial^2 f}{\partial x\partial x^t}.$$

For the function $f(x,y)=y^2\log x$, the Hessian is given as

$$H_f(x,y)=\begin{bmatrix}-\frac{y^2}{x^2} & \frac{2y}{x}\\ \frac{2y}{x} & 2\log x\end{bmatrix}.$$

It is clear that the Hessian is a symmetric matrix. Moreover, as a generalization of the univariate result, the Hessian of a convex function is positive semi-definite, i.e., for all $x\in\mathbb{R}^p$ we have $x^tHx\geq 0$, whereas the Hessian of a concave function is negative semi-definite, i.e., for all $x\in\mathbb{R}^p$ we have $x^tHx\leq 0$.

2.4.2 Taylor series

An important numerical application of derivatives is the Taylor expansion. The idea of the Taylor expansion is to approximate an infinitely differentiable function f by another function g, such that at one particular point x_0 the function values correspond, the first derivatives correspond, the second derivatives correspond, etc. An English mathematician, Brook Taylor (1685–1731), derived in 1715 a general method for obtaining such "approximation."

He defined the infinite series

$$\begin{aligned} g(x) &= f(x_0) + (x - x_0) f^{(1)}(x_0) + \frac{(x_0 - x_0)^2}{2!} f^{(2)}(x_0) \\ &\quad + \frac{(x_0 - x_0)^3}{3!} f^{(3)}(x_0) + \dots \\ &= \sum_{k=0}^{\infty} (x - x_0)^k f^{(k)}(x_0). \end{aligned} \tag{2.9}$$

The expression in (2.9) is called the **Taylor series**. By taking derivatives and plugging in x_0 for x, we can see that the derivatives of f and g indeed match for the point x_0:

$$\begin{aligned} g(x_0) &= f(x_0) \\ g^{(1)}(x_0) &= f^{(1)}(x_0) + (x_0 - x_0) f^{(2)}(x_0) + \frac{(x_0 - x_0)^2}{2!} f^{(3)}(x_0) + \dots \\ &= f^{(1)}(x_0) \\ g^{(2)}(x_0) &= f^{(2)}(x_0) + (x_0 - x_0) f^{(3)}(x_0) + \dots \\ &= f^{(2)}(x_0) \\ \vdots &\quad \vdots \end{aligned}$$

It can even be shown that for certain functions, such as the exponential, the sine and cosine, the functions f and g are identical on their entire domain. They are examples of *analytic functions*, for which this property holds. The infinite series can also be truncated for some finite power. This is typically referred to as the **Taylor polynomial** or **Taylor approximation**. The function $f(x) = \sin(x)$ can be approximated for values of x near $x_0 = -1$ by a cubic Taylor polynomial as

$$\begin{aligned} \sin(x) &\approx \sin(-1) + (x+1)\cos(-1) - \frac{(x+1)^2}{2}\sin(-1) - \frac{(x+1)^3}{6}\cos(-1) \\ &= -0.84 + 0.54(x+1) + 0.42(x+1)^2 - 0.09(x+1)^3 \end{aligned}$$

For example, for $x = -0.5$, the approximation gives -0.483, whereas $\sin(-0.5) = -0.479$. In Figure 2.6, it is clear that the truncated series better approximates $f(x) = \sin(x)$ the more terms are included in the series. In practice this means that the Taylor expansion can be used to approximate functions, particularly for values x close to x_0.

The Taylor expansion can also be generalized to *multivariate functions*. For example, for a twice differentiable function $f : \mathbb{R}^p \to \mathbb{R}$, the quadratic Taylor approximation for values near x_0 involves the gradient and the Hessian and is given by

$$f(x) \approx f(x_0) + (x - x_0)^t \nabla f(x_0) + \frac{1}{2}(x - x_0)^t H_f(x_0)(x - x_0).$$

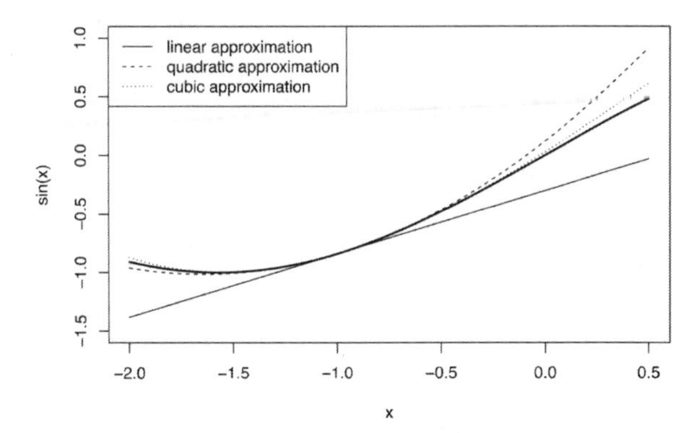

FIGURE 2.6
Three Taylor approximations of the function $f(x) = \sin(x)$.

In fact, the function $f(x, y) = y^2 \log x$ can be approximated for (x, y) close to, say, $(x_0, y_0) = (1, 1)$, as

$$\begin{aligned} y^2 \log x &\approx \begin{bmatrix} x-1 & y-1 \end{bmatrix} \begin{bmatrix} 1 \\ 0 \end{bmatrix} + \frac{1}{2} \begin{bmatrix} x-1 & y-1 \end{bmatrix} \begin{bmatrix} -1 & 2 \\ 2 & 0 \end{bmatrix} \begin{bmatrix} x-1 \\ y-1 \end{bmatrix} \\ &= (x-1) - \frac{1}{2}(x-1)^2 + 2(x-1)(y-1) \\ &= -\frac{1}{2}x^2 + 2y(x-1) + \frac{1}{2} \end{aligned}$$

For example, for the point $\begin{bmatrix} 1.5 & 1.5 \end{bmatrix}^t$ the approximation gives 0.875, whereas the true value is $1.5^2 \log 1.5 = 0.912$.

2.5 Maxima and Minima

A common task in practical problems is to find the minimum or a maximum of a particular function. For example, one may be interested in finding out for which input values a certain objective function is optimized, such as how much of a certain catalyst is needed to maximize the speed of an important chemical reaction. Consider a function $f : \mathbb{R}^2 \longrightarrow \mathbb{R}$, then the **minimum** is defined as the point or set of points (x_0, y_0), such that

$$x_0 = \operatorname*{argmin}_{x \in \mathbb{R}^p} f(x)$$

and $y_0 = f(x_0)$. The **maximum** is defined in an analogous way. Consider the quadratic function,

$$f(x) = x^2.$$

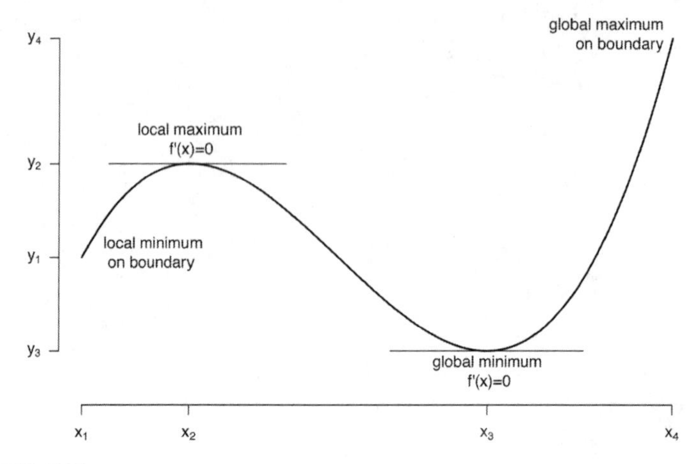

FIGURE 2.7
The local and global minima and maxima of function f on a bounded interval $[x_1, x_4]$.

From Figure 2.2, it is clear that it achieves its minimum at $(x_0, y_0) = (0, 0)$ and that this minimum is unique. The function does not have a maximum as for $x \to \pm\infty$ we have that $f(x) \to \infty$. In fact, for any quadratic function $f(x) = ax^2 + bx + c$, the optimum can be found at

$$x_0 = -b/2a \tag{2.10}$$

To explore the concept of optimization in more detail, consider the function $f : [x_1, x_4] \longrightarrow \mathbb{R}$ in Figure 2.7. As can be seen from the figure, f has a minimum at (x_3, y_3) and a maximum at (x_4, y_4). In addition, one can see a local minimum of f at (x_1, y_1) and a local maximum of f at (x_2, y_2) A **local minimum** is defined as a point or sets of points (x_0, y_0) such that in the direct neighborhood of x_0 it holds that $f(x) > y_0$. A **local maximum** is defined analogously.

A special class of optimization problems are the **convex optimization problems**. This requires finding an optimum of a convex function f over a convex domain. A convex domain is a set such that for every x, y in that set also $wx+(1-w)y$ belongs to that set for each value of $w \in [0, 1]$. Many practical objective functions that we will encounter in this book are convex or locally convex, and therefore, we will often be able to make use of the very useful properties of convex optimization: (i) the optimum is always a minimum, due to the cup shape nature of convex functions; (ii) every local minimum in a convex optimization problem is automatically a global minimum, (iii) if the minimum is not unique, then the set of minima is itself convex; (iv) finally, if the function is strictly convex, i.e., if for all x, y we have $f(wx + (1-w)y) < wf(x) + (1-w)f(y)$, then the minimum is unique.

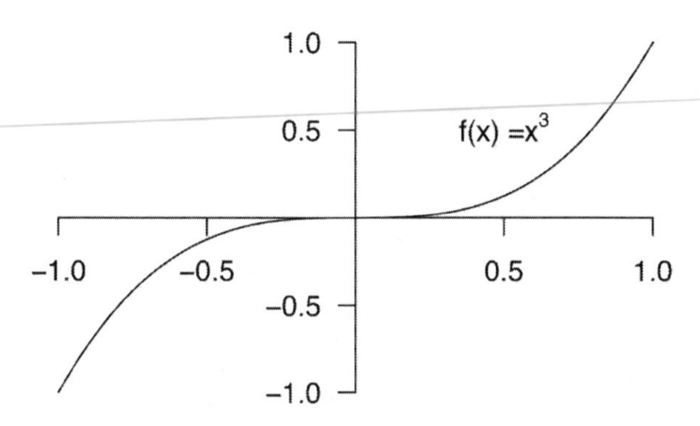

FIGURE 2.8
The cubic function $f(x) = x^3$ has both a zero first derivative and second derivative at $x = 0$, $f'(0) = f''(0) = 0$. This is an indication of a saddle point.

Figure 2.7 also gives a hint on how to find minima and maxima of a function. Besides the boundary points, the optima of a differentiable function are found where the first derivative is equal to zero. In fact, the function in this figure is given as $f(x) = x(x - 0.5)(x - 1)$, which has derivative $f'(x) = 3x^2 - 3x + 1/2$. The solution of $f'(x) = 0$ can be found using the general solution of a quadratic equation given in Equation (2.5). Therefore, the solution is given as $x = \frac{3 \pm \sqrt{3}}{6}$, i.e., $x_2 = 0.21$ and $x_3 = 0.79$.

2.5.1 Second derivative test, saddle points, and inflection points

If for some x_0 we have that $f'(x_0) = 0$, is it possible without drawing the function to say whether it has a (local) minimum or maximum? As long as f has a second derivative in x_0, then the answer is "yes." If $f''(x_0) < 0$, then the first derivative was going down in x_0. As $f'(x_0) = 0$, this means that for $x < x_0$ we have $f'(x) > 0$, i.e., the function f was going up before x_0. And on the other side, for $x > x_0$ we have $f'(x) < 0$, i.e., the function f was going down after x_0. This means that f must have a maximum at x_0. An analogous argument can be used to conclude that if $f''(x_0) > 0$, then the optimum is a minimum. This is known as the **second derivative test**.

What happens if the second derivative is zero? Let's consider an example where that happens. Consider the function $f(x) = x^3$, shown in Figure 2.8. Setting the first derivative to zero $f'(x) = 3x^2 = 0$, we find $x = 0$ as a potential optimum. However, the second derivative at $x = 0$ is also zero, i.e., $f''(x)|_{x=0} = 6x|_{x=0} = 0$ and therefore, $(0, f(0))$ is neither a minimum nor a maximum, but a so-called **saddle point**. In the graph this means that the function has a saddle-like shape near $x = 0$.

To summarize, for a twice-differentiable function with $f'(x_0) = 0$, we can distinguish the following three cases:

1. x_0 is a (local) minimum, if $f''(x_0) > 0$;
2. x_0 is a (local) maximum, if $f''(x_0) < 0$;
3. x_0 is a saddle point, if $f''(x_0) = 0$.

This can also be generalized to multivariate settings. If $H_f(x_0)$ is positive-definite, then the function has a (local) minimum at x_0. If $H_f(x_0)$ is negative-definite, then the function has a (local) maximum at x_0. Whereas if $H_f(x_0)$ has both positive and negative eigenvalues, then the function has a saddle point at x_0. A $p \times p$ matrix A is said to be **positive-definite**, if for all non-zero $x\mathbb{R}^p$, we have $x^t Ax > 0$. Reversely, it is said to be **negative-definite** if for all non-zero x in $\mathbb{R}^p$, we have $x^t Ax < 0$.

As in interesting curiosity, we could ask ourselves the question what would happen if $f''(x_0) = 0$, but $f'(x_0) \neq 0$. This is called an **inflection point**: it describes the point, where the function f changes from being locally convex to being locally concave, or vice versa. Consider for example, the function

$$f(x) = \frac{1}{\sigma\sqrt{2\pi}} e^{-\frac{1}{2\sigma^2}x^2}.$$

Its second derivative,

$$f''(x) = \frac{1}{\sigma\sqrt{2\pi}} \left(\frac{1}{\sigma^2} - \frac{x^2}{\sigma^4}\right) e^{-\frac{1}{2\sigma^2}x^2}$$

is zero at $x = -\sigma$ and $x = \sigma$, but in both cases $f'(\pm\sigma) \neq 0$. Therefore, the function f, shown in Figure 2.9, has two inflection points at $\pm\sigma$. We will come across this function later in the book. It is also known as the density function of a normal distribution with zero mean and standard deviation σ.

2.5.2 Newton-Raphson algorithm for finding optima

Sometimes, it is not possible to find the optimum of a function explicitly, as there are no analytical solutions to the problem $f'(x) = 0$. Instead of finding an explicit analytical solution, it is sometimes possible and sufficient to find a numerical solution through some algorithm. There are several numerical methods for finding minima and maxima of a function, but many of them are based on the Newton-Raphson algorithm, which we describe here.

A forerunner of the method was described by Isaac Newton (1642–1726/7) and Joseph Raphson (1648–1715) in the late 17th century, and it was derived in its modern form by Thomas Simpson (1710–1761) in 1740. Consider a twice differentiable function f with its second-order Taylor approximation g in some

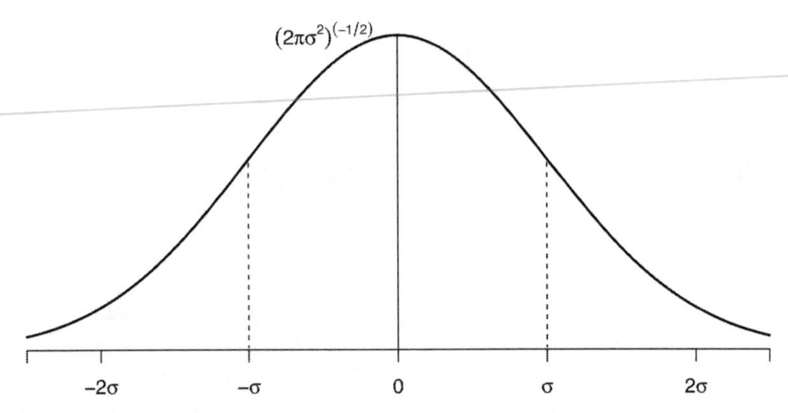

FIGURE 2.9
The function $f(x) = \frac{1}{\sigma\sqrt{2\pi}}e^{-\frac{1}{2\sigma^2}x^2}$ has two inflection points at $x = \pm\sigma$.

initial value x_0, as shown in Figure 2.10,

$$\begin{aligned} g(x) &= f(x_0) + (x - x_0)f'(x_0) + (x - x_0)^2 f''(x_0)/2 \\ &= \frac{1}{2}f''(x_0)x^2 + (f'(x_0) - x_0 f''(x_0))x + f(x_0) - x_0 f'(x_0) + \frac{1}{2}x_0^2 f''(x_0). \end{aligned}$$

Given that g approximates f, the idea is that the optimum of g will approximate the optimum of f. Given the explicit form of the optimum of a quadratic function in (2.10), a first approximation of the optimizer of f is given as

$$x_1 = -\frac{f'(x_0) - x_0 f''(x_0)}{f''(x_0)} = x_0 - \frac{f'(x_0)}{f''(x_0)}.$$

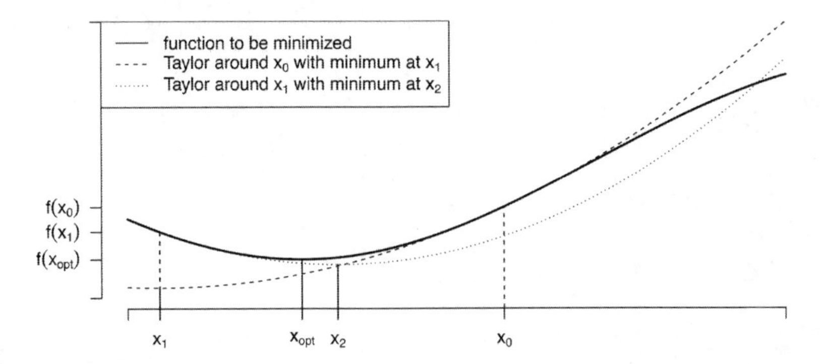

FIGURE 2.10
Two steps of a Newton-Raphson minimization procedure starting at x_0 to obtain x_2 as approximation of the true minimum x_{opt}.

This can be iterated for $k = 1, 2, \ldots$,

$$x_k = x_{k-1} - \frac{f'(x_{k-1})}{f''(x_{k-1})},$$

and if the sequence converges, it tends to converge quickly to the optimum x_{opt}, as shown in Figure 2.10. If a function has multiple (local) optima, such as in Figure 2.7, then the algorithm converges to any of these values depending on the starting value x_0. Each optimum that is characterized by $f'(x_{\text{opt}}) = 0$ has its own **basin of attraction**, typically a set of values close to x_{opt}. It may also be that for certain starting values, the algorithm does not converge. In this case the algorithm can be restarted with a different value.

The method can be generalized in a straightforward way to higher dimensions. For any twice differentiable function $f : \mathbb{R}^p \longrightarrow \mathbb{R}$, its optimum x_{opt} can be found iteratively if the following sequence is converging for $k = 1, 2, \ldots$

$$x_k = x_{k-1} - H_f^{-1}(x_{k-1}) \nabla_f (x_{k-1}),$$

where $x_0 \in \mathbb{R}^p$ is some initial value, H_f is the Hessian, and ∇_f is the gradient of f.

2.6 Integration

Integration of a function is related to the question about the area underneath the curve of that function. Let's consider our trusted parabola, shown in Figure 2.11. We indicate the area between the parabola and the x-axis between the bounds a and b by the symbolic representation for a **definite integral**,

$$\int_a^b x^2 \, dx = \text{area between x-axis, } x^2 \text{ and the bounds } x = a \text{ and } x = b.$$

If we want to calculate the area under the curve between a and b, we can subdivide the region $[a, b]$ into n equally sized sections, calculate the partial sum, and then let n go to infinity. This can be done with, e.g., the limit type of the techniques we learned in Section 2.2,

$$\begin{aligned}\int_a^b x^2 \, dx &= \lim_{n\to\infty} \sum_{k=1}^{n} \frac{b-a}{n} \left(\frac{(a + (k-1)\frac{b-a}{n})^2 + (a + k\frac{b-a}{n})^2}{2} \right) \\ &= \lim_{n\to\infty} \left(2a^2 n + 2a\frac{b-a}{n} \underbrace{\sum_{k=1}^{n}(2k-1)}_{n^2} + \left(\frac{b-1}{n}\right)^2 \underbrace{\sum_{k=1}^{n}(k^2 + (k-1)^2)}_{4n^3+n} \right)\end{aligned}$$

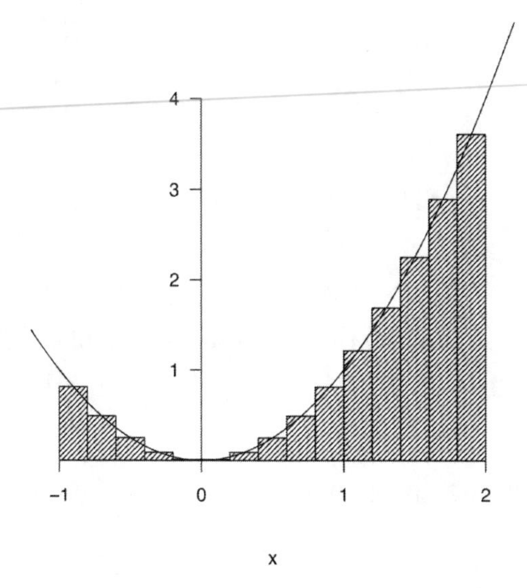

FIGURE 2.11
The area under the curve can be calculated via series, representing sums of infinitesimally small rectangles.

$$= (b-1)ab + \lim_{n\to\infty} \frac{(b-a)^3}{3}\left(1+\frac{1}{4n^2}\right)$$
$$= \frac{1}{3}(b^3 - a^3)$$

Clearly, this calculation is quite involved and it would be very inconvenient if we had to do this for every function we wanted to integrate. However, by closer inspection of the answer we have just obtained, we can learn about a close connection between integration an differentiation:

$$\int_a^b x^2\, dx = \frac{1}{3}x^3\,|_{x=b} - \frac{1}{3}x^3\,|_{x=a} = \left[\frac{1}{3}x^3\right]_a^b,$$

where the derivative of the answer $x^3/3$ is exactly equal to the input function x^2. In fact, this leads us to the following rule that integration is the reverse of differentiation:

The **definite integral** of f between a and b can be calculated as

$$\int_a^b f(x)\,dx = F(b) - F(a),$$

most of the time, where the derivative of F is f,

$$\frac{dF}{dx}(x) = f(x).$$

The function F is referred to the **indefinite integral** of f,

$$F(x) =: \int f(x)\,dx.$$

N.B. The indefinite integral is not unique, as you can always add a constant:

$$(F + c)' = f.$$

For example, we can now calculate the following integral,

$$\int_0^\pi \sin x\,dx = [-\cos\pi] - [-\cos 0] = 1 + 1 = 2,$$

because clearly $\frac{d}{dx}[-\cos x] = \sin x$.

2.6.1 Improper integrals

We have modified the integration rule by stating that it is true *most of the time*, because originally the German mathematician Riemann had shown to hold only for functions of f that are finite in the interval $[a, b]$ (Riemann, 1867). However, there are a number of useful extensions for the rule to apply more generally. Although these extensions are referred to **improper integrals**, they are perfectly sound:

- **Unbounded f in the interior.** If the function f is unbounded near an interior point c of $[a, b]$, then the integral can be split into two:

$$\int_a^b f(x)\,dx = \int_a^c f(x)\,dx + \int_c^b f(x)\,dx.$$

  For example, $\int_{-1}^4 \frac{1}{\sqrt{|x|}}\,dx = \int_{-1}^0 \frac{1}{\sqrt{|x|}}\,dx + \int_0^4 \frac{1}{\sqrt{|x|}}\,dx.$

- **Unbounded f near the boundary.** If the previous scenario occurred, we should now be able to solve an integral of a function that is unbounded on its boundary. This is done via a limit:

$$\int_a^b f(x)\, dx := \lim_{t \downarrow a} \int_t^b f(x)\, dx \text{ if } f \text{ unbounded in } a \tag{2.11}$$

$$\int_a^b f(x)\, dx := \lim_{t \uparrow b} \int_a^t f(x)\, dx \text{ if } f \text{ unbounded in } b \tag{2.12}$$

For example,

$$\int_0^4 \frac{1}{\sqrt{|x|}}\, dx = 2\sqrt{|4|} - \lim_{t \downarrow 0} 2\sqrt{|t|} = 4 - 0 = 4.$$

For the inverse function, we find

$$\int_0^2 \frac{1}{x}\, dx = \log|2| - \lim_{t \downarrow 0} \log|t| = \infty.$$

Therefore, this integral *diverges*. By the way, this is closely connected with the observation in section 2.2 that the harmonic series $\sum_{k=1}^{\infty} 1/k$ diverges.

- **Infinite integration bounds.** Although originally intended for functions on a finite interval, the definition of the integral can be extended to function on infinite intervals:

$$\int_{-\infty}^{\infty} f(x)\, dx = \lim_{a \to -\infty, b \to \infty} \int_a^b f(x)\, dx.$$

For example, probability density function of an exponential distribution with rate $\lambda > 0$ integrates to one,

$$\int_0^{\infty} \lambda e^{-\lambda x}\, dx = \lim_{t \to \infty} -e^{-\lambda t} + e^0 = 1.$$

2.6.2 Practical integration rules

Given that integration is just "reverse differentiation," the table with standard derivatives is equally useful for integration. In fact, Table 2.1 in section 2.4 contains the standard integrals for many standard functions. In practice, we often find that we need to integrate functions that are combinations of other functions. Just as there are rules for derivatives of combinations of functions, so there are such combination rules for integrals, although they typically require more work.

- **Sums and scalar multiplication.** The first rule is a simple one and says that addition and multiplication by a scalar simply distributes:

$$\int_a^b cf(x) + dg(x)\, dx = c \int_a^b f(x)\, dx + d \int_a^b g(x)\, dx.$$

So, for example,

$$\int_0^1 3e^x + 2\cos x \; dx = 3\int_0^1 e^x \; dx + 2\int_0^1 \cos x \; dx.$$

- **Integration by parts.** By reversing the product rule for differentiation, it is possible to obtain the following *integration by parts* rule:

$$\int_a^b F(x)g(x) \; dx = [F(x)G(x)]_a^b - \int_a^b f(x)G(x) \; dx,$$

where F is the integral of f and G the integral of g. For example,

$$\begin{aligned}\int_0^1 xe^x \; dx &\overset{F(x)=x}{\underset{g(x)=e^x}{=}} [xe^x]_0^1 - \int_0^1 e^x \; dx \\ &= e - [e^x]_0^1 \\ &= 1.\end{aligned}$$

- **Substitution rule.** Whereas integration by parts is the product rule for integration, the substitution rule is the composition rule for integration. Let $g : [a,b] \longrightarrow D \subset \mathbb{R}$ a continuous, differentiable function, such that D is an interval. Let $f : D \longrightarrow \mathbb{R}$ continuous, then

$$\int_a^b f(g(x))g'(x) \; dx = \int_{g(a)}^{g(b)} f(u) \; du.$$

For example,

$$\begin{aligned}\int_0^\pi \sin(x)e^{\cos x} \; dx &\overset{u=\cos x}{\underset{du=-\sin(x)dx}{=}} \int_{\cos 0}^{\cos \pi} -e^u \; du \\ &= [-e^u]_1^{-1} \\ &= e - e^{-1}\end{aligned}$$

2.6.3 Multiple integrals

If instead of an area under a curve, we are interested in calculating some volume, we need to use multiple integrals. Fortunately, the definition of integrals of univariate functions can be extended easily to integrals of multivariate functions by iteratively solving univariate integrals.

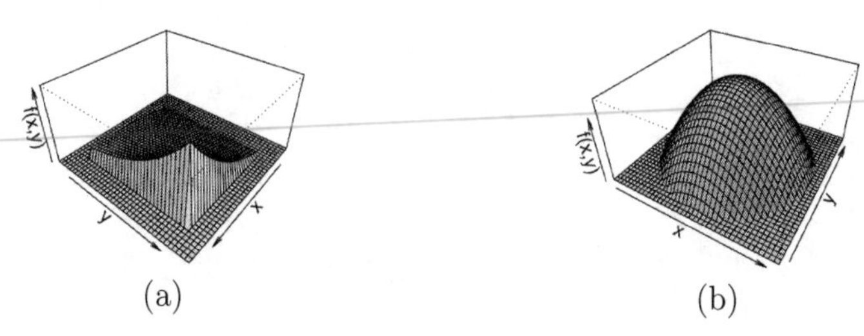

FIGURE 2.12
(a) $f(x,y) = xe^{xy}$ and (b) a two-dimensional parabola.

Consider $f : D \longrightarrow \mathbb{R}$, where $D \subset \mathbb{R}^p$ can be written as

$$D = \{x \in \mathbb{R}^p | a < x_1 < b, l_2(x_1) \leq x_2 \leq u_2(x_1), \dots, l_p(x_{-p}) \leq x_p \leq u_p(x_{-p})\}$$

then the $p+1$ dimensional volume under f on D is calculated as

$$\int_D f(x)\ dx = \int_a^b \left[\int_{l_2(x_1)}^{u_2(x_1)} \left[\cdots \left[\int_{l_p(x_{-p})}^{u_p(x_{-p})} f(x)\ dx_p \right] \cdots \right] dx_2 \right] dx_1.$$

If the order of the x's in defining D would be different, then the integral can be calculated accordingly.

Example. Consider the two-dimensional curve shown in Figure 2.12a on the rectangular domain D,

$$D = \{(x,y) | 0 \leq x \leq 1, -1 \leq y \leq 1\}.$$

Our aim is to determine the volume below the function $f(x,y) : D \longrightarrow \mathbb{R}$,

$$f(x,y) = xe^{xy}.$$

This volume is indicated by means of the multivariate integral,

$$\int_D f(x,y)\ dxdy := \text{volume under the curve } f(x,y) \text{ for } (x,y) \in D$$

The integral can be written as two nested univariate integrals.

$$\begin{aligned}
\int_D xe^{xy}\ dxdy &= \int_0^1 \left[\int_{-1}^1 xe^{xy}\ dy \right] dx \\
&= \int_0^1 \left[e^{xy} \right]_{y=-1}^{y=1}\ dx \\
&= \int_0^1 e^x - e^{-x}\ dx \\
&= e - 1 + e^{-1} - 1
\end{aligned}$$

Notice that it is easier to first integrate inside with respect to y and only then with respect to x, although either way would yield the same answer.

A more challenging example, bringing together a lot of different aspects of integration is given below.

Example. Consider the two-dimensional parabola shown in Figure 2.12b. For values (x, y) inside the unit circle D, where

$$D = \left\{(x, y) | -1 \leq x \leq 1, -\sqrt{1-x^2} \leq y \leq \sqrt{1-x^2}\right\},$$

our aim is to determine the volume below the parabola, given by $f(x, y) : D \longrightarrow \mathbb{R}$,

$$f(x, y) = 1 - x^2 - y^2.$$

By using the iterative definition of D in terms of x and y, this integral can be written as two nested univariate integrals,

$$\begin{aligned}
\int_D 1 - x^2 - y^2 \, dxdy &= \int_{-1}^{1} \left[\int_{-\sqrt{1-x^2}}^{\sqrt{1-x^2}} 1 - x^2 - y^2 \, dy \right] dx \\
&= \int_{-1}^{1} 2\sqrt{1-x^2}(1-x^2) - \frac{2}{3}(1-x^2)^{3/2} \, dx \\
&= \frac{4}{3} \int_{-1}^{1} (1-x^2)^{3/2} \, dx \\
&\overset{x=\sin u}{\underset{dx=\cos(u)du}{=}} \frac{4}{3} \int_{\arcsin(-1)}^{\arcsin(1)} (1 - \sin^2 u)^{3/2} \cos u \, du \\
&= \frac{4}{3} \int_{-\pi/2}^{\pi/2} \cos^4 u \, du \\
&\overset{\cos^2 u=}{\underset{(1+\cos 2u)/2}{=}} \frac{1}{3} \int_{-\pi/2}^{\pi/2} [1 + 2\cos 2u + \cos^2 2u] \, du \\
&= \frac{1}{3} \int_{-\pi/2}^{\pi/2} [1 + 2\cos 2u + \frac{1}{2}(1 + \cos 4u)] \, du \\
&= \frac{1}{3} \left[u + \sin 2u + \frac{1}{2}(u + \frac{1}{4} \sin 4u) \right]_{-\pi/2}^{\pi/2} \\
&= \frac{1}{3} \left(\pi + 0 + \frac{1}{2}(\pi + 0) \right) \\
&= \pi/2
\end{aligned}$$

2.6.4 Interchange integration and differentiation

Given that differentiation and integration are fundamental tools for any engineer, it is not uncommon to see them show up in the same problem together, as, for example,

$$\frac{\partial}{\partial t}\int_0^1 x^2 \log t \; \partial x = \frac{\partial}{\partial t}\frac{1}{3}(\log t - 0) = \frac{1}{3t}.$$

Interestingly, if the integral sign and the differentiation sign had been interchanged, we would have obtained the same result,

$$\int_0^1 \frac{\partial}{\partial t} x^2 \log t \; \partial x = \int_0^1 \frac{\partial}{\partial t}\frac{x^2}{t} \; \partial x = \left(\frac{1^3}{3t} - \frac{0}{3t}\right) = \frac{1}{3t}.$$

This poses the question, under what conditions do we have the equality $\frac{\partial}{\partial t}\int_a^b f(x,t) \; \partial x = \int_a^b \frac{\partial f(x,t)}{\partial t} \; \partial x$? The answer was discovered by the 17th century mathematician, Gottfried Leibniz (1646–1716).

Theorem 2.1 Leibniz integration rule. *Consider a function $f(x,t)$ for which both the function and its partial derivative $\frac{\partial f(x,t)}{\partial t}$ are continuous on $[a,b] \times t_0$, then*

$$\left.\frac{\partial}{\partial t}\int_a^b f(x,t) \; \partial x\right|_{t=t_0} = \left.\int_a^b \frac{\partial f(x,t)}{\partial t} \; \partial x\right|_{t=t_0}.$$

The theorem is particularly useful in neat ways to calculate difficult integrals.

2.7 Differential Equations

When Heraclitus, a Greek philosopher, declared in the 6th century BC that "All is Flux," he hit upon an important principle in the physicial sciences that when a system changes it typically does not do so according to some general, central plan, but more locally, relative to its current state. However, it was only until the 17th century with the development of calculus by Newton and Leibniz that this idea was formalized by means of differential equations.

A differential equation relates the change of a system, i.e., a derivative of a function, to the state of the system, i.e., the function itself. In short, it consists of three elements: (1) a function $y(t)$, (2) its derivative $y'(t)$, and (3) a relationship between the function and its derivative f. An example of an ordinary differential equation[1] is

$$y'(t) = f(y(t), t).$$

[1] There are also other differential equations, such as partial differential equations and stochastic differential equations, but they will not be considered here.

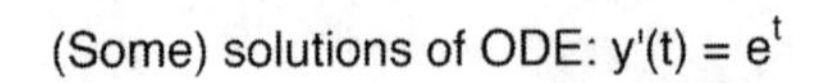

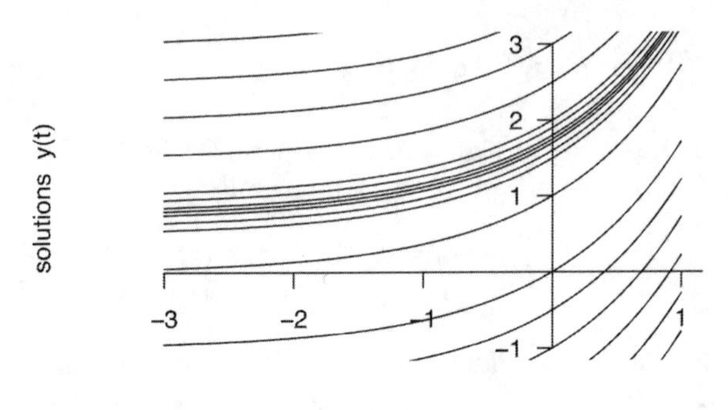

FIGURE 2.13
The solution y of an ordinary differential equation is a collection of functions that satisfy the condition $y'(t) = f(y(t), t)$ for some known function f.

We think of y as an *unknown* function of, say, time t and of f as a *known* function of y and t. It is called a differential *equation* in the sense that the aim of the equation is finding the function or functions y that satisfy it. This is called *solving the differential equation.*

It is instructive to see that differential equations can be solved with the tools we already have available. For example, consider the function $f(y, t) = e^t$, then we can solve the differential equation in the following steps.

$$\begin{aligned} y'(t) &= e^t \\ y(t) &= c + \int_{-\infty}^{t} e^s \, ds \\ y(t) &= c + e^t, \end{aligned} \tag{2.13}$$

where $c \in \mathbb{R}$ is an arbitrary constant. To see that we need this constant c, go from the second line back to the first line: by differentiating the second equation, we exactly obtain the first because the derivative of a constant is zero. So rather than a single solution, the differential equation (2.13) has an infinite number of solutions. Figure 2.13 shows some of the solutions of the differential equation, where each line represents a function $y(t)$ that satisfies $y'(t) = e^t$.

Although it has been quite easy to solve the above differential equation, in general this is not so easy. In fact, it may actually be impossible to solve an arbitrary equation. However, there are several classes of differential equations that can be solved easily. The next sections are dedicated to those methods. For those that cannot be solved analytically, we will encounter in Chapter 4 methods to solve them numerically.

2.7.1 Equilibrium solutions of differential equations

We start with a very special solution that is of particular interest in many applications. A differential equation shows how a "system" y changes over time as a function of the underlying state y itself. For example, how does a cup of water heat up when some amount of energy has been added to it? Or, how do the number of prey change, when some number of predators enter a region? It is often interesting for practitioners to know what the system eventually, i.e., after the dust has settled, looks like. This eventual state $\lim_{t\to\infty} y(t)$ is often an equilibrium solution of the differential equation.

An *equilibrium solution* of a differential equation $y'(t) = f(y(t), t)$ is defined as a constant function

$$y(t) = c,$$

such that for all t

$$f(c, t) = 0.$$

Consider the following example,

$$y'(t) = t^2(y(t) - 2)(y(t) - 4).$$

Then by solving for c

$$t(c - 2)(c - 4) = 0,$$

we find that the constant functions

$$y(t) = 2 \text{ and } y(t) = 4$$

are the equilibrium solutions of the differential equation. These equilibrium solutions are legitimate solutions of the differential equation. Furthermore, they are typically quite easy to find and of practical importance, but only if they are stable. On their own equilibrium solutions are not enough, and therefore, we need to introduce the concept of stable and unstable equilibria.

An equilibrium solution $y(t) = c$ of a differential equation $y'(t) = f(y(t), t)$ is *stable* if $\forall t$,

$$f(y, t) > 0 \text{ for } y \text{ just below } c \text{ and} \tag{2.14}$$

$$f(y, t) < 0 \text{ for } y \text{ just above } c \tag{2.15}$$

If $f(y, t)$ is entirely positive or negative in the neighbourhood of $y = c$, then the solution is semi-stable. Otherwise, the solution is *unstable*.

If the initial value at $t = 0$ is a slightly perturbed value of an unstable equilibrium, then the process will tend to move away from that value, whereas

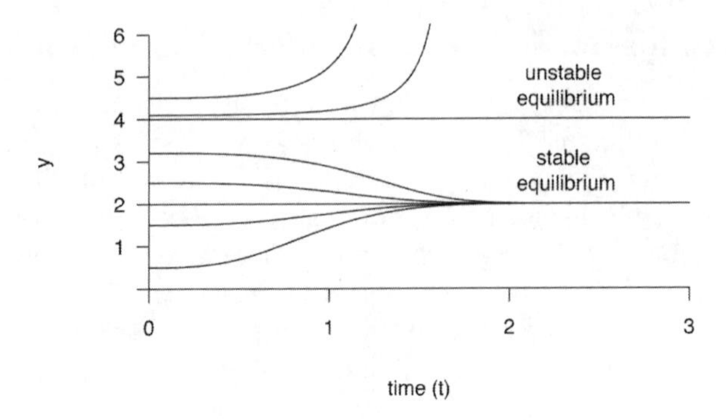

FIGURE 2.14
The solution $y(t) = 2$ is a stable equilibrium solution of the differential equation $y'(t) = t^2(y(t) - 2)(y(t) - 4)$, whereas $y(t) = 4$ is an unstable equilibrium solution.

for a stable equilibrium, the process will tend to move towards it over time. For the above example, for values of y close to 4, the function f is greater than zero above 4 and less than zero for values below 4. According to the definition, this is an *unstable equilibrium.* In Figure 2.14 it is clear that the solutions of the differential equation move away from $y = 0$. On the other hand, for values of y close to 2, the function f is negative above 2 and positive beneath 2. This means that this is a stable equilibrium solution. In the same figure, solutions close to $y = 2$ tend to converge to $y = 2$.

2.7.2 First-order equations with separable variables

In this section we focus on a particular form of a possible differential equation that has a, more or less, explicit solution.

If the differential equation can be written in the following form,

$$y'(t) = \frac{f_1(t)}{f_2(y)}, \tag{2.16}$$

then we call the expression a first-order differential equation with separable variables. We call a differential equation of first order if only the first derivative appears in the expression. The differential equation has an explicit solution,

$$y(t) = F_2^{-1}(F_1(t) + C), \tag{2.17}$$

as long as f_1 and f_2 are integrable functions with integrals F_1 and F_2.

We present an intuitive derivation,

$$
\begin{aligned}
\frac{dy}{dt} &= \frac{f_1(t)}{f_2(y)} \\
f_2(y)dy &= f_1(t)dt && \text{Moving } t \text{ and } y \text{ terms to either side} \\
\int f_2(y)dy &= \int f_1(t)dt + C && \text{Integrating both sides} \\
F_2(y) &= F_1(t) + C && \text{Which then leads to the required result} \quad (2.17)
\end{aligned}
$$

From this derivation, it now also becomes clear why this differential equation has *separable variables*. The ordinary differential equation in (2.16) is in fact often written in the following form,

$$f_2(y)dy - f_1(t)dt = 0,$$

which shows explicitly how it separates the function $f_2(y)$ and its differential dy from the function $f_1(t)$ and its differential dt.

Example 2.3 *Consider the following differential equation,*

$$\frac{dy}{dt} = ty(t) \text{ with initial value } y(0) = 2.$$

We first separate the variables, writing

$$\frac{1}{y}dy - tdt = 0.$$

By integrating both sides, we find

- *For the y variable,*

$$
\begin{aligned}
F_2(y) &= \int \frac{1}{y}dy \\
&= \log|y|.
\end{aligned}
$$

- *For the t variable,*

$$
\begin{aligned}
F_1(t) &= \int t \, dt \\
&= t^2/2
\end{aligned}
$$

- *For the right-hand side,*

$$\int 0 = C,$$

for any constant C

- *The general solution of the differential equation is given as,*

$$\log|y| = t^2/2 + C$$
$$y(t) = e^{t^2/2+C}$$

for any C.

- *Given the initial value $y(0) = 2$, we can find the value for C as*

$$2 = e^{0+C}$$
$$C = \log 2$$

For good measure, we consider another example, just to show that elegant explicit solutions are the exception, rather than the rule, even for separable first-order ordinary differential equations.

Example 2.4 *Consider the differential equation we studied before for its equilibrium solutions,*

$$\frac{dy}{dt} = t^2(y(t) - 2)(y(t) - 4),$$

that can be rewritten in separable form,

$$\frac{1}{(y-2)(y-4)}dy - t^2 dt = 0.$$

By integrating both sides, we find

- *For the y variable,*

$$\begin{aligned} F_2(y) &= \int \frac{1}{(y-2)(y-4)}dy \\ &= \int \frac{1}{2(y-4)} - \frac{1}{2(y-2)}dy \\ &= 0.5\log|y-4| - 0.5\log|y-2| \\ &= \frac{1}{2}\log\frac{|y-4|}{|y-2|}. \end{aligned}$$

- *For the t variable,*

$$\begin{aligned} F_1(t) &= \int t^2\, dt \\ &= t^3/3 \end{aligned}$$

- *For the right-hand side,*

$$\int 0 = C,$$

for any constant C

- *The general solution of the differential equation is given as,*

$$y(t) = F_2^{-1}(F_1(t) + C),$$

for any C*. Unfortunately, this solution is not explicit as the inverse of* $F_2(y) = \frac{1}{2}\log\frac{|y-4|}{|y-2|}$ *is not an explicit function.*

In many physical problems differential equations naturally appear. We will consider a number of them throughout the book. Here is a typical chemical engineering problem.

Example 2.5 *A tank contains 125 liters of brine whose salt concentration is 0.2kg per liter. Two liters of fresh water flow into the tank each minute, and at the same time 3 liters of the solution flow out of the tank. Assuming that the tank is stirred constantly to keep the mixture uniform, then calculate the salt content of the brine as a function of time.*

It is clear that the amount of solution in the tank is $(150 - t)kg$ *after* t *minutes. If* y *is the amount of salt in solution* t*, then* $3/(125 - t)$ *is the fraction of brine that is flowing out. Therefore, the change in salt content* dy *is reduced by* $y \times 3/(125 - t)$ *per time frame* dt*. This leads to the following differential equation,*

$$dy = -\frac{3ydt}{125 - t},$$

with $y(0) = 0.2 \times 150 = 25kg$*. By separating the variables, we obtain*

$$\frac{1}{3}\int \frac{1}{y}dy = \int \frac{1}{125 - t}dt,$$

which results into

$$y(t) = 5(125 - t)^{1/3}.$$

2.8 Complex Numbers and Functions

In this section we introduce the complex numbers and complex functions. Although complex numbers were originally introduced by Gerolamo Cardano (1501–1576) in his *Ars Magma* (1545) to simplify solving cubic equations, they have a much wider impact and application. They are used in spectroscopy and in nuclear chemistry, where the decay of instable nuclei can be modeled by a complex Hamiltonian. They also appear in chemical reaction theory, where a transition structure can be characterized by the presence of imaginary frequencies. But we begin with the basics.

Fundamentally, complex numbers are needed to be able to solve the "problem,"

$$x^2 = -1,$$

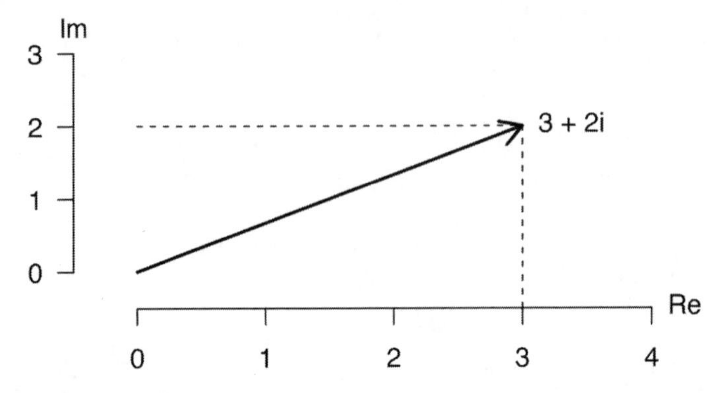

FIGURE 2.15
The complex numbers $\mathbb{C}$ can be represented by a two-dimensional plane, in which the x-axis represents the real part and the y-axis represents the imaginary part. A complex number is a vector in this plane.

for some number x. For the longest time, this was not considered a problem. It was simply impossible, because if you multiply a number with itself, you get something non-negative. However, as Cardano discovered, by pretending that this problem has a solution, it was possible to simplify the solutions of a cubic equation dramatically. It was Leonard Euler (1707–1783) who introduced the notation,

$$i = \sqrt{-1},$$

which still forms the basis for complex numbers.

Definition 2.7 *A* **complex number** z *is defined as,*

$$z = a + bi,$$

with $a, b \in \mathbb{R}$ *and* $i^2 = -1$. *The collection of all complex numbers is indicated by* $\mathbb{C}$. *A complex number consists of*

- *a real part:* $\Re z = a$, *and*
- *an imaginary part:* $\Im z = b$.

A common way to represent the complex numbers $\mathbb{C}$ is by means of the complex plane, which is a two-dimensional plane with a real x-axis and an imaginary y-axis, as shown in Figure 2.15.

In section 2.3.4 we saw how we could transform vectors from their Euclidean representation into the polar coordinate representation. The same idea is used to write complex numbers in another, convenient form. If $\theta = \arctan(b/a)$ is the angle of the complex number z and $r = \sqrt{a^2 + b^2}$ its length, then $z = a + bi$ can be written as

$$z = r(\cos\theta + i\sin\theta),$$

which will be represented as the complex exponential

$$z = re^{i\theta}.$$

This representation is particularly convenient when multiplying two complex numbers $z_j = r_j e^{i\theta_j}$ $(j = 1, 2)$,

$$z_1 z_2 = r_1 r_2 e^{i(\theta_1 + \theta_2)}.$$

It is fairly straightforward to see that if $f(x) = e^{tx}$ for x real and t complex, then $f'(x) = te^{tx}$ and $\int f(x)dx = e^{tx}/t$ just as for t real.

2.9 Exercises

1. Consider the vectors $x_1 = (1, \sqrt{3})^t$, $x_2 = (\sqrt{3}, 1)^t$ and $x_1 = (-1, \sqrt{3})^t$.
   (a) Calculate the sum of x_1 and x_2. What kind of object is the sum? In what space does it lie?
   (b) What is the norm of x_1?
   (c) Calculate the inner product between x_1 and x_2 and between x_2 and x_3. Determine the angle between x_1 and x_2 and the angle between x_2 and x_3.
   (d) For x_1 and x_2 check that the triangle inequality holds.
   (e) Calculate the matrix by multiplying x_1 with the transpose of x_2, i.e., $x_1 x_2^t$. What happens if instead you multiply the transpose of x_1 with x_2, i.e., $x_1^t x_2$.
2. Consider the function $f(x) = \sqrt{x}$ on the domain $D_f = [0, 4]$.
   (a) Determine the range Im_f of f.
   (b) Is the function f convex, concave or neither of the two?
   (c) Determine whether the function f is continuous.
   (d) What function would you obtain if you were to compose f with itself? Determine its domain and its range.
   (e) Determine the inverse function f^{-1} of f. Use the inverse function to "solve" the equation $f(x) = \sqrt{3}$.
3. Give an example of a real-world function that requires more than one input.
4. Give the domain and range of the following multivariate functions:
   (a) $f(x, y) = x^2 + 2y^2 + 3$
   (b) $f(x, y) = 1 - 2y$

(c) $f(x, y) = \frac{2}{x-y}$
(d) $f(x, y) = \sin^2 xy + \cos^2 xy$
(e) $f(x, y) = \begin{bmatrix} 2x + y \\ y - x \end{bmatrix}$
(f) $f(x, y, z) = \begin{bmatrix} 2x + y - z \\ y - x \\ x + z \end{bmatrix}$

5. The last two functions given in the previous exercise are special cases of linear transformations,

$$f(x) = Ax.$$

(a) Determine the matrices A for both functions.
(b) Determine the standard 2-by-2 inverse for the matrix A_1 belonging to the function $f(x, y) = \begin{bmatrix} 2x + y \\ y - x \end{bmatrix}$
(c) Use Gauss-Jordan elimination to determine the inverse for the matrix A_2 that belongs to the function $f(x, y, z) = \begin{bmatrix} 2x + y - z \\ y - x \\ x + z \end{bmatrix}$
(d) Determine the eigendecomposition of A_1 and A_2 by using your software of choice.

6. **Eigendecomposition.** Show that for a symmetric matrix $p \times p$ A with p linearly independent eigenvectors $u_1, \ldots, u_p$ of length 1, the following holds for the eigenvector matrix $U = [u_1, \ldots, u_p]$,

$$U^{-1} = U^t.$$

7. Determine whether the following sequences converge or diverge and if they converge to what value.

(a) $\lim_{n \to \infty}(1 + \log x/n)^n$

8. Use the sequence $(1 + x/n)^n \longrightarrow e^x$ to prove with the help of the definition of a derivative that the derivative of e^x is itself.

9. **Differentiation.** Determine the derivatives of the following functions

(a) $f(x) = 2x \log(\cos(x))$.
(b) $g(x) = \log(\sin(x)) - x \tan(x - 2) + \frac{4}{\sqrt{x}}$.
(c) $h(t) = \frac{t^2 \cos(x)}{te^{2t}}$.

10. **Gradients and Hessians.**

    (a) Calculate the gradient and the Hessian of the function $f(x, y) = fracsin\, y1 + x^2$.

11. **Taylor series.**

    (a) Consider the function $f(x) = \sin(x^2)$.

    i. Determine the first two derivatives of f.
    ii. Determine the second-order Taylor approximation of the function f around the point $x = \sqrt{\pi}$.
    iii. Determine the limit $\lim_{x \to \sqrt{(\pi)}} \frac{\sin(x^2)}{x - \sqrt{\pi}}$.

    (b) Determine the second-order Taylor series of the function $f(x) = \log(x^2 + 1)$ around the point $x = 0$ and find the limit $\lim_{x \to 0} \frac{\log(x^2+1)}{x^2}$.

12. **Extremal points.** Determine the location of the maximum of the function $f(x) = \frac{x^2 - x + 3}{x^2 + 1}$.

13. **Integration.**

    (a) Calculate the following definite integrals

    i. $\int_{-1}^{0} \frac{2}{3x-2}\, dx$.
    ii. $\int_{0}^{\pi/8} \frac{4}{\cos^2(2x)} dx$.

    (b) Determinate the following indefinite integrals

    i. $\int x^2 e^{3x}\, dx$.
    ii. $\int x^2 \ln(x)\, dx$.
    iii. $\int \frac{1}{x(x+1)}\, dx$.

14. **Bivariate function.** Consider the following bivariate function, functie,

$$f(x, y) = x + y^2.$$

    (a) Calculate the gradient of f in the point $(1, 1)$.
    (b) Calculate the tangent plane to the function f in the point $(1, 1, 2)$.
    (c) Calculate the volume under the function that lies above the following triangle in the (x, y) plane.

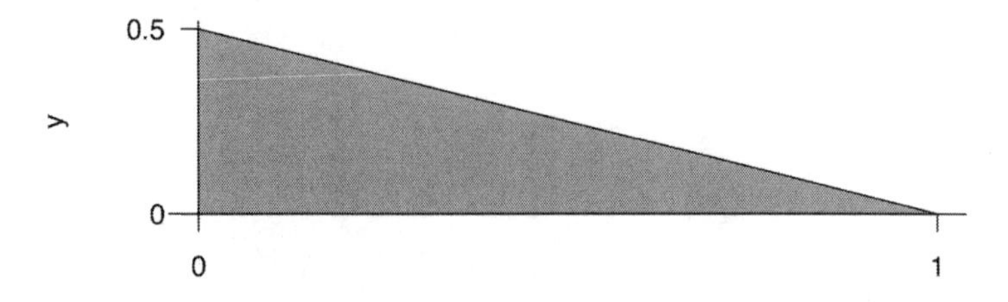

15. **Stationary point of a function with two variables.** Given the function,

$$f(x, y) = -x^2 + 2xy - 2y^2 - 4x - 10y - 4.$$

calculate the stationary point of f and determine whether it is a minimum, a maximum or a saddle point.

16. **Stability of a differential equation**. Consider the following differential equation.

$$y'(t) = 2y(t) + 4$$

(a) Determine the equilibrium for the above differential equation.
(b) Solve the differential equation with initial value $y(0) = -2$.
(c) Solve the differential equation with initial value $y(0) = 1$.

17. **Differential equation**. Consider the following ordinary differential equation

$$y'(t) = -2ty^2(t).$$

Solve the differential equation with the initial value $y(0) = \frac{1}{2}$.

18. **Differential equation**. Solve the following initial value problem:

$$\begin{cases} y'(t) = -\frac{y(t)}{t^2} - \frac{1}{t^3} \\ y(1) = 0 \end{cases}$$

19. **Second-order differential equation**. Give de real valued solution of the following differential equation with initial values,

$$2x''(t) + 4x'(t) + 4x(t) = 0, \quad x(0) = 2, \quad x'(0) = 1.$$

3

Probability Essentials

This book will make use of probability calculus throughout, as statistics is sometimes described as inverse probability. In order to make the text self-contained, we will introduce in this chapter the fundamental probability concepts and notation, as well as the standard techniques to calculate with probabilities.

We begin with describing the fundamental properties of sets, which in probability theory are called events. A probability is a way to measure the size of these event sets. By defining a function on the elements of the sets, it is possible to define random variables, which are the working horses of probability theory and particularly useful for the applied scientist. For example, measurements of chemical quantities in an experiment can be represented by random variables that attain values with a certain probability.

In this chapter you learn

- In Section 3.1 about events and probability of events.
- In Section 3.2 about random variables, and their properties, such as means and variances.
- In Section 3.2.10 about an important class of random variables, called the exponential family, which has many favorable properties.
- In Section 3.3 about generating pseudo-random measurements that form the basis for simulations of chemical experiments.

3.1 Probability of Events

Events happen or do not happen. However, the event that a die lands on an even number is more likely than that it lands on a six. Probability is a way to measure events in terms of how likely it is to happen. To make this precise, we need to define *events* mathematically. We do this by means of set theory.

DOI: 10.1201/9781003178194-3

3.1.1 Basic set theory

A (random) experiment can have a number of outcomes. The set of all possible outcomes Ω is defined as the **sample space**. For example, when rolling a die, the sample space is given as $\Omega_{\text{die}} = \{1, 2, 3, 4, 5, 6\}$, whereas when testing the concentration of sodium hypochloride (NAOCl) in water, the outcome of the test is a value inside $\Omega_{\text{NaOCl}} = [0, 1]$.

An element $\omega \in \Omega$ is typically referred to as an **elementary outcome**. For example, household bleach typically has a sodium hypochlorite concentration around 0.0525, but still safe levels of NAOCl in water are around hundred times lower, at 0.0005. Both of these values are possible elementary outcomes in a safety test of drinking water. An **event** A is defined as a subset of the sample space, i.e., $A \subset \Omega$. For example, safe levels of NAOCl in drinking water correspond to the event $A = [0, 0.0005]$.

If the event of interest A has no elements, then we write $A = \emptyset$. If A has only two elements 1/2 and 1/3, then we write $A = \{1/2, 1/3\}$. The **complement** of A^c of A consists of all ω not in A, i.e., $A^c = \{w \in \Omega \mid \omega \notin A\}$. The **intersection** $A \cap B$ is defined as the set with elements both in A and in B. The **union** $A \cup B$ is defined as the set with elements either in A or in B, or in both. Taking the complement, intersection, or union are the basic set operations. These operations can be combined via the **Distributive Law**

$$(A \cup B) \cap C = (A \cap C) \cup (B \cap C)$$

$$(A \cap B) \cup C = (A \cup C) \cap (B \cup C)$$

and **DeMorgan's Law** :

$$(A \cup B)^c = A^c \cap B^c$$

$$(A \cap B)^c = A^c \cup B^c.$$

Example 3.1 *Let $\Omega = [0, 1]$ denote all real numbers between zero and one. Two subsets that represent segments are $A_1 = (0, 1/2)$, $A_2 = (1/2, 1]$. The first segment is open to the left and to the right and the second is open to the left and closed to the right. Then,*

$$\begin{aligned} A_1 \cap A_2 &= \emptyset \\ 0 &\in A_1^c,\ 1/2 \in A_1^c,\ \{0, 1/2\} \subset A_1^c \\ (A_1 \cup A_2)^c &= \{0, 1/2\} \end{aligned}$$

Example 3.2 *Let $\Omega = [0, 1]$, $A_1 = (0, 1/2)$, and $A_2 = [1/3, 2/3]$. Then, we have*

$$A_1^c = \{0\} \cup [1/2, 1], \quad A_2^c = [0, 1/3) \cup (2/3, 1].$$

$$A_1 \cap A_2 = [1/3, 1/2), \quad A_1 \cup A_2 = (0, 2/3].$$

$$(A_1 \cup A_2)^c = A_1^c \cap A_2^c = \{0\} \cup (2/3, 1]$$

$$(A_1 \cap A_2)^c = A_1^c \cup A_2^c = [0, 1/3) \cup [1/2, 1]$$

A **partition** of $\Omega = [0, 1]$ is a collection of sets $A_1, \cdots, A_n$ that have empty intersections $A_i \bigcap A_j = \emptyset$ for $i \neq j$, and Ω as its union; $\bigcup_{i=1}^{n} A_i = \Omega$.

Example 3.3 *If* $A = [0, 1/2)$, *then* $A^c = [1/2, 1]$, *so that* $A \cup A^c = \Omega$ *and the collection* $\{A, A^c\}$ *is a partition of* $\Omega = [0, 1]$.

3.1.2 Laplace's definition of probability

Marquis Pierre-Simon Laplace (1749–1827) was a polymath who also contributed to probability and statistics. In 1779, he gives an important definition of a probability of an event as

> *the ratio of the number of cases favorable to it, to the number of all cases possible when nothing leads us to expect that any one of these cases should occur more than any other, which renders them, for us, equally possible.*

or in a more concise mathematical formulation,

$$P(A) = \frac{|A|}{|\Omega|} = \frac{\text{no. of elementary outcomes in } A}{\text{no. of elementary outcomes in } \Omega}.$$

The definition crucially assumes that each of the elementary outcomes is equally likely. Although this is assumption sounds restrictive, it is sometimes possible to define the sample space and its elementary outcomes in such a way that they are. Nevertheless, the definition is not fully general and next section considers a more general definition. Despite its drawbacks, Laplace definition is the most intuitive introduction into the concept of probability.

Example 3.4 Laplace's definition of probability. *Consider the sufficiently chaotic roll of a perfectly symmetrical die. In this case it seems reasonable to assume that the elementary outcomes are more or less equally likely. Therefore, the probability that the roll lands even is*

$$P(\textit{die lands even}) = \frac{3}{6} = 0.5.$$

It is clear that Laplace definition of probability requires us to be able to count the number of elementary outcomes in the event and in the sample space. Only in the most trivial cases, such as the die example above, it is possible to use simple enumeration. In most cases we need to use *counting principles* to calculate the number of elementary outcomes. We distinguish three counting rules, whereby each rule is a specific application of the previous one.

The **multiplication rule** states that if an experiment consists of k subexperiments, each with n_i mutually compatible outcomes ($i = 1, \ldots, k$), then the total number of outcomes N is

$$N = n_1 \times n_2 \times \ldots \times n_k.$$

Example 3.5 Multiplication rule. *Consider game in which a blindfolded player roles a die, flips a coin, and picks a card from a deck of cards. The player will win a prize if the die lands even, the coin shows tails, and the card is an ace. What is the probability that the player wins the prize?*

It is clear that both the sample space Ω *and the event* A *can be seen as a concatination of three subspaces* $\Omega = \Omega_1 \times \Omega_2 \times \Omega_3$ *and subevents* $A = A_1 \times A_2 \times A_3$*, respectively, where*

$$\begin{aligned}
\Omega_1 &= \{1, 2, 3, 4, 5, 6\} \\
A_1 &= \{2, 4, 6\} \\
\Omega_2 &= \{H, T\} \\
A_2 &= \{T\} \\
\Omega_3 &= \{H2, \ldots, HA, C2, \ldots, CA, D2, \ldots, SA\} \\
A_3 &= \{HA, CA, DA, SA\}
\end{aligned}$$

Using the multiplication rule both for the numerator and the denominator, we obtain

$$P(A) = \frac{3 \times 1 \times 4}{6 \times 2 \times 52} = \frac{12}{624} = 0.019.$$

Example 3.6 Birthday paradox (1). *Consider a class with 60 students. Assume that all of the students were born in a non-leap year. How many arrangements of the birthdays are possible?*

One way to think about this problem is by seeing this as the repitition of the same experiment 60 times. Given that each experiment has 365 outcomes, the total number of outcomes N *is given as*

$$N = 365^{60} = 5.46 \times 10^{153}.$$

This brings us to the second rule. The rule of **permutations** states that if one considers shuffling n distinct objects, then there are

$$n! = n \times (n - 1) \times \ldots \times 2 \times 1$$

possible arrangements. One can think of a permutation as n experiments in which the choice of the number of possible arrangements each time decreases by one. The permutation rule is therefore an application of the multiplication rule.

Example 3.7 Permutation. *A system consists of seven molecules operating in some hierarchical order, numbered from 1 to 7. A catalyst is added to the system and a chemical reaction sets in. The molecules line up. If in the line-up two or more molecules of consecutive increasing order are next to each other, then they bond together and become one. What is the probability that after only one addition of the catalyst all molecules bond together?*

The sample space consists of all possible permutations of the hierarchical orders 1 to 7. The event of interest consists of the unique ordering 1-2-3-4-5-6-7. Therefore,

$$P(\textit{all molecules bond}) = \frac{1}{7!} = 0.0002.$$

Example 3.8 Birthday paradox (2). *A class of 60 students were all born in non-leap years. What is the probability that they all have different birthdays?*

The event that all 60 students have different birthdays consists of 60 subevents, where the first has 365 possibilities, the second 364, and so on. Combined with the size of the sample space calculated in example 3.6, we get a probability of

$$P(\textit{60 different birthdays}) = \frac{365!/305!}{365^{60}} = 0.0059.$$

Interestingly, it is extremely likely to have at least one matching birthday in a class of 60 students. This is sometimes known as the birthday paradox.

This brings us at the third counting principle, **combinations**. Consider a number of n objects, consisting of two groups of identical types, of size k and $n-k$, respectively. The total number N of different configurations is given by,

$$N = \frac{n!}{k!(n-k)!} = \binom{n}{k}.$$

Example 3.9 Combinations. *An important chemical reaction that takes place in plants is called photosynthesis. It converts carbon dioxide and water into plant food, i.e., glucose and oxygen, according to the following reaction,*

$$6CO_2 + 6H_2O + \textit{light} \longrightarrow C_6H_{12}O_6 + 6O_2.$$

The speed of a reaction is proportional to the number of ways the reaction can occur. If a certain part of a leave exposed to light contains 1000 molecules of carbon dioxide and 200 molecules of water, then in how many ways can the photosynthetic reaction occur?

Among the 1000 molecules of carbon dioxide and 200 molecules of water, there are $\binom{1000}{6}$ resp. $\binom{200}{6}$ possible ways of selecting different sets of six carbon dioxide and water molecules, according to the number of combinations. Then using the multiplication rule, there are in total

$$N = \binom{1000}{6}\binom{200}{6} = 1.13 \times 10^{26}$$

different ways a photosynthesis reaction can occur.

We conclude this section by returning to the Laplace's definition of probability. There is another way of looking at this definition. Although it is not particularly helpful for calculating practical probabilities – and, therefore, we have not used it so far – it is helpful to for defining an extension of the definition of probability in the section below.

Definition 3.1 Reformulation of Laplace's definition of probability. *Consider a sample space Ω with equally likely outcomes, then the probability of an event $A \subset \Omega$ is given as*

$$P(A) = \sum_{\omega \in \Omega} I_A(\omega) \times \frac{1}{|\Omega|},$$

where $|\Omega|$ is the number of elementary outcomes in the set Ω, and I_A is the indicator function on A,

$$I_A(\omega) = \begin{cases} 1 \text{ if } \omega \in A \\ 0 \text{ if } \omega \notin A \end{cases}$$

The indicator function will play an important role throughout the book.

3.1.3 General definition of probability

Unfortunately, the concept of probability of Laplace has its limitations. Most crucially, it is only possible to define a probability on finite sample spaces and only with respect to examples in which some kind of physical symmetry exists. For more general scenarios, we require a more careful mapping P from an event to a probability value in $[0, 1]$.

First, it turns out that, in general, it is impossible to define a consistent probability for an arbitrary subset of a the sample space. Instead, we focus on certain collections $\mathcal{F}$ of events, called *sigma algebra* or *sigma field*, that satisfy certain conditions.

Definition 3.2 *$\mathcal{F}$ is a* **sigma field** *in Ω if it satisfies the following conditions*

1. *$\Omega \in \mathcal{F}$*
2. *If $A \in \mathcal{F}$, then its complement $A^c \in \mathcal{F}$*
3. *If $A_n \in \mathcal{F}$, for all $n = 1, \cdots, \infty$, then this also holds for its union $\bigcup_{n=1}^{\infty} A_n \in \mathcal{F}$.*

Note that any $A \in \mathcal{F}$ is a subset of the largest set Ω in $\mathcal{F}$.

Example 3.10 *The sigma field associated with a set A is the collection that follows from its complement, the union $A \cup A^c = \Omega$ and $\Omega^c = \emptyset$. This yields the sigma field generated by the set A as the collection*

$$\sigma(A) = \{A, A^c, \Omega, \emptyset\}$$

Example 3.11 *If $A, B \in \mathcal{F}$, then $A^c, B^c \in \mathcal{F}$. The latter implies by DeMorgan's Law that*

$$A \cap B = (A^c \cup B^c)^c \in \mathcal{F}.$$

Indeed the above requirement 3 can be replaced by: If $A_n \in \mathcal{F}$, for all $n = 1, \cdots, \infty$, then this also holds for its intersection $\bigcap_{n=1}^{\infty} A_n \in \mathcal{F}$.

If we consider $\mathcal{I}$ the collection of all intervals from $\Omega = [0,1]$, then the sigma field $\mathcal{B} = \sigma(\mathcal{I})$ generates the so-called **Borel Sets**. These include, e.g., $(0,x]$ for all $x \in [0,1]$ but also singletons, since

$$\{x\} = \cap_{n=1}^{\infty}(x - 1/n, x]$$

The Borel sets are particularly useful in practice, and we will return to them in section 3.2. Now that we have defined the concept of a sigma field, it is now possible to extend Laplace's definition of probability to the general case.

As a formal basis, we need the concept of a probability space $(\Omega, \mathcal{F}, P)$, where $\mathcal{F}$ is a sigma field of sets which are in the sample space Ω. Sets in the sigma field are conceived as events that occur with a certain probability.

Definition 3.3 *A* **probability space** $(\Omega, \mathcal{F}, P)$ *consists of a sample space* Ω*, an associated sigma field* $\mathcal{F}$ *and a set function* $P : \mathcal{F} \to [0,1]$ *that satisfies the following conditions*

1. $0 \leq P(A) \leq P(\Omega) = 1$ *for any* $A \in \mathcal{F}$,
2. *If* $A_1, A_2, \ldots$ *are disjoint set in* $\mathcal{F}$*, then*

$$P\left(\bigcup_{n=1}^{\infty} A_n\right) = \sum_{n=1}^{\infty} P(A_n)$$

We refer to P *as a* **probability measure**.

This definition, due to the Russian mathematician Kolmogorov (1903–1987), is less specific than Laplace's definition. In fact, by focusing on the essentials, it can be extended to infinite sample spaces, in which the elementary outcomes are not equally probable. The previous definition has a number of consequences, which shows that it indeed behaves like a probability.

Example 3.12 *If* $A \in \mathcal{F}$*, then* $A^c \in \mathcal{F}$ *and, therefore,*

$$1 = P(\Omega) = P(A \cup A^c) = P(A) + P(A^c).$$

Consequently, we have that $P(A^c) = 1 - P(A)$.

Example 3.13 *For any two events* $A, B \in \mathcal{F}$ *we have*

$$P(A \cup B) = P(A) + P(B) - P(A \cap B).$$

Since $A \cup B$ *can be written as the disjoint union of* $A \cap B^c$, $A \cap B$*, and* $A^c \cap B$*, we have*

$$P(A \cup B) = P(A \cap B^c) + P(A \cap B) + P(A^c \cap B).$$

Combining this with $P(A) = P(A \cap B^c) + P(A \cap B)$ *and* $P(B) = P(A \cap B) + P(A^c \cap B)$*, we get the required result.*

Example 3.14 Uniform(0,1) distribution. *The concepts of the size of a set, probability of an event, and the integral become exactly equal in the following example, which is generally referred to as a* uniform *distribution. Suppose that* $A = [0, x) \subset \Omega = [0, 1]$ *and that each* $\omega \in \Omega$ *is "equally likely." One way to translate the concept of equally likely for an infinitely uncountable set of values into a mathematically rigorous definition, by defining the probability of* A *as its size relative to that of* Ω. *Since the size of the unit interval equals one, we obtain* $P([0, x)) = x/1 = x$. *More formally we define the probability of event* A *as the integral over the indicator function*

$$P(A) = \int_{\Omega} I_A(\omega) d\omega = \int_A d\omega = \int_0^x d\omega = x.$$

Similarly, for $B = [a, b] \subset \Omega = [0, 1]$, *we have*

$$P(B) = \int_{\Omega} I_B(\omega) d\omega = \int_a^b d\omega = b - a.$$

Note that for $B = (a, b)$, *we would have obtained the same probability. The existence of a function that integrates to the probability of the set is particular useful. We refer to it as the* probability density function.

The Uniform(0,1) distribution is very useful for simulation. If somehow are able to draw samples from a Uniform distribution, it is possible to draw samples from any other distribution.

Example 3.15 *If* $\Omega = [0, 1]$, $A_i = [i/n, (i + 1)/n)$, *for* $i = 0, \cdots, n - 1$, *then* $P(A_i) = 1/n$ *for all* i. *For example, if we take* $n = 6$, *then we obtain the probabilities of a fair dice.*

The concept of probability that we have defined so far is somewhat static. In chemical engineering we are often interested in being able to say how the probability of a particular outcome changes, when we change the experimental setting. This means that we want to define the concept of a *conditional probability.*

Definition 3.4 *Consider a probability space* $(\Omega, \mathcal{F}, P)$. *Then for any* $B \in \mathcal{F}$ *the triplet* $(\Omega, \mathcal{F}, P(\ |B))$ *is a valid probability space, where*

$$P(A|B) = \frac{P(A \cap B)}{P(B)}$$

is referred to as the **conditional probability** *of* A *given* B.

Example 3.16 *Suppose that* $A_1 \cdots, A_n$ *is a partition of* Ω, *and that all* $P(A_i)$ *and all* $P(B|A_i)$ *are known, then the unknown* $P(B)$ *can be computed from*

$$P(B) = \sum_{i=1}^{n} P(A_i) \cdot P(B|A_i)$$

Theorem 3.1 Bayes theorem. *If for $i = 1, \ldots, n$ the probabilities $P(B|A_i)$ and $P(A_i)$ are known, then the unknown $P(B|A)$ can be computed from*

$$P(A_i|B) = \frac{P(B|A_i) \cdot P(A_i)}{\sum_{i=1}^{n} P(B|A_i)P(A_i)}.$$

This is referred to as Bayes Theorem, named after the 18th century English Presbyterian minister Thomas Bayes (1701–1761).

Bayes theorem is practical tool for calculating probabilities in complex settings, by breaking the problem down in often simpler probabilities. Moreover, it has become the corner stone of modern epistemological theory about how to continuously incorporate evidence in a coherent manner.

Example 3.17 Bayes theorem. *Consider a particular rare disease D that is not easily diagnosable. Molecular diagnostic research aims to devise a test to detect the disease. When given to a person, the result can either be positive $+$, indicating the presumed presence of D, or negative $-$ indicating presumed absence of D. The effectiveness of the test is given in terms of two conditional probabilities. The* **sensitivity** *$P(+|D)$ is the probability of a positive test, given that the person has the disease. The* **specificity** *$P(-|D^c)$ is the probability of a negative test, given that the person is healthy. These values are typically ascertained under laboratory conditions, say,*

$$P(+|D) = 0.99, \quad P(-|D^c) = 0.95.$$

From the perspective of the patient, however, the predictive probability is important. The **negative predictive value** *$P(H|-)$ is the probability of being healthy, given a negative test. The* **positive predictive value** *$P(D|+)$ refers to the probability of having the disease, given a positive test. Assume that the test is given to a particular population and that within that population the proportion of people having the disease at a certain moment is given by its prevalence $P(D)$, say*

$$P(D) = 0.01,$$

as it is a rare disease. Then Bayes theorem can be used to calculate the positive predictive probability,

$$P(D|+) = \frac{P(+|D)P(D)}{P(+|D)P(D) + P(+|D^c)P(D^c)} = \frac{0.99 \times 0.01}{0.99 \times 0.01 + 0.05 \times 0.99} = 0.17,$$

only 17%! This is the reason why population-wide scanning programs for rare diseases can result in many false-positives.

3.1.4 Independence

Often we can simplify the calculation of the probability of an event by making use of the concept of independence. Typically, two events are independent if they are defined on two physically separated systems. For example, if we draw a card from a well-shuffled deck of card and flip a coin, then the events $A = \{\text{Ace}\}$ and $B = \{\text{Head}\}$ are physically non-interfering and therefore can assumed to be independent. This idea is a bit vague, although often useful in practice, and so we provide a formal definition of independence of two events.

Definition 3.5 *Two events A and B are* **independent** *if*

$$P(A \cap B) = P(A) \cdot P(B) \tag{3.1}$$

If A and B are independent, then the definition holds for all sets generated by the sigma field of $\{A, B\}$, that is, we have

$$\begin{aligned} P(A \cap B^c) &= P(A) \cdot P(B^c) \\ P(A^c \cap B^c) &= P(A^c) \cdot P(B^c) \\ P(A^c \cap B^c) &= P(A^c) \cdot P(B^c) \end{aligned}$$

If A and B are independent, then conditioning on event B makes no difference for the probability of event A to happen. In particular, if $P(B) > 0$, then

$$P(A|B) = \frac{P(A \cap B)}{P(B)} = \frac{P(A) \cdot P(B)}{P(B)} = P(A),$$

i.e., event B has no influence on the probability for A.

Example 3.18 Physical separation. *Consider drawing a card from a deck and flipping a fair coin, then the events $A = \{Ace\}$ and $B = \{Head\}$ can, under normal circumstances (i.e., no magician is involved or any other connection between the deck and the coin), assumed to be independent. Therefore,*

$$P(A \cap B) = P(A)P(B) = \frac{4}{52}\frac{1}{2} = 0.038$$

Although we will often invoke independence for situations of physically separated systems, like the one considered in the previous example, it is interesting to note that the concept of independence is broader than that.

Example 3.19 No physical separation. *Consider the events $A = \{even\}$ and $B = \{more\ than\ 2\}$ when rolling a die. Using Laplace's definition of probability, it is easy to check that*

$$\begin{aligned} P(A \cap B) &= P(\{4, 6\}) \\ &= \frac{2}{6} \\ &= \frac{1}{2}\frac{4}{6} \\ &= P(A)P(B). \end{aligned}$$

So, apparently, events A and B are independent, even though they are not physically separated.

Finally, we want to sound a word of warning as the concept of independent events is sometimes confused with that of disjoint events. Note that in the two examples we considered above the intersections of the independent events were not empty. Furthermore, if two events are disjoint and they have non-zero probability, then they are not independent: if one of the event happens, the other one can certainly not happen.

The definition of **independence** can be generalized to collections of events. That is, the events $A_1, A_2, \ldots$ in $\mathcal{F}$ are independent if for every finite subset $j_1, \cdots, j_n$ we have

$$P\left(\cap_{i=1}^{n} A_{j_i}\right) = \prod_{i=1}^{n} P(A_{j_i}).$$

3.2 Random Variables

In this section we introduce the work-horse of probability: the random variable. Random variable are used to select or define what part of the random experiment is interesting for the chemical engineer. We derive various useful properties of random variables and introduce various methods for practical computation.

3.2.1 Definition of random variables

Definition 3.6 *A* **random variable** X *is a real valued function defined on the sample space,*

$$X : \Omega \to \mathbb{R},$$

such that its inverse image $X^{-1}(B) \in \mathcal{F}$ *for all Borel sets* B *in* $\mathbb{R}$.

Example 3.20 Bernoulli random variable. *Let* $\Omega = [0, 1]$ *and* $A \subset \Omega$. *Then a simple random variable is defined by the indicator function*

$$X(\omega) = I_A(\omega).$$

It attains the value one with probability $P(X = 1) = P(A)$ *and the value zero with probability* $P(X = 0) = P(A^c) = 1 - P(A)$. *This is called the* **Bernoulli** *random variable .*

We can generalize this idea considerably. The following example shows a generalization of a Bernoulli random variable to an arbitrary discrete distribution.

Example 3.21 Simple random variable. *Suppose that* $A_1, \cdots, A_n$ *is a partition of* $\Omega = [0, 1]$ *in the sense that the intersections* $A_i \bigcap A_j = \emptyset$ *for* $i \neq j$

and $\bigcup_{i=1}^{n} A_i = [0,1]$, *where* n *is infinite is allowed. Then for any sequence of distinct* $\{x_1, \cdots, x_n\}$ *of real numbers, we have a random variable*

$$X(\omega) = \sum_{i=1}^{n} x_i I_{A_i}(\omega).$$

This defines all discrete random variables that attain integer values x_i *with probability*

$$P(X(\omega) = x_i) = P(A_i),$$

for all $i = 1, \cdots, n$. *When* n *is finite, then the random variable* X *is called simple.*

Practically speaking, random variables are often defined on a specific sample space with a specific purpose in mind.

Example 3.22 Sum of two four-sided dice. *Consider rolling two four-sided dice. The sample space consists of 16 equally likely outcomes,*

$$\Omega = \{(\omega_1, \omega_2) | \omega_1, \omega_2 = 1, 2, 3, 4\}.$$

Consider the random variable defined as the sum of the two dice,

$$X(\omega) = \omega_1 + \omega_2.$$

The distribution of X *can be read off from the table below.*

X		ω_2			
		1	2	3	4
ω_1	1	2	3	4	5
	2	3	4	5	6
	3	4	5	6	7
	4	5	6	7	8

In particular, $P(X = 2) = P(X = 8) = 1/16$, $P(X = 3) = P(X = 7) = 2/16$, $P(X = 4) = P(X = 6) = 3/16$ *and finally* $P(X = 5) = 4/16$.

3.2.2 Distribution functions

In order to work efficiently with random variables, we typically do not need to refer to the underlying sample space. Instead, we will typically make use of two functions in most practical calculations. These are the cumulative distribution function and the probability density function.

Definition 3.7 *The* **cumulative distribution function** *(CDF)* F_X *indicates the probability of the event that* X *attains smaller values than* x,

$$\begin{aligned} F_X(x) &= P(\{\omega \in \Omega : X(\omega) \leq x\}) \\ &= P(X \leq x). \end{aligned}$$

We often drop the subscript X in the notation of the CDF, when it is sufficiently clear from the context which random variable we are considering. It follows from the definition of probability that the distribution function F is non-decreasing

$$F(y) \leq F(x), \quad \text{for any } y < x,$$

that it has a left limit

$$\lim_{y \uparrow x} F(y) = F(x-),$$

and that it is continuous from the right

$$\lim_{y \downarrow x} F(y) = F(x).$$

Definition 3.8 *If the derivative of* F_X *exists, then we define the* **probability density function** f_X *as the non-negative function*

$$f_X(x) = F'_X(x).$$

In this case we call the random variable absolutely continuous.

If the random variable X *can take on countably many values, then the derivative does not exist and the* **probability mass function** *is defined as,*

$$f_X(x) = F_X(x) - F_X(x-).$$

In this case we call the random variable discrete.

The previous enumeration is not exhaustive. It could be possible for a random variable to be partially discrete and partially continuous. We will not consider these cases in this book.

Example 3.23 Uniform distribution. *For* $\Omega = [0,1]$ *we may define* $X(\omega) = I_{[0,x)}(\omega)$, *so that for* $x \in [0,1]$

$$F_X(x) = P(\omega : X(\omega) \leq x) = P(\omega \in [0,x)) = x.$$

This X *is the uniformly distributed random variable on domain* $[0,1]$, *where any outcome* x *is equally likely. From* $F(x) = x$, *we immediately find the density function*

$$f_X(x) = \begin{cases} 1 & 0 \leq x \leq 1 \\ 0 & \textit{elsewhere} \end{cases}$$

Note that Ω does not play an explicit role in the previous two examples. Once the distribution function of a random variable is given, there is no need to refer back to the original samples space anymore.

There are some useful properties of the probability density and mass functions f_X. In particular, for any set $B \subset \mathbb{R}$, the probability that the random variable X lies inside B is equal to the integral or sum, respectively, over the set B of f_X,

$$P(X \in B) = \begin{cases} \int_B f_X(x) & \text{if } X \text{ continuous} \\ \sum_{x \in B} f_X(x)\; dx & \text{if } X \text{ discrete,} \end{cases}.$$

From this it follows naturally that $P(X \in \mathbb{R}) = 1$. The density function can also be used to derive the distributions of transformations $g(X)$ of the original random variable X.

Theorem 3.2 Change of variable formula. *Consider an absolute continuous random variable X with density X and a monotone, differentiable function $g : \mathbb{R} \longrightarrow \mathbb{R}$. Then the density of $g(X)$ is given by*

$$f_{g(X)}(t) = f_X(g^{-1}(t)) \left| \frac{dg^{-1}}{dt}(t) \right|.$$

Example 3.24 Transforming uniform into exponential. *Consider a Uniform(0,1) distributed random variable $U \sim U(0,1)$ and transformation $g(u) = -\log(1-u)$. We are interested in the distribution of $T = g(U)$. In order to apply the above theorem, note that*

$$\begin{aligned} g^{-1}(t) &= 1 - e^{-t} \\ \frac{dg^{-1}}{dt}(t) &= e^{-t} \end{aligned}$$

and, therefore, the probability density of T on $t \in [0, \infty)$ is given by

$$f_T(t) = 1 \times |e^{-t}| = e^{-t}.$$

This distribution is called the exponential distribution with parameter 1,

$$T \sim Exp(1).$$

3.2.3 Moments of a random variable

Random variables take on numeric values. Often it is interesting to know what this value is on average or how it varies between different experiments. In this section we define the concept of an expected value of a random variable in order to make these ideas precise.

Definition 3.9 *The expected value or* **expectation** *of a random variable X is defined as*

$$EX = \begin{cases} \int_{-\infty}^{\infty} x \cdot f(x)\; dx & \text{if } X \text{ absolutely continuous,} \\ \sum_x x \cdot f(x) & \text{if } X \text{ discrete,} \end{cases} \tag{3.2}$$

under the assumption that integral and sum, respectively, exist.

Example 3.25 Mean of sum of two four-sided dice. *In example 3.22 we encountered the sum X of two four-sided dice, which ranged from 2 until 8. We find that the average value of the random variable X is given as,*

$$\begin{aligned} EX &= 2f_X(2) + 3f_X(3) + \ldots + 8f_X(8) \\ &= \frac{2}{16} + \frac{6}{16} + \frac{12}{16} + \frac{20}{16} + \frac{18}{16} + \frac{14}{16} + \frac{8}{16} \\ &= 5, \end{aligned}$$

which is not surprising as we saw above that the probability mass function was symmetric around 5.

Example 3.26 Mean of a simple random variable. *For a simple random variable $X(\omega) = \sum_{i=1}^{n} x_i I_{A_i}(\omega)$, where $A_1, \cdots, A_n$ partition Ω, we find*

$$E[X] = \sum_{i=1}^{n} x_i E[I_{A_i}(\omega)] = \sum_{i=1}^{n} x_i P(A_i).$$

Consider a function $g : \mathbb{R} \longrightarrow \mathbb{R}$, then it can be shown that the expectation of the random variable $g(X)$ can be calculated efficiently, via

$$E[g(X)] = \begin{cases} \int_{-\infty}^{\infty} g(x) \cdot f(x) dx & \text{if } X \text{ continuous} \\ \sum_x g(x) \cdot f(x) & \text{if } X \text{ discrete,} \end{cases} \tag{3.3}$$

There are various functions g that are of particular interest. For $g(x) = (x - \mu)^2$, we obtain

$$\text{Var}(X) = E\left[(X - \mu)^2\right],$$

called the **variance** of X. The variance stands for the *average squared deviation from the mean.* Given that the variance does not have the same units of the original random variable, it is common for practical purposes to consider its square root, which is called the standard deviation,

$$SD(X) = \sqrt{\text{Var}(X)}.$$

Example 3.27 Standard deviation of sum of 2 four-sided dice. *Using the results from examples 3.22 and 3.25, we calculate the variance as*

$$\begin{aligned} Var(X) &= (2-5)^2 f_X(2) + (3-5)^2 f_X(3) + \ldots + (8-5)^2 f_X(8) \\ &= \frac{9}{16} + \frac{8}{16} + \frac{3}{16} + \frac{0}{16} + \frac{3}{16} + \frac{8}{16} + \frac{9}{16} \\ &= 2.5, \end{aligned}$$

Therefore, the standard deviation is given as

$$SD(X) = \sqrt{2.5} = 1.58.$$

Given that the sum extends by 3 on either side of the mean 5, namely from 2 to 8, this value of 1.58 is not unreasonable as the average deviation from the mean – or, more precisely, the square root of the average square deviation of the mean.

It is often convenient to use the relation,

$$\mathrm{Var}(X) = E(X^2) - (EX)^2,$$

which writes the variance as functions of EX^n with $n = 1, 2$. We give these quantities a special name, so-called *moments.*

Definition 3.10 Moments of a random variable. *Consider the function $g(x) = x^n$, then the quantity*

$$E[X^n],$$

is known as the n-th moment of X.

It turns out that the moments can be calculated efficiently, using a special function, appropriately named the *moment generating function.*

Definition 3.11 Moment generating function. *For $g(x) = e^{tx}$, the moment generating function of X is given as*

$$M_X(t) = E[e^{tX}].$$

If $M_X(t) < \infty$ for t in an open interval around zero, then it generates the n-th moment from evaluating the n-th derivative at zero

$$E[X^n] = M_X^{(n)}(0) = \frac{d^n}{dt^n} M_X(t)\bigg|_{t=0},$$

since $Ee^{tX} = \sum_{n=0}^{\infty} \frac{E(X^n)}{n!} t^n$.

Closely related to the moment generating function is the complex function called the *characteristic function,*

Definition 3.12 Characteristic function. *For the complex function $g(x) = e^{itx}$, the characteristic function is defined as*

$$\varphi_X(t) = E[e^{itX}],$$

also known as the **Fourier transform** *for absolutely continuous random variables.*

Just like the moment generating function, the characteristic function also has the moment generating property

$$\varphi_X^{(n)}(0) = \frac{d^n}{dt^n} \varphi_X(t)\bigg|_{t=0} = i^n E[X^n].$$

It is, in fact, the moment generating function evaluated at it, that is $M_Y(it) = \varphi_Y(t)$. The main advantage of the characteristic function over the moment generating function is that it always exists, is uniformly continuous, and that it is bounded by 1 $|\varphi(t)| \leq \varphi(0) = 1$. Moreover, it shares with the moment generating function the property that it uniquely determines the distribution, in the sense that X and Y are equally distributed if and only if $\varphi_X = \varphi_Y$. In fact, it is possible to recover the probability density function from the characteristic function by means of the inversion formula; if $\int_{-\infty}^{\infty} |\varphi_X(t)|dt < \infty$, then X has an absolutely continuous and bounded density f_X, given by

$$f_X(x) = \frac{1}{2\pi} \int_{-\infty}^{\infty} e^{itx} \varphi_X(t)dt.$$

The moment generating and characteristic functions are not often practically useful, but are most often used to verify and proof important properties. One particularly useful property in that respect is given by

$$\varphi_{aY+b}(t) = e^{ibt} \varphi_Y(at).$$

We encounter this property when proving the central limit theorem.

3.2.4 Some standard probability distributions

In this section we define a number of standard probability distributions that will come in handy in the rest of the book.

Example 3.28 Bernoulli distribution. *Consider a Bernoulli random variable $X \sim$ Bern(p) with probability mass function*

$$f_X(x) = \begin{cases} 1-p & \textit{if } x = 0 \\ p & \textit{if } x = 1. \end{cases}$$

The event $X = 1$ is often labeled as a "success" and the event $X = 0$ as a "failure," or as "true" and "false," respectively. The moments of X now follow directly. That is,

$$E(X) = 0 \cdot P(X = 0) + 1 \cdot P(X = 1) = p$$

and, upon noting that $X = X^2$, we have

$$\text{Var}(X) = E(X^2) - (EX)^2 = E(X) - p^2 = p - p^2 = p(1-p).$$

The variance of a Bernoulli random variable equals the probability of success times that of failure. The moment generating function is given by

$$\begin{aligned} M_X(t) &= f_X(0)e^{0t} + f_X(1)e^{1t} \\ &= 1 - p + pe^t. \end{aligned}$$

and, therefore, its characteristic function as

$$\varphi_X(t) = 1 + p + e^{it}.$$

Example 3.29 Uniform distribution. *A fundamental continuous distribution is the Uniform over an interval* $[a, b]$, *i.e.,*

$$U \sim U(a, b).$$

Its density function is given by

$$f_U(x) = \frac{1}{b-a} I_{[a,b]}(x).$$

It is easy to check that $0 \leq f_U(x)$ *and* $\int_{-\infty}^{\infty} f_U(x)\, dx = 1$. *The expected value follows from*

$$E(U) = \int_a^b x f_U(x)\, dx = \int_a^b \frac{x}{b-a}\, dx = \frac{a+b}{2}.$$

The variance follows from

$$\text{Var}(X) = E(X - EX)^2 = \int_a^b \left(x - \frac{a+b}{2}\right)^2 \frac{1}{b-a}\, dx = \frac{(b-a)^2}{12}.$$

The moment generating function is given by $M_X(t) = \frac{e^{bt} - e^{at}}{(b-a)t}$. *In the important special case where* $[a, b] = [0, 1]$, *we have* $E(X) = 1/2$, *and* $\text{Var}(X) = 1/12$.

Example 3.30 Exponential distribution. *If* T *is has an exponential distribution with parameter* λ,

$$T \sim Exp(\lambda),$$

then its density $f_T(t) = \lambda e^{-\lambda t}$ *for* t *in* $[0, \infty)$. *It follows immediately that the distribution function*

$$F(t) = P(X \leq t) = \int_0^t \lambda e^{-\lambda x}\, dx = 1 - \lambda e^{-\lambda t}.$$

Using integration by parts the expected value of T *can easily be found*

$$\begin{aligned} E(T) &= \int_0^\infty x \cdot f_T(x) dx = \int_0^\infty \lambda x e^{-\lambda x}\, dx \\ &= \lambda x \frac{e^{-\lambda x}}{-\lambda}\bigg|_0^\infty - \int_0^\infty \lambda \frac{e^{-\lambda x}}{-\lambda}\, dx \\ &= 0 + \frac{1}{\lambda} \int_0^\infty \lambda e^{-\lambda x}\, dx = \frac{1}{\lambda}. \end{aligned}$$

We may compute the expected value of T^n, *by*

$$\begin{aligned} E[T^n] &= \int_0^\infty x^n \cdot f_T(x)dx = \int_0^\infty \lambda x^n e^{-\lambda x}\, dx \\ &= \lambda x^n \frac{e^{-\lambda x}}{-\lambda}\Big|_0^\infty - \int_0^\infty n x^{n-1} \lambda \frac{e^{-\lambda x}}{-\lambda}\, dx \\ &= 0 + \frac{n}{\lambda}\int_0^\infty x^{n-1}\lambda e^{-\lambda x}\, dx \\ &= \frac{n}{\lambda}\cdot E[X^{n-1}] \end{aligned}$$

Repeating these steps $n-1$ *times it follows that* $E[T^n] = n!\lambda^{-n}$. *We conclude that the expected value of* $E[T^n]$ *exists for all* n, *the moments of* T *of all orders exist. The moment generating function for* T *is given as*

$$\begin{aligned} M_T(t) &= \int_0^\infty e^{tx}\lambda e^{-\lambda x}\, dx = \int_0^\infty \lambda e^{-(\lambda - t)x}\, dx \\ &= \frac{\lambda e^{-(\lambda-t)x}}{-(\lambda - t)}\Big|_0^\infty = \frac{\lambda}{\lambda - t} \end{aligned}$$

For the expected value of, say, X^2 *we find*

$$M_X^{(2)}(t) = \frac{d^2}{dt^2}M_X(t) = \frac{d^2}{dt^2}\lambda(\lambda-t)^{-1} = \frac{d}{dt}\lambda(\lambda-t)^{-2} = 2\lambda(\lambda-t)^{-3},$$

so that $E(X^2) = M_X^{(2)}(0) = 2\lambda^{-2}$.

Example 3.31 Normal distribution *The random variable* X *is normally distributed with parameters* μ *and* σ^2,

$$X \sim N(\mu, \sigma^2),$$

if it has density function

$$f_X(x) = \frac{1}{\sqrt{2\pi\sigma^2}} \exp\left[-\frac{1}{2}(x-\mu)^2/\sigma^2\right]. \tag{3.4}$$

It can be verified in several ways that $E[X] = \mu$ *and* $\mathrm{Var}[X] = \sigma^2$. *The moment generating function follows from the transformation* $y = x - \mu$,

$$\begin{aligned} M_X(t) &= \int_{-\infty}^\infty e^{tx}\frac{1}{\sqrt{2\pi\sigma^2}}e^{-(x-\mu)^2/2\sigma^2}dx && (3.5)\\ &= e^{\mu t + \sigma^2 t^2/2}\cdot \frac{1}{\sqrt{2\pi\sigma^2}}\int_{-\infty}^\infty e^{-(y-t\sigma^2)^2/2\sigma^2}dy && (3.6)\\ &= e^{\mu t + \sigma^2 t^2/2}. && (3.7) \end{aligned}$$

Observe that it can be differentiated n*-times, so that the moments of any order exists. Its characteristic function follows from*

$$\varphi_X(t) = M_X(it) = \exp(i\mu t - \sigma^2 t^2/2).$$

3.2.5 Joint and marginal distribution functions

Two random variables X and Y have a joint distribution, if they are defined on the same sample space. In this case, they jointly vary as a result of the outcome on the underlying sample space. The **joint cumulative distribution function** F_{XY} describes how the variables X and Y are related and is given by

$$F_{XY}(x, y) = P(X \leq x \cap Y \leq y)$$

for all $(x, y) \in \mathbb{R}^2$. If both X and Y are absolutely continuous, then the **joint density function** follows from the mixed second-order partial derivative

$$f_{XY}(x, y) = \frac{\partial^2 F_{XY}(x, y)}{\partial x \partial y},$$

where the order of differentiation is immaterial. The concept of the **joint probability mass function** for discrete random variables can be defined analogously by replacing differentiation by differencing, as was done in the univariate case. Conversely, the joint distribution function can be written as

$$F_{XY}(x, y) = \int_{-\infty}^{x} \int_{-\infty}^{y} f_{XY}(s, t)\, dsdt,$$

for absolutely continuous distributions. By replacing the integrals with sums, we get the analogous discrete version. Whereas the joint density and distribution functions give the precise overall relation between two random variables, there is an often-used one-number summary, called the *correlation.*

Definition 3.13 *The* **covariance** *between X and Y is defined as*

$$Cov(X, Y) = E(X - EX)(Y - EY).$$

The **correlation coefficient** ρ *between X and Y is defined as*

$$\rho = \frac{\text{Cov}(X, Y)}{\sqrt{\text{Var}(X) \cdot \text{Var}(Y)}} \tag{3.8}$$

It can be shown that a correlation lies between -1 and 1, where

- $\rho = -1$ indicates a perfect negative linear relationship;
- $\rho = 0$ indicates no linear relationship;
- $\rho = 1$ indicates a perfect positive linear relationship.

If two random variables have a joint distribution, then it is common to write them as a vector $\mathbf{Y} = (Y_1, Y_2)^T$. The expectation of $\mathbf{Y}$ is an element of $\mathbf{R}^2$ so that their expected value becomes

$$E[\mathbf{Y}] = E\begin{bmatrix} Y_1 \\ Y_2 \end{bmatrix} = \begin{bmatrix} E(Y_1) \\ E(Y_2) \end{bmatrix} = \begin{bmatrix} \mu_1 \\ \mu_2 \end{bmatrix} = \boldsymbol{\mu}. \tag{3.9}$$

For the variance of $\mathbf{Y}$ obtain a **variance matrix** $\mathbf{\Sigma}$ defined as

$$\text{Var}(\mathbf{Y}) = E(\mathbf{Y} - \boldsymbol{\mu})(\mathbf{Y} - \boldsymbol{\mu})^T = \mathbf{\Sigma} = \begin{bmatrix} \sigma_{11} & \sigma_{12} \\ \sigma_{21} & \sigma_{22} \end{bmatrix},$$

with its ij-th element

$$\sigma_{ij} = E(Y_i - \mu_i)(Y_j - \mu_j) = \text{Cov}(Y_i, Y_j),$$

for $i = 1, 2$, $j = 1, 2$. When $i = j = 1$ we have $\sigma_{11} = E(Y_1 - \mu_1)^2 = \sigma_1^2$ the variance of Y_1. Similarly, when $i = j = 2$ we have $\sigma_{22} = E(Y_2 - \mu_2)^2 = \sigma_2^2$ the variance of Y_2. When i, and j differ, then we have

$$E(Y_i - \mu_i)(Y_j - \mu_j) = E(Y_j - \mu_j)(Y_i - \mu_i),$$

so that we see that co-variance is symmetric concept for which $\sigma_{ij} = \sigma_{ji}$, so that the variance matrix $\mathbf{\Sigma}$ is symmetric. The correlation coefficient ρ between Y_i and Y_j is then given as

$$\rho = \frac{\sigma_{12}}{\sigma_1 \cdot \sigma_2}$$

It is possible to recover the univariate distribution from the multivariate distribution. We refer to this univariate distribution as the marginal distribution. The **marginal distribution** of X can be found by integrating out y, that is

$$\begin{aligned} f_X(x) &= \int_{-\infty}^{\infty} f_{XY}(x, y)\, dy \\ F_X(x) &= \int_{-\infty}^{x} \int_{-\infty}^{\infty} f_{XY}(s, y)\, dyds = \int_{-\infty}^{x} f_X(s)\, ds. \end{aligned}$$

Analogously, we obtain the marginal distribution $F_Y(y)$ and marginal density $f_Y(y)$ of Y.

Example 3.32 Bivariate normal distribution. *The density function of the bivariate normal distribution may be written in matrix terms as*

$$f(\boldsymbol{y}) = (2\pi)^{-2/2} |\det \mathbf{\Sigma}|^{-1/2} \exp\{-(\boldsymbol{y} - \boldsymbol{\mu})^T \mathbf{\Sigma}^{-1} (\boldsymbol{y} - \boldsymbol{\mu})/2\}.$$

To compute the marginal distributions, we need to simplify this formula. Using that $\sigma_{12} = \rho\sigma_1\sigma_2$, *we find for the determinant of the variance matrix*

$$\det \mathbf{\Sigma} = \sigma_{11}\sigma_{22} - \sigma_{12}^2 = \sigma_1^2 \sigma_2^2 (1 - \rho^2)$$

and the inverse of the variance matrix

$$\mathbf{\Sigma}^{-1} = \frac{1}{\det \mathbf{\Sigma}} \begin{bmatrix} \sigma_{22} & -\sigma_{12} \\ -\sigma_{12} & \sigma_{11} \end{bmatrix}.$$

After a few simplifications the density can be written as $f_{Y_1,Y_2}(y_1, y_2) =$

$$\frac{1}{2\pi\sigma_1\sigma_2\sqrt{1-\rho^2}} \cdot \exp\left[-\frac{1}{2(1-\rho^2)}\left\{\left(\frac{y_1-\mu_1}{\sigma_1}\right)^2 - 2\rho\left(\frac{y_1-\mu_1}{\sigma_1}\right)\left(\frac{y_2-\mu_2}{\sigma_2}\right) + \left(\frac{y_2-\mu_2}{\sigma_2}\right)^2\right\}\right]$$

To compute the marginal distribution of Y_1*, we use the substitutions* $w = (y_1 - \mu_1)/\sigma_1$*,* $z = (y_2 - \mu_2)/\sigma_2$*, and* $dy = \sigma_2 dz$ *to integrate out variable* y *from the joint density as follows*

$$\begin{aligned}
f_{Y_1}(y_1) &= \int_{-\infty}^{\infty} f_{Y_1,Y_2}(y_1, y_2)dy \\
&= \int_{-\infty}^{\infty} \frac{1}{2\pi\sigma_1\sigma_1\sqrt{1-\rho^2}} \cdot \exp\left[-\frac{1}{2(1-\rho^2)}\left\{\left(w^2 - 2\rho wz + z^2\right)\right\}\right]\sigma_2 dz \\
&= \frac{e^{-w^2/2}}{\sqrt{2\pi}\sigma_1} \cdot \int_{-\infty}^{\infty} \frac{1}{\sqrt{2\pi}\sqrt{1-\rho^2}} \cdot \exp\left[-\frac{1}{2(1-\rho^2)}\left\{\left((z-\rho w)^2\right)\right\}\right] dz \\
&= \frac{1}{\sqrt{2\pi}\sigma_1} \cdot \exp\left[-\frac{1}{2}\left(y_1-\mu_1\right)^2/\sigma_1^2\right].
\end{aligned}$$

We find that Y_1 *is* $N(\mu_1, \sigma_1^2)$ *distributed.*

3.2.6 Independent random variables

In this section we consider the important special case in which two random variables do not have any relationship, which we call *independence.* We encountered independent events before and we will define independent random variables in a similar way.

Definition 3.14 *Two random variables* X_1 *and* X_2 **independent** *if the joint distribution is the product of the marginal distributions*

$$F_{X_1X_2}(x_1, x_2) = F_{X_1}(x_1) \cdot F_{X_2}(x_2).$$

for all real values x_1 *and* x_2*.*

Since the density and mass functions are just the partial derivatives and differences, respectively, of the distribution function, we have that X_1 and X_2 are independent if and only if their joint density or mass function equals the product of marginal density or mass functions

$$f_{X_1X_2}(x_1, x_2) = f_{X_1}(x_1) \cdot f_{X_2}(x_2).$$

for all x_i and x_j. If two random variables are independent, then this also holds for any two functions of these. That is, if X_1 and X_2 are independent, then $g_1(X_1)$ and $g_2(X_2)$ are independent.

Example 3.33 Independence implies zero covariance. *Note that the covariance is defined as the expected values of a product of two functions. If X_1 and X_2 are independent, then the covariance*

$$\begin{aligned}\mathrm{Cov}(X_1, X_2) &= E\left[(X_1 - EX_1)(X_2 - EX_2)\right] \\ &= E\left[X_1 - EX_1\right] \cdot E\left[X_2 - EX_2\right] = 0\end{aligned}$$

For the sum of two independent random variables, we obtain

$$\begin{aligned}\mathrm{Var}(X_1 + X_2) &= E\left((X_1 - EX_1) + (X_2 - EX_2)\right)^2 \\ &= E(X_1 - EX_1)^2 + 2E\left((X_1 - EX_1) \cdot (X_2 - EX_2)\right)^2 \\ &\quad + E(X_2 - EX_2)^2 \\ &= \mathrm{Var}(X_1) + 2\mathrm{Cov}(X_1, X_2) + \mathrm{Var}(X_2) \\ &= \mathrm{Var}(X_1) + \mathrm{Var}(X_2) \qquad (3.10)\end{aligned}$$

The variance of the sum equals the sum of the variances. Obviously, this also holds for all finite sums of independent random variables.

The concept of independence can be generalized from situations with two random variables to cases with n random variables.

Definition 3.15 *The n random variables $X_1, \cdots, X_n$ are* **independent**, *if for all $x \in \mathbb{R}^n$*

$$f_{X_1,\ldots,X_n}(x) = \prod_{i=1}^{n} f_{X_i}(x_i).$$

As a consequence, the expectation of any product of functions of independent random variables can be calculated in a marginal way. That is, if $X_1, \cdots, X_n$ are independent, then

$$E\left[\prod_{i=1}^{n} g_i(X_i)\right] = \prod_{i=1}^{n} E\left[g_i(X_i)\right].$$

Furthermore, for a collection of identically and independently distributed random variables $X_1, \cdots, X_n$, the moment generating function of its sum becomes

$$M_{\sum_{i=1}^n X_i}(t) = E[e^{t\sum_{i=1}^n X_i}] = \prod_{i=1}^{n} E[e^{tX_i}] = M_{X_1}(t)^n,$$

and in a completely similar manner we find for characteristic functions that

$$\varphi_{\sum_{i=1}^n X_i}(t) = \varphi_{X_1}(t)^n.$$

These properties facilitate the computation of moments for sums of independent random variables.

Example 3.34 Binomial as sum of Bernoulli. *Let p be the probability that an insect dies if a certain amount of insecticide is given. Let Y_i be a Bernoulli random variable with $Y_i = 1$ representing the event that i-th insect dies and $Y_i = 0$ if not. Suppose that all $Y_1, \cdots, Y_n$ are independent and are identically distributed. The latter simply comes down to $P(Y_i = 1) = p$ for all i with $p \in (0, 1)$. Given a certain dose of a chemical compound, we may count the number of successes out of n by $Y = \sum_{i=1}^{n} Y_i$ and, next, compute the probability of y successes out of n. This comes down to*

$$P(Y = y) = \binom{n}{y} p^y (1-p)^{n-y} = \frac{n!}{x!(n-y)!} p^y (1-p)^{n-y}$$

That is, there are $\binom{n}{y}$ number of ways (combinations) of y successes out of n and $p^y(1-p)^{n-y}$ is probability of y consecutive successes and $n - y$ failures. The mean and the variance of the number of successes can be computed as follows

$$E(Y) = E(\sum_{i=1}^{n} Y_i) = \sum_{i=1}^{n} E(Y_i) = \sum_{i=1}^{n} p = np.$$

Using the independence between the elementary events $Y_1, \cdots, Y_n$, we find

$$\text{Var}(Y) = \text{Var}\left(\sum_{i=1}^{n} Y_i\right) = \sum_{i=1}^{n} \text{Var}(Y_i) = \sum_{i=1}^{n} p(1-p) = np(1-p).$$

Here we conveniently used the fact that, for independent random variables, the variance of a sum equals the sum of the variances. The characteristic function for Y follows from

$$\begin{aligned}
\varphi_Y(t) &= E[e^{itY}] = \sum_{y=0}^{n} e^{ity} \binom{n}{y} p^y (1-p)^{n-y} \\
&= \sum_{y=0}^{n} \binom{n}{y} (pe^{it})^y (1-p)^{n-y} \\
&= \left(1 + p(e^{it} - 1)\right)^n
\end{aligned}$$

Example 3.35 Poisson as limit of Binomial distribution. *Consider we want to model relatively rare counts in a very large population, such as the number of deaths as a result of some low level chemical pollution. Assume that each person in the population of size n has a small probability p_n of experiencing the event, then the total number of counts X_n has a Binomial(n, p_n) distribution. Assuming that $np_n \to \lambda$ as $n \to \infty$, then the binomial theorem implies that*

$$\begin{aligned}
\varphi_{X_n}(t) &= \left(1 + p_n(e^{it} - 1)\right)^n \\
&= \left[\left(1 + p_n(e^{it} - 1)\right)^{\frac{1}{p_n(e^{it}-1)}}\right]^{np_n(e^{it}-1)} \\
&\to e^{\lambda(e^{it}-1)}.
\end{aligned}$$

This is the characteristic function of a so-called **Poisson distribution**, *with probability mass function*

$$f(y) = \frac{e^{-\lambda}\lambda^y}{y!},$$

for all non-negative integers y. So a binomial distribution with a small success probability can be approximated by the Poisson. In order to model counts in chemical and many other processes, the Poisson distribution is an ideal candidate.

3.2.7 Conditional distributions

In this section we define the concept of a conditional distribution, together with its conditional expectation and conditional variance.

Definition 3.16 *Consider the random variables X and Y and a value x for which $f_X(x) > 0$. Then the* **conditional density** *$f_{Y|X=x}$ of Y given $X = x$ is defined as*

$$f_{Y|X=x}(y) = \frac{f(x, y)}{f_X(x)}.$$

This yields a proper density function which is non-negative and integrates to one. It can be illuminating to see Y as the outcome of a chemical process depending on a predictive variable X such as amount of catalyst, pressure, or temperature. The **conditional expectation** of $g(Y)|X = x$, that is of a function g of outcome Y given that X attains value x, is simply the integral over y weighted by the conditional density

$$E[g(Y)|X = x] = \int_{-\infty}^{\infty} g(y) \cdot f_{Y|X=x}(y)dy.$$

From this $E[Y|X = x]$ and $E[Y^2|X = x]$ are immediate, so that we can compute the conditional variance of Y as

$$\text{Var}[Y|X = x] = E[Y^2|X = x] - (E[Y|X = x])^2.$$

Example 3.36 *Suppose we have the continuous random vector X, Y with density*

$$f(x, y) = e^{-y} I_{\{(0,\infty),(x,\infty)\}}(x, y),$$

that is $0 < x < y < \infty$, and we want to compute its conditional variance. The marginal density of X becomes

$$f_X(x) = \int_{-\infty}^{\infty} f(x, y)dy = \int_{x}^{\infty} e^{-y}dy = e^{-x},$$

so that we see that X is exponentially distributed. The conditional density now becomes

$$f_{Y|X=x}(y) = \frac{f(x, y)}{f_X(x)} = e^{x-y} I_{\{(0,\infty),(x,\infty)\}}(x, y).$$

The conditional expectation of Y*, given* $X = x$*, can be computed by integration by parts*

$$\begin{aligned} E[Y|X=x] &= \int_{-\infty}^{\infty} y \cdot f_{Y|X=x}(y)dy = \int_{0}^{\infty} y \cdot e^{x-y} I_{\{(0,\infty),(x,\infty)\}}(x,y)dy \\ &= e^{x} \int_{x}^{\infty} y \cdot e^{-y} dy \\ &= x+1. \end{aligned}$$

Given that this is true for all x*, we can write* $E[Y|X] = X + 1$*. Now the* **conditional variance** *follows from*

$$\begin{aligned} \mathrm{Var}[Y|X=x] &= E[Y^2|X=x] - (E[Y|X=x])^2 \\ &= \int_{x}^{\infty} y^2 f_{Y|X=x}(y)dy - \left(\int_{x}^{\infty} y f_{Y|X=x}(y)dy\right)^2 \\ &= x^2 + 2x + 2 - (x+1)^2 \\ &= 1, \end{aligned}$$

which as is smaller than $\mathrm{Var}[Y] = 2$.

The latter finding holds indeed more general. That is, if X and Y are correlated, then

$$\mathrm{Var}[Y|X=x] < \mathrm{Var}[Y].$$

In the analysis of chemical data, we often want to find the "best" predictive function g of some outcome Y with some input X, i.e., we wish to minimize the squared discrepancy between the outcome Y, such as the concentration of a chemical substance, and a function g of a predictive variable X, such as the amount of light absorption. For this we have the fundamental property

$$\min_{g} E[(Y - g(X))^2] = E[(Y - E[Y|X])^2],$$

where g ranges over all functions. This globally minimizing function $g(x) = E[Y|X = x]$ is called the regression of Y on X.

Example 3.37 Some ODE regression functions. *By the qualitative analysis of the solution of an ordinary differential equation (ODE), we often conjecture a model formulation such as*

$$Y = f(X) + \varepsilon,$$

where the error term ε *represents error in the measurements. Obviously, the latter is small when the* $\mathrm{Var}[\varepsilon]$ *is small. Without loss of generality we may assume the error to have mean zero* $E(\varepsilon) = 0$ *and that the error is stochastically independent to predictive variable* X*. Under these model assumptions, it follows that* $E[Y|X] = f(X)$*.*

Depending on the form of the ODE, we find a different function f. Given the form $y'(x) = \beta$, we would find the linear model $f(x) = \alpha + \beta x$. Whereas $y'(t) = (b/t - K)y$ with initial condition $y(0) = 0$ we find, for the concentration of a chemical quantity in time t, the class of concentration curves is given as $f(t) = at^b e^{-K \cdot t}$.

Example 3.38 Conditional bivariate normal. *Consider the bivariate normal distribution (X, Y). In Example 3.32 we found expression for $f(x, y)$ and $f(x)$. Now to compute the condition density of $Y|X = x$ we use*

$$
\begin{aligned}
f_{Y|X=x}(y) &= \frac{f(x,y)}{f_X(x)} \\
&= \frac{1}{2\pi\sigma_X\sigma_Y\sqrt{1-\rho^2}} \\
&\quad \cdot \exp\left[-\frac{1}{2(1-\rho^2)}\left\{\left(w^2 - 2\rho wz + z^2\right)\right\}\right] \cdot \sqrt{2\pi}\sigma_X e^{w^2/2} \\
&= \frac{1}{\sqrt{2\pi}\sigma_Y\sqrt{1-\rho^2}} \cdot \exp\left[-\frac{1}{2(1-\rho^2)}\left\{\left(\rho^2 w^2 - 2\rho wz + z^2\right)\right\}\right] \\
&= \frac{1}{\sqrt{2\pi}\sigma_Y\sqrt{1-\rho^2}} \\
&\quad \cdot \exp\left[-\frac{1}{2(1-\rho^2)\sigma_Y^2}\left\{(y-\mu_Y) - \left(\rho\frac{\sigma_X}{\sigma_Y}(x-\mu_X)\right)\right\}^2\right]
\end{aligned}
$$

We conclude that $Y|X = x$ has a normal distribution with expected value

$$\mu_Y + \rho \cdot \frac{\sigma_X}{\sigma_Y} \cdot (x - \mu_X)$$

and variance $(1-\rho^2)\sigma_Y^2$. Note that the expected value is indeed a function of x. In fact, it happens to be a linear function with intercept $\mu_Y - \mu_X \cdot \rho \cdot \sigma_Y/\sigma_X$ and slope $\rho \cdot \sigma_Y/\sigma_X$. For the variance we have

$$\text{Var}[Y|X = x] = (1-\rho^2)\sigma_Y^2 < \sigma_Y^2 = \text{Var}[Y],$$

if X and Y have a non-zero correlation ρ.

3.2.8 Random variables related to the normal

Among all distributions, the normal distribution is historically, philosophically, and scientifically the most important distribution. As can be seen in Chapter 6, also the inference of models for chemical processes relies heavily on the normal distribution. A normally distributed random variable X with the mean μ and variance σ^2 has density function

$$f_X(x) = \frac{1}{\sqrt{2\pi\sigma^2}} e^{-(x-\mu)^2/2\sigma^2}$$

for any real x and is denoted by $N(\mu, \sigma^2)$. Several other distributions derived from the normal are of great importance for testing chemical hypotheses. These are the standard normal $N(0,1)$, χ^2, t, and F-distributions.

Example 3.39 Standard normal distribution. *The linear transform $Z = (X-\mu)/\sigma$ of any normal distribution $X \sim N(\mu, \sigma^2)$ yields the standard normal distribution with density function*

$$f_Z(z) = \frac{1}{\sqrt{2\pi}} e^{-z^2/2}, \quad \text{for any real } z.$$

We will often refer to the quantiles z_α of this distribution, where

$$F_Z(z_\alpha) = \alpha.$$

The standard normal distribution shows up in standard maximum likelihood inference theory (Section 6.5.1).

Example 3.40 Chi-square distribution. *For a standard normal distribution Z, the random variable Z^2 has a chi-square $\chi^2(1)$ distribution with density*

$$f_{Z^2}(x) = \frac{1}{\sqrt{2\pi}} x^{-1/2} e^{-x/2}, \quad \text{for } x > 0.$$

It has expected value 1 and variance 2. A sum of n independent chi-square distributed random variables $U = \sum_{i=1}^{n} Z_i^2$ has a chi-square distribution with parameter n with density function

$$f_U(u) = \frac{1}{\Gamma(n/2)2^{n/2}} u^{(n/2)-1} e^{-u/2}, \quad \text{for } u > 0.$$

It has mean n and variance $2n$.

The chi-square distribution appears as the distribution of the likelihood ratio statistic in section 6.5.6.

Example 3.41 t-distribution. *If the standard normal Z is independent from a $\chi^2(n)$ random variable U, then*

$$T = \frac{Z}{\sqrt{U/n}}$$

has a $t(n)$-distribution with density

$$f_T(t) = \frac{\Gamma((n+1)/2)}{\Gamma(n/2)\sqrt{n\pi}} \left(1 + \frac{t^2}{n}\right)^{-\frac{n+1}{2}} \quad \text{for any real } t.$$

The t-distribution is useful for testing hypotheses about means (Chapter 6) and regression coefficients (Chapter 7 and following).

Example 3.42 F-distribution. *If $U \sim \chi^2(m)$ is independent from $V \sim \chi^2(n)$, then the random variable*

$$X = \frac{U/m}{V/n}$$

has an $F(m,n)$ distribution with density

$$f_X(x) = \frac{\Gamma((m+n)/2)}{\Gamma(m/2)\Gamma(n/2)} \left(\frac{m}{n}\right)^{m/2} \frac{x^{(m-2)/2}}{[1+(m/n)x]^{(m+n)/2}} \quad \text{for} \quad x > 0.$$

The F-distribution is useful for testing a smaller linear model against a larger model in Chapters 6 and 11.

Example 3.43 Characteristic function of normal distribution. *The characteristic function of the standard normal random variable Y can be derived as*

$$\begin{aligned}
\varphi_Y(t) &= E[e^{itY}] = \int_{-\infty}^{\infty} e^{ity} dF_Y(y) = \int_{-\infty}^{\infty} e^{ity} f_Y(y)dy \\
&= \int_{-\infty}^{\infty} e^{ity} \frac{1}{\sqrt{2\pi}} e^{-y^2/2} dy \\
&= \frac{1}{\sqrt{2\pi}} \int_{-\infty}^{\infty} \exp\left\{-\frac{1}{2}\left(y^2 - 2yit + (it)^2 - (it)^2\right)\right\} dy \\
&= e^{-t^2/2} \cdot \frac{1}{\sqrt{2\pi}} \int_{-\infty}^{\infty} \exp\left\{-\frac{1}{2}(y-it)^2\right\} dy \\
&= e^{-t^2/2}.
\end{aligned}$$

If X has a normal distribution with mean μ and standard deviation σ, then we find from $X = \mu + \sigma Y$ that

$$\begin{aligned}
\varphi_X(t) &= \varphi_{\mu+\sigma Y}(t) = e^{i\mu t}\varphi_{\sigma Y}(t) \\
&= e^{i\mu t}\varphi_Y(\sigma t) = e^{i\mu t} e^{-\sigma^2 t^2/2} \\
&= \exp\left\{i\mu t - \sigma^2 t^2/2\right\}
\end{aligned}$$

In chemical applications we very often work with a **linear combination** of normally distributed measurements. An important property is given by the follow example.

Example 3.44 Sums of normal are normal. *Suppose that X_i are independently normally distributed random variables with mean μ_i and standard* *deviation* σ_i, $i = 1, \cdots, n$. *To derive the distribution of a weighted sum we use*

$$\begin{aligned}\varphi_{\sum_i w_i X_i}(t) &= \prod_{i=1}^{n} \varphi_{w_i X_i}(t) = \prod_{i=1}^{n} \varphi_{X_i}(w_i t) \\ &= \prod_{i=1}^{n} \exp\left\{w_i \mu_i t - w_i^2 \sigma_i^2 t^2/2\right\} \\ &= \exp\left\{\sum_{i=1}^{n} w_i \mu_i t - \sum_{i=1}^{n} w_i^2 \sigma_i^2 t^2/2\right\}.\end{aligned}$$

We see that $\sum_{i=1}^{n} w_i X_i$ *has a normal distribution with mean* $\sum_{i=1}^{n} w_i \mu_i$ *and variance* $\sum_{i=1}^{n} w_i^2 \sigma_i^2$. *The mnemonic here is that (weighted) sums of normally distributed random variables are normally distributed.*

Example 3.45 Replication. *If* n *independent random variables* X_i *have the same normal distribution* $N(\mu, \sigma^2)$, *then for* $w_i = 1/n$ *we find*

$$\frac{1}{n}\sum_{i=1}^{n} X_i \sim N(\mu, \sigma^2/n).$$

If we have n *independent measurements that follow the same normal distribution, then their average is normally distributed with the same mean and variance* σ^2/n, *tending to zero with* n.

3.2.9 Multivariate normal distribution

In example 3.32 we came across the bivariate normal distribution. This distribution can be extended from $n = 2$ to any size. Given its importance, we dedicate an entire section to it.

The random vector $\mathbf{Y}$ has a multivariate normal distribution $N(\boldsymbol{\mu}, \boldsymbol{\Sigma})$ with mean $\boldsymbol{\mu}$ and variance matrix $\boldsymbol{\Sigma}$. The matrix $\boldsymbol{\Sigma}$ is the **variance matrix** defined as

$$\text{Var}(\mathbf{Y}) = E(\boldsymbol{Y} - \boldsymbol{\mu})(\boldsymbol{Y} - \boldsymbol{\mu})^T,$$

with its ij-th element

$$\sigma_{ij} = E(Y_i - \mu_i)(Y_j - \mu_j) = \text{Cov}(Y_i, Y_j), \text{ for } i, j = 1, \cdots, n.$$

To see that $\boldsymbol{\Sigma}$ is positive definite, we write

$$\begin{aligned}\mathbf{a}^T \boldsymbol{\Sigma} \mathbf{a} &= \text{Var}(\mathbf{a}^T \mathbf{Y}) = E\left[\mathbf{a}^T (\boldsymbol{Y} - \boldsymbol{\mu})(\boldsymbol{Y} - \boldsymbol{\mu})^T \mathbf{a}\right] \\ &= E\left[\mathbf{a}^T (\boldsymbol{Y} - \boldsymbol{\mu})\right]^2 \\ &> 0 \end{aligned} \tag{3.11}$$

for all non-zero real vectors $\mathbf{a}$.

The density function of the multivariate normal is

$$f(\boldsymbol{y}) = (2\pi)^{-n/2}|\det \boldsymbol{\Sigma}|^{-1/2} \exp\{-(\boldsymbol{y}-\boldsymbol{\mu})^T\boldsymbol{\Sigma}^{-1}(\boldsymbol{y}-\boldsymbol{\mu})/2\}.$$

If all covariances are zero, $\text{Cov}[Y_i, Y_j] = \sigma_{ij} = 0$ for all $i \neq j$, then the matrix $\boldsymbol{\Sigma}$ is diagonal. Then the $\det \boldsymbol{\Sigma} = \prod_{i=1}^{n} \sigma_{ii}$ so that

$$\begin{aligned}
P(\mathbf{Y} \leq \mathbf{z}) &= \int_{\mathbf{Y} \leq \mathbf{z}} f(\boldsymbol{y}) dy_1 \cdots dy_n \\
&= \int_{\mathbf{Y} \leq \mathbf{z}} (2\pi)^{-n/2}|\det \boldsymbol{\Sigma}|^{-1/2} \exp\{-(\boldsymbol{y}-\boldsymbol{\mu})^T\boldsymbol{\Sigma}^{-1}(\boldsymbol{y}-\boldsymbol{\mu})/2\} dy_1 \cdots dy_n \\
&= \int_{Y_1 \leq z_1} \frac{1}{\sqrt{2\pi\sigma_{ii}}} e^{-\frac{(y_1-\mu_1)^2}{2\sigma_{11}}} dy_1 \cdots \int_{Y_n \leq z_n} \frac{1}{\sqrt{2\pi\sigma_{nn}}} e^{-\frac{(y_n-\mu_n)^2}{2\sigma_{nn}}} dy_n \\
&= P(Y_1 \leq z_1) \cdots P(Y_n \leq z_n).
\end{aligned}$$

Hence, the random variables $Y_1, \cdots, Y_n$ are independent. Whereas it is always true that for independent $Y_1, \cdots, Y_n$ all covariances are zero, the reverse is only true for normally distributed data.

The **characteristic function** of the multivariate normal $N(\boldsymbol{\mu}, \boldsymbol{\Sigma})$ equals

$$\varphi_{\mathbf{Y}}(\mathbf{t}) = \exp\{i\boldsymbol{t}^T\boldsymbol{\mu} - \boldsymbol{t}^T\boldsymbol{\Sigma}\boldsymbol{t}/2\}$$

From this it is immediate that

$$\varphi_{\mathbf{AY}}(\mathbf{t}) = \exp\{i\boldsymbol{t}^T\boldsymbol{A}\boldsymbol{\mu} - \boldsymbol{t}^T\boldsymbol{A}\boldsymbol{\Sigma}\boldsymbol{A}^T\boldsymbol{t}/2\}$$

which shows that $\mathbf{AY}$ has distribution $N(\boldsymbol{A}\boldsymbol{\mu}, \boldsymbol{A}\boldsymbol{\Sigma}\boldsymbol{A}^T)$ for any m by n matrix $\mathbf{A}$.

In Chapter 7 we will use two properties of quadratic forms of normal distributions. Let $\boldsymbol{y} \sim N(\boldsymbol{\mu}, \boldsymbol{\Sigma})$ and $a = \text{Rank}(\boldsymbol{A}\boldsymbol{\Sigma})$, then

1. $(\boldsymbol{y}-\boldsymbol{\mu})^T\boldsymbol{A}(\boldsymbol{y}-\boldsymbol{\mu}) \sim \chi^2(a)$ if and only if $\boldsymbol{\Sigma}\boldsymbol{A}\boldsymbol{\Sigma}\boldsymbol{A}\boldsymbol{\Sigma} = \boldsymbol{\Sigma}\boldsymbol{A}\boldsymbol{\Sigma}$.
2. $\boldsymbol{P}^T\boldsymbol{y}$ is independent to $(\boldsymbol{y}-\boldsymbol{\mu})^T\boldsymbol{A}(\boldsymbol{y}-\boldsymbol{\mu})$ if and only if $\boldsymbol{\Sigma}\boldsymbol{A}\boldsymbol{\Sigma}\boldsymbol{P} = \mathbf{0}$.

3.2.10 Exponential family of distributions

Most distributions important for modeling of chemical processes, such as the Bernoulli, binomial, Poisson, normal and exponential, are members of the exponential family of distributions.

Definition 3.17 *A random variable Y is a member of the **exponential family** of distributions if its density can be written as*

$$0 \leq f(y, \theta) = \exp\{a(y)b(\theta) + c(\theta) + d(y)\},$$

for a parameter θ from a set Θ and certain functions $a, b, c,$ and d that depend on the type of distribution. The set of y values for which $f(y, \theta) > 0$ is not allowed to depend on θ.

For members of the exponential family many useful properties hold, such as that interchanging differentiation and integration is allowed. Such properties simplify several derivations of expected value and properties of maximum likelihood estimation in Chapter 6. To see this we derive an extra method to compute the expected value and the variance for Y. Without loss of generality we assume that $b'(\theta) \neq 0$ for all θ. Using, the fact the density f integrates to one, interchanging integration and differentiation, and the chain rule we find

$$\begin{aligned} 0 &= \frac{d}{d\theta}1 = \frac{d}{d\theta}\int_{-\infty}^{\infty} f(y,\theta)dy = \int_{-\infty}^{\infty} \frac{\partial}{\partial\theta} f(y,\theta)dy \\ &= \int_{-\infty}^{\infty} \left[a(y)b'(\theta) + c'(\theta)\right] f(y,\theta)dy \\ &= E\left[a(Y)\right] b'(\theta) + c'(\theta) \end{aligned} \tag{3.12}$$

So that

$$E[a(Y)] = -\frac{c'(\theta)}{b'(\theta)}.$$

Similarly, using the previous and the product rule, we find

$$\begin{aligned} 0 &= \frac{d^2}{d\theta^2}\int_{-\infty}^{\infty} f(y,\theta)dy = \int_{-\infty}^{\infty} \frac{\partial}{\partial\theta}\frac{\partial}{\partial\theta} f(y,\theta)dy \\ &= \int_{-\infty}^{\infty} \frac{\partial}{\partial\theta}\left[a(y)b'(\theta) + c'(\theta)\right] f(y,\theta)dy \\ &= \int_{-\infty}^{\infty} \left[a(y)b''(\theta) + c''(\theta)\right] f(y,\theta) + \left[a(y)b'(\theta) + c'(\theta)\right]\frac{\partial}{\partial\theta} f(y,\theta)dy \\ &= b''(\theta)E[a(Y)] + c''(\theta) + \int_{-\infty}^{\infty} b'(\theta)^2(a(y) - E[a(Y)])^2 f(y,\theta)dy \\ &= b''(\theta)E[a(Y)] + c''(\theta) + b'(\theta)^2\text{Var}[a(Y)] \end{aligned} \tag{3.13}$$

So that

$$\text{Var}[a(Y)] = \frac{b''(\theta)c'(\theta) - b'(\theta)c''(\theta)}{(b'(\theta))^3}.$$

For discrete members of the exponential family, the derivation is very similar.

Example 3.46 Poisson distribution. *If Y is Poisson distributed, then the density can be written as*

$$f(y,\theta) = \frac{\theta^y}{y!}\exp(-\theta) = \exp\left[y\log(\theta) - \theta - \log y!\right]$$

so that we may take $a(y) = y$, $b(\theta) = \log(\theta)$, and $d(y) = -\log y!$. Hence, the Poisson is a member of the exponential family and we find

$$E(Y) = -\frac{c'(\theta)}{b'(\theta)} = -\frac{-1}{1/\theta} = \theta,$$

and

$$\mathrm{Var}[a(Y)] = \frac{b''(\theta)c'(\theta) - b'(\theta)c''(\theta)}{(b'(\theta))^3} = \frac{-\frac{1}{\theta^2} \cdot -1 - 0 \cdot \frac{1}{\theta}}{\left(\frac{1}{\theta}\right)^3} = \theta.$$

Example 3.47 Binomial distribution. *If Y has a binomial distribution, then the density can be written as*

$$f(y, \pi) = \binom{n}{y} \pi^k (1-\pi)^{n-k} = \exp\left[y \log\left(\frac{\pi}{1-\pi}\right) + n\log(1-\pi) + \log\binom{n}{y}\right]$$

so that we may take $a(y) = y$, $b(\pi) = \log(\pi) - \log(1-\pi)$, and $d(y) = \log\binom{n}{y}$. Hence, the binomial is a member of the exponential family for $0 < \pi < 1$. Note that now have $b'(\pi) = 1/(\pi(1-\pi))$ and $c'(\pi) = -n/(1-\pi)$, so that

$$E(Y) = -\frac{c'(\theta)}{b'(\theta)} = n\pi.$$

Similarly, the reader can check that $\mathrm{Var}[Y] = n\pi(1-\pi)$.

3.3 Pseudo Random Number Generation

In example 3.24 we showed that it is possible to transform any sample from a Uniform(0,1) distribution into any other distribution. So, the aim of this section is to describe a method for sampling a random element from $[0, 1]$, being equiprobable compared to any other element from the unit interval. Any algorithm to simulate a number within a computer would in principle be deterministic. For practical purposes, this philosophical insight need not bother us, when it is possible to produce **pseudo random numbers** that pass all reasonable randomness tests.

To give an example of a pseudo random number generator, we consider the Wichmann-Hill I Generator (W.H. Press and Vettering, 2007) defined by

$$\begin{aligned}
w_i &= a_1 w_{i-1} \mod m_1 \\
x_i &= a_2 x_{i-1} \mod m_2 \\
y_i &= a_3 y_{i-1} \mod m_3 \\
z_i &= a_4 z_{i-1} \mod m_4 \\
u_i &= \left(\frac{w_i}{m_1} + \frac{x_i}{m_2} + \frac{y_i}{m_3} + \frac{z_i}{m_i}\right) \mod 1
\end{aligned}$$

where the u_i, for $i = 1, 2, \cdots$ form the required sequence. The NAG Library implementation includes 273 sets of parameters, a_j, m_j, for $j = 1, 2, 3, 4$, to choose from. The constants a_i are in the range 112 to 127, and the constants

m_j are prime numbers in the range 16718909 to 16776971, which are close to $2^{24} = 16777216$. These constants have been chosen so that each of the resulting 273 generators are essentially independent. All calculations can be carried out in 32-bit integer arithmetic and the generators give good results with the spectral test (Knuth, 1997). The period of each of these generators would be at least 2^{92} if it were not for common factors between $(m_1 - 1)$, $(m_2 - 1)$, $(m_3 - 1)$, $(m_4 - 1)$. However, each generator should still have a period of at least 2^{80}. Further discussion of the properties of these generators can be found in Maclaren (1989).

If we have a random number $\omega \in (0, 1)$ and the distribution function $F(x)$ is given, then the inverse transformation gives the solution to $\omega = F(x)$ as

$$x = F^{-1}(\omega).$$

By repeating this process n times, a sample $\{x_i\}_{i=1}^n$ of independent pseudo-random numbers from the distribution F can be generated.

Suppose U is uniformly distributed on $(0, 1)$ and that g is a monotonically increasing function of $u \in (0, 1)$ onto the range of random variable Y. If we now take $y = g(u)$, then we have $g(u) \le y$ if and only if $u \le g^{-1}(y)$. From this it follows that

$$F_Y(y) = P(Y \le y) = P(U \le g^{-1}(y)) = F_U(g^{-1}(y)) = g^{-1}(y),$$

using the fact that $F_U(u) = u$. For purposes of conducting a simulation study, this leads to straightforward methods for generating random numbers from a distribution.

Example 3.48 *Suppose that*

$$y = g(u) = \log(\frac{u}{1-u}),$$

then we find the inverse to be equal to

$$u = g^{-1}(y) = (1 + e^{-y})^{-1}.$$

From this we deduce that

$$F_Y(y) = F_U(g^{-1}(y)) = (1 + e^{-y})^{-1},$$

which is the distribution function of the standard logistic random variable

Similarly if U is uniformly distributed on $(0, 1)$ and g is a monotonically decreasing and we take $y = g(u)$, then we have

$$F_Y(y) = 1 - F_U(g^{-1}(y)).$$

Example 3.49 *Suppose that* $y = g(u) = -\log(u)/\theta$ *for a positive* θ*, then* $u = g^{-1}(y) = e^{-\theta y}$ *and we find that*

$$F_Y(y) = 1 - F_U(g^{-1}(y)) = 1 - e^{-\theta y},$$

in which we recognize the exponential distribution with parameter θ*. This is useful for chemodynamics and stoichiometry.*

Example 3.50 *To generate two independently distributed standard normal random variables one may use the Box-Muller method. Let* U_1 *and* U_2 *be independent uniform random variables. Take*

$$R = \sqrt{-2\log U_2} \text{ and } \theta = 2\pi U_1$$

then

$$X = Rcos(\theta) \text{ and } X = Rsin(\theta)$$

are independently standard normal distributed.

3.4 Notes and Comments

The general expected value definition (3.9) can indeed also be written as

$$E[g(X)] = \int_\Omega g(X(\omega))dP(\omega)$$

keeping the randomness explicit in the formula.

Note that all applications are from the probability space $(\Omega, \mathcal{F}, P)$, however, the latter is often not made explicit, in particular when the distribution function is known. It is custom to suppress the little omega in the notation of a random variable. Its randomness, however, plays a role throughout, be it on measurement precision, hypothesis testing, or simulation studies.

Probabilistic thinking plays a key role in all chapters that follow. It allows for clear model formulations which increases the clarity of thinking. For example, the linear model in regression analysis can be formulated by a single multivariate normal distribution. The latter is useful for defining measurement models for chemical quantities, it forms the basis for statistical testing, and it is the limiting distribution of averages due to the Central Limit theorem.

All models below are formulated in terms of random variables. The latter can be simulated by a random number generator, producing data that conform to the model. The analysis of these data gives estimates of the model which can be evaluated with respect to their precision. Such is often of great help in designing experiments.

After an experiment is properly designed, it is often actually conducted. This yields the data from which the parameters of the model are estimated. From the evaluation of the precision of the estimated parameters, we make inferences about the true values of the parameters that rest in nature. For good book length introductions to probability theory for engineers, we refer to Bain and Engelhardt (1992) as well as to Montgomery and Runger (2018).

3.5 Notes on Using R

The above Wichmann-Hill I Generator is given for illustration. Further information on random number generation, including several references to the literature, can be obtained from R its help on `RNGkind`. The basic functions for computations on distributions are available from R using `p` for probability, `d` for density, `q` for quantile, and `r` for random number generation. In particular we have for the normal distribution the functions `pnorm`, `dnorm`, `qnorm`, and `rnorm`. By the function `sample`, any discrete random variable can be simulated, but functions specific for the Binomial, Poisson, or, e.g., the Exponential may be useful and can easily be found by `?Distributions`.

3.6 Exercises

1. **Rolling a die.** A die is rolled and a coin is flipped simultaneously.
   (a) Describe the sample space S of this "experiment."
   (b) Count the number of outcomes in E={Die is Even} and F={Coin shows Tail}.
   (c) Count the number of outcomes in the events $E \cup F$ and $E \cap F$.
   (d) Calculate the probability of E, F, $E \cup F$, and $E \cap F$.
2. **Running.** Dwain Chambers and Maurice Greene are running a 100m dash against each other. *Both* their finishing times are recorded.
   (a) Draw the sample space (Hint: First imagine 1 possible outcome).
   (b) Draw the event that Chambers beats Greene by a full second.
3. **Number of heads.** Three coins are flipped and the number of heads are recorded as X.
   (a) Describe the Range of X, R_X.

(b) Find the probability mass function of X, p_X.
(c) Find the cumulative distribution function of X, F_X
(d) Calculate the expectation and standard deviation of X.

4. **A random variable.** A random variable Y has a range $R_Y = \{-1, 2, 4, 5\}$. The cumulative distribution function of Y is given as

$$F_Y(-1) = 0.2, \quad F_Y(3) = 0.45, \quad F_Y(4) = 0.90.$$

(a) Calculate the probability mass function of Y, p_Y.
(b) Calculate the mean and the variance of Y.
(c) Calculate the Moment Generating Function of Y.
(d) Calculate the probability that Y is -1 given that it is less than 3.

5. **Choosing balls.** Two balls are randomly chosen without replacement from an urn containing 8 white, 4 black, and 2 orange balls. Suppose that we win £5 for each white ball, lose £1 for each black ball and stay even for each orange ball. Let X denote our winnings.

(a) What are the possible values of X?
(b) What is the probability mass function of X?

6. **Partial information.** For random variable Z we know that $p_Z(5) = 0.2$ and some partial information about the cdf.

$$F_Z(2) = 0.3, \quad F_Z(5) = 0.85$$

(a) Calculate $P(Z > 5)$ and $P(Z \geq 5)$.
(b) Calculate $P(2 < Z \leq 5)$ and $P(2 < Z < 5)$.

7. **Uniform distribution**. Use a random number generator to simulate from a uniformly distribution over the interval $[0, 1]$.

(a) Generate 10000 pseudo-random deviates and construct a histogram consisting of ten bars.
(b) Repeat the previous using the option to have probabilities on the vertical axis. Increase the number of pseudo random deviates to 100000. Do the bars have the same height?
(c) Use the same random number generator to simulate a product failure with probability 0.05. Generate 1000 products and construct a frequency table.

8. **Binomial distribution**. Suppose that a sample of water has 5 percent probability to contain a pollutant. Assume 100 samples obtained to be independent and equal with respect to presence of pollutant.

(a) What is the probability of 4 polluted samples?

(b) What is the probability of 4 or more polluted samples?
(c) What is the probability of at least for 4, but strictly less than 10 polluted samples?

9. **Poisson distribution**. Suppose that Y has a Poisson distribution with probability mass function

$$P(Y = y) = \frac{\lambda^y e^{-\lambda}}{y!}, \quad y = 0, 1, 2, \cdots,$$

where λ is the mean number of events during a given unit of time, area, or volume.

(a) Show that $E[Y] = \text{Var}(Y) = \lambda$ and that its characteristic function $\varphi_Y(t) = e^{\lambda(e^{it}-1)}$.
Hint: Conveniently compute $E[Y(Y-1)]$ for determining the variance.
(b) Suppose that the number Y of cracks per concrete specimen for a particular type of cement mix has a Poisson distribution with $\lambda = 2$. Use a visualization of probability mass functions to check the above approximation of the binomial by the Poisson. Is the approximation relatively precise if $n = 100$ and $p = 0.02$?
(c) Generate 10,000 realizations of the Poisson distribution with $\lambda = 2$ and compute its average. Is the average close to 2?
Remark: Note that this is an application of the law of large numbers, a subject of Chapter 5.

10. **Nitrate ion concentration**. Suppose that the nitrate ion concentration in μg ml^{-1} has a normal distribution with mean 0.50 and standard deviation 0.02.

(a) What is the probability of a concentration larger than 0.48? Why is this equal to the probability of a concentration smaller than 0.52?
(b) What is the probability that the concentration is within $(0.47, 0.53)$.
(c) What is the probability that the concentration is more than two standard deviations away from the mean?
(d) What is the probability that the concentration is more that five standard deviations away from the mean?

11. **Regression of concentration on absorbance.** Suppose that X is light absorbance and Y is the concentration of a chemical substance and that these are bivariate normally distributed according to

$$\begin{aligned} N(\boldsymbol{\mu}, \boldsymbol{\Sigma}) &= N\left(\begin{pmatrix} \mu_X \\ \mu_Y \end{pmatrix}, \begin{bmatrix} \sigma_X^2 & \sigma_{XY} \\ \sigma_{XY} & \sigma_Y^2 \end{bmatrix}\right) \\ &= N\left(\begin{pmatrix} 114.2 \\ 0.6 \end{pmatrix}, \begin{bmatrix} 13,004 & 60.5 \\ 60.5 & 0.285 \end{bmatrix}\right) \end{aligned}$$

(a) Compute the correlation coefficient $\rho = \sigma_{XY}/(\sigma_X \cdot \sigma_Y)$.

(b) Compute the regression intercept and its slope.

(c) Compute the conditional variance of Y given $X = x$.

12. **Standard normal distribution.** Consider the standard normal distribution with density function for any real z equal to

$$f_Z(z) = \frac{1}{\sqrt{2\pi}} e^{-z^2/2}.$$

(a) Show that random variable Z has expected value zero.

(b) Show that the normal density attains its maximum at $z = 0$.

(c) Show that the normal density has inflection points at -1 and 1. One standard deviation to the left and to the right relative to its mean.

13. **t-distribution converges to the standard normal.** It can be show by the Stirling approximation that the density of the t-distribution tends to that of the standard normal. In many applied books on statistics, it is stated that when the degrees of freedom is larger than 40, the two distributions are equal. A more direct way to observe the similarity between the two types of distributions is by comparing the density functions.

(a) Plot the density function of the standard normal in a plot together with that of the t-distribution with degrees of freedom equal to 10, and 40. What do you conclude?

14. **Singular normal distribution.** Suppose we have the tri-variate normal distribution with mean zero and variance matrix

$$\mathbf{\Sigma} = \begin{bmatrix} 1 & -1/2 & -1/2 \\ -1/2 & 1 & -1/2 \\ -1/2 & -1/2 & 1 \end{bmatrix}$$

(a) Explain why that the variance matrix is singular.

(b) Is the density function for this normal distribution properly defined?

(c) Note, however, that it is possible to sample from this distribution. Show how this can be accomplished mathematically and program the steps in R checking your derivations.
Hint 1: Consider the eigen decomposition of $\mathbf{\Sigma}$ and a suitable linear transformation.
Hint 2: It is handy to use the library `mvtnorm` to sample from a bivariate normal distribution.

Hint 3: Check your computations by computing the variance matrix for, say, 10^6 random deviates.

15. **Exponential distribution**. Suppose that X is the time between detections of a particle by a Geiger counter and that it has an Exponential distribution with $\lambda = 1.4$ seconds.

    (a) What is the probability to detect a particle within half a second.

    (b) Compute the conditional probability $P(X < 3.5|X > 3)$ and compare with $P(X < 0.5)$.

    (c) Simulate 10^6 observations of times between detections and use these to make a comparison with the probabilities just computed.

16. **Lack of memory**. Suppose that lifetime X of an object is exponentially distributed, with density function $f_X(x) = \lambda e^{-\lambda x}$ for $x \in [0, \infty)$. Show that this distribution has the lack of memory property

$$P(X > s|X > t) = P(X > s - t)$$

    for $0 \leq t < s$. That is, given that the probability that the life time is longer than s, given that it has lived longer than t, depends only on the difference between s and t. Interestingly, it is the only continuous distribution with this property Billingsley (2012).

17. **Cauchy distribution**. A standard Cauchy distributed random variable X has density $f_X(x) = 1/(\pi(1 + x^2))$ for any real x. It may also be defined as the ratio of two independently standard normally distributed random variables.

    (a) Compute its distribution function $F(x) = \frac{1}{\pi}\left(\arctan(x) + \frac{2}{\pi}\right)$.

    (b) Show that it $E|X| = \infty$. That is, its tails to too thick for the expected value to be properly defined.

18. **Simulating distributions.** We may verify theoretical properties by simulation.

    (a) Suppose that n independent random variables X_i have the same normal distribution $N(\mu, \sigma^2)$, then

$$\frac{1}{n}\sum_{i=1}^{n} X_i \sim N(\mu, \sigma^2/n).$$

    Take $\mu = 10$, $\sigma = 2$, $n = 10$, and generate 10^5 times these n pseudo random variables and their average. Use these 10^5 averages to compute a histogram with 100 bars and add to this the proper density curve. Are the two almost equal?

(b) Suppose that n independent random variables Z_i have the standard normal distribution $N(0, 1)$, then

$$\sum_{i=1}^{n} X_i^2 \sim \chi^2(df = 10).$$

Take $n = 10$, and generate 10^5 times these n pseudo random variables and compute their sum of squares. Use these 10^5 sums to compute a histogram with 100 bars and add to this the proper density curve. Are the two almost equal?

Part II

Numerics and Error Propagation

4

Introduction to Numerical Methods

This chapter gives an introduction into numerical methods for finding a zero or an optimum of a given function. The latter is relevant if we wish to maximize a function that represents the yield of a chemical process. To find a root of a function we start with the basic, but fundamental, method of fixed point iteration. It is based on the idea of generating a sequence of numbers by applying a function repeatedly on the previous outcome. A special case is the Newton iterative method, which often converges faster. The Newton method can be applied to find an optimum of a function numerically.

Numerical methods can also be used for finding the solution of a differential equation. Many chemical processes can be completely characterized by a single differential equation involving the rates of the reactions and an initial condition. However, many differential equations that arise in chemical practice are too complicated to solve analytically, so that numerical algorithms are needed to approximate their solution. We start by explaining the basic, but fundamental, Euler method and how it can be generalized to the more precise and stable Runge-Kutta method. The latter can be further generalized to solve several equations simultaneously.

In this chapter you learn

- In Section 4.1 what a fixed point iteration is and which type of problems it solves.
- In Section 4.2 the basic Euler method and the popular Runge-Kutta iterative methods are explained and applied to solve problems from chemistry.
- In Section 4.3 about a famous example about three chemical species involving differential algebraic equations.

4.1 Fixed Point Problems

Many non-linear numerical problems are, in fact, special cases of the problem to find an x such that $x = g(x)$.

DOI: 10.1201/9781003178194-4

Example 4.1 *The problem of finding a solution to* $x^2 - 3x + 1 = 0$, *is equivalent to finding the fixed point* $x = (x^2 + 1)/3$, *i.e., finding a fixed point for the function* $g(x) = (x^2 + 1)/3$.

Solving the equation $f(x) = 0$, *is equivalent to finding a fixed point for* $g(x)$ *defined as* $g(x) = x - \frac{f(x)}{f'(x)}$, *at least if the first-order derivative is not equal to zero near the solution.*

If we wish to maximize a function f *via setting its first order derivative equal to zero, then this is equivalent to finding a fixed point for* $g(x)$ *defined as* $g(x) = x - \frac{f'(x)}{f''(x)}$, *if the second-order derivative is nonzero near the point of maximization.*

If A *is a matrix, then an eigenvector* v *is defined as* $Av = \lambda v$. *Hence, we immediately recognize the fixed point problem* $v = \lambda^{-1} Av$.

4.1.1 Fixed point iteration

Suppose that our aim is to solve

$$f(x) = 0,$$

that is, to find the x that zeros the function f. A point x solving $f(x) = 0$ is called a zero or a **root** of the function. It is often quite useful to study the closely related problem of finding an x for which $x = g(x)$. A point p for which $p = g(p)$ is called a **fixed point**. A very simple example allows us to illustrate several properties of fixed point iteration.

Example 4.2 *Finding a root of*

$$f(x) = x^2 - 3x + 1$$

is equivalent to finding the fixed point $x = (x^2+1)/3$. *That is,* x *is a fixed point of the function* $g_1(x) = (x^2 + 1)/3$. *Note, however, that equivalent functions with fixed points are* $g_2(x) = 3 - 1/x$ *and* $g_3(x) = x^2 - 2x + 1$. *Such flexibility in choosing the function corresponding to the fixed point equation may be an advantage in solving the problem.*

An algorithm to solve the fixed point problem is remarkably simple in its definition and nicely illustrates the idea behind iteration: We start with some number x_0 and compute $x_1 = g(x_0)$ for a known continuous function g. Next, compute $x_2 = g(x_1)$, $x_3 = g(x_2), \cdots$. In this way we generate a sequence of numbers $\{x_0, x_1, x_2, \cdots\}$ by $g(x_{n-1}) = x_n$, for $n = 1, 2, 3, \cdots$. Each x_n is defined recursively by applying the function g to its predecessor x_{n-1}. If the sequence converges to a real number x, that is $x_n \to x$, as $n \to \infty$, then $g(x_n) \to x = g(x)$, and we have obtained a solution to the fixed point problem.

Example 4.3 *To illustrate the properties of the fixed point approach, we compare it with the exact solution from the well-known abc-formula. The complete*

TABLE 4.1
Fixed point iteration on the functions $g_1(x) = (x^2+1)/3$ and $g_2(x) = 3 - 1/x$ with starting points $x_0 = 1$ and $x_0 = 3$.

	$g_1(x) = \frac{1}{3}(x^2+1)$		$g_2(x) = 3 - 1/x$	
n	x_n	x_n	x_n	x_n
0	1.000000	3.000000	1.000000	3.000000
1	0.666667	3.333333	2.000000	2.666667
2	0.481481	4.037037	2.500000	2.625000
3	0.410608	5.765889	2.600000	2.619048
4	0.389533	11.415160	2.615385	2.618182
5	0.383912	43.768626	2.617647	2.618056
6	0.382463	638.897535	2.617978	2.618037
7	0.382093	136063.686907	2.618026	2.618034
8	0.381998	...	2.618033	2.618034
9	0.381974	...	2.618034	2.618034
10	0.381968	...	2.618034	2.618034

solution to the root-finding problem in Example 4.2 is $x = \frac{3}{2} \pm \sqrt{\frac{5}{4}}$, *which is approximately equal to* $x \approx 0.381968$ *or* $x \approx 2.618034$. *The sequences generated by fixed point iteration on the functions* $g_1(x) = (x^2+1)/3$ *and* $g_2(x) = 3 - 1/x$ *using the starting points* $x_0 = 1$ *and* $x_0 = 3$, *respectively, are given in Table 4.1. The sequence generated by* g_1 *from* $x_0 = 1$ *converges to solution* 0.381968, *whereas that from* $x_0 = 3$ *diverges. The sequence generated by* g_2 *from* $x_0 = 1$, *as well as that from* $x_0 = 3$ *both converge to the solution* 2.618034.

Properties of generated sequences are given by the convergence theorem for fixed point iteration:

Theorem 4.1 *If* $x = p$ *is a solution to* $x = g(x)$, *and* g *has a continuous derivative on an interval* I *around* p, *and* $|g'(x)| \leq K < 1$ *for all* x *in* I, *then the sequence generated by fixed point iteration converges.*

This can be seen as follows: from the mean value theorem (Stewart, 2007; Apostol, 2007a), it follows that there exists a t between x and p such that

$$g(x) - g(p) = g'(t) \cdot (x - p).$$

Using that $p = g(p)$ it follows that

$$\begin{aligned} |x_n - p| &= |g(x_{n-1}) - g(p)| = |g'(t) \cdot (x_{n-1} - p)| = |g'(t)| \cdot |x_{n-1} - p| \\ &\leq K|x_{n-1} - p|. \end{aligned}$$

Repeating this $n - 1$ times yields

$$|x_n - p| \leq K^n |x_0 - p|.$$

Since $K \in (0,1)$, it follows that $K^n \to 0$, as $n \to \infty$, so that the right-hand side tends to zero. Hence, $|x_n - p| \to 0$ or $x_n \to p$, as $n \to \infty$.

Example 4.4 *Let's re-consider example 4.3 with the convergence theorem. From $g_1'(x) = \frac{2}{3}x$, we have $|g_1'(x)| < 1$ for all $x \in [0,1]$, so that the conditions hold and the sequence generated from $x_0 = 1$ converges. However, for $x_0 = 3$, we have $g_1'(x_0) = 2 > 1$ so that convergence cannot be guaranteed. For g_2, we have $g_2'(x) = 1/x^2$ and we obtain $g_2(x_0) = g_2(1) = 2$ after the first step. Then we obtain $g_2'(2) < 1$, and this holds for all values of x_n larger than one, so that the sequence converges. Similarly, for $x_0 = 3$, we have $g_2'(3) < 1$, so that convergence is guaranteed. Note that g_1 and g_2 yield a single but different solution generated from different starting points.*

The order of convergence can be obtained from a second-order Taylor approximation around p. That is,

$$x_{n+1} = g(x_n) = g(p) + g'(p)(x_n - p) + \frac{1}{2}g''(p)(x_n - p)^2 + \cdots \tag{4.1}$$

Let $\epsilon_n = x_n - p$ be the error at iterative step n. Then $g(p) = p$ implies

$$\epsilon_{n+1} = g'(p)\epsilon_n + \frac{1}{2}g''(p)\epsilon_n^2 + \cdots$$

So if $g'(p) \neq 0$, then the error ϵ_{n+1} at step $n+1$ linearly depends on error ϵ_n at step n. When this holds the convergence is called linear. If $g'(p) = 0$ and $g''(p) \neq 0$, then the convergence is called quadratic.

4.1.2 Newton iteration

The Newton Iterative method aims to solve the equation $f(x) = 0$, that is to find the x that zeroes the function f. The method can be derived from a first-order Taylor-approximation around x_n, that is

$$f(x_{n+1}) = f(x_n) + (x_{n+1} - x_n)f'(x_n) + \cdots$$

Setting this to zero

$$f(x_{n+1}) = f(x_n) + (x_{n+1} - x_n)f'(x_n) = 0$$

and solving for x_{n+1} we find

$$x_{n+1} = x_n - \frac{f(x_n)}{f'(x_n)}. \tag{4.2}$$

If $f'(x) \neq 0$, and the generated sequence converges ($x_n \to x$), then we obtain in the limit the equation $x = x - f(x)/f'(x)$, which gives the solution x that zeros the function $f(x) = 0$.

Example 4.5 *A classical example is to numerically approximate the root* $\sqrt{2}$ *by solving* $f(x) = x^2 - 2$. *From the derivative* $f'(x) = 2x$, *we find the iterative steps*

$$x_{n+1} = x_n - \frac{f(x_n)}{f'(x_n)} = x_n - \frac{x_n^2 - 2}{2x_n} = \frac{1}{2}\left(x_n + \frac{2}{x_n}\right),$$

for $x_n \neq 0$. *The first 4 steps yield the sequence* $x_0 = 1.000000, x_1 = 1.500000, x_2 = 1.416667, x_3 = 1.414216, x_4 = 1.414214$, *the latter of which equals* $\sqrt{2}$ *in 6 decimal places.*

Example 4.6 *Suppose we have the following function* f *and its first order derivative* f':

$$f(x) = x^3 + 2x^2 + 10x - 20 \Rightarrow f'(x) = 3x^2 + 4x + 10.$$

In order to find a zero for the function f *we program Newton's iterative procedure. Note, however, that we do not know beforehand the number of iterative steps to take in order to reach a certain level of precision. Such is nicely handled by a stopping criterion in a while loop. The R-code to implement Newton's algorithm reads*

```
f  <- function(x){x^3+2*x^2+10*x-20}
df <- function(x){3*x^2+4*x+10}
g  <- function(x){x-f(x)/df(x)}              # updating function
n  <- 1                                      # set counter to 1
x  <- c(10,g(10))                            # start
while (n<100 & abs(x[n+1]-x[n])>.0001) {
   n <- n+1
   x[n+1] <- g(x[n])
}
x
[1] 10.000000 6.342857 3.913918 2.381589 1.597070 1.383032 1.368866
[8]  1.368808
```

By plotting the function and adding a vertical line $x = 1.368808$ *it can be verified that the Newton procedures gives a precise approximation to the zero of the function* f, *see Figure 4.1.*

To analyze the order of convergence of Newton's algorithm, we note that it is in fact a fixed point method, since Equation (4.2) implies $x_{n+1} = g(x_n)$, where

$$g(x) = x - \frac{f(x)}{f'(x)}.$$

The latter has the first-order derivative

$$g'(x) = 1 - \frac{f'(x)f'(x) - f(x)f''(x)}{(f'(x))^2} = \frac{f(x)f''(x)}{(f'(x))^2}. \tag{4.3}$$

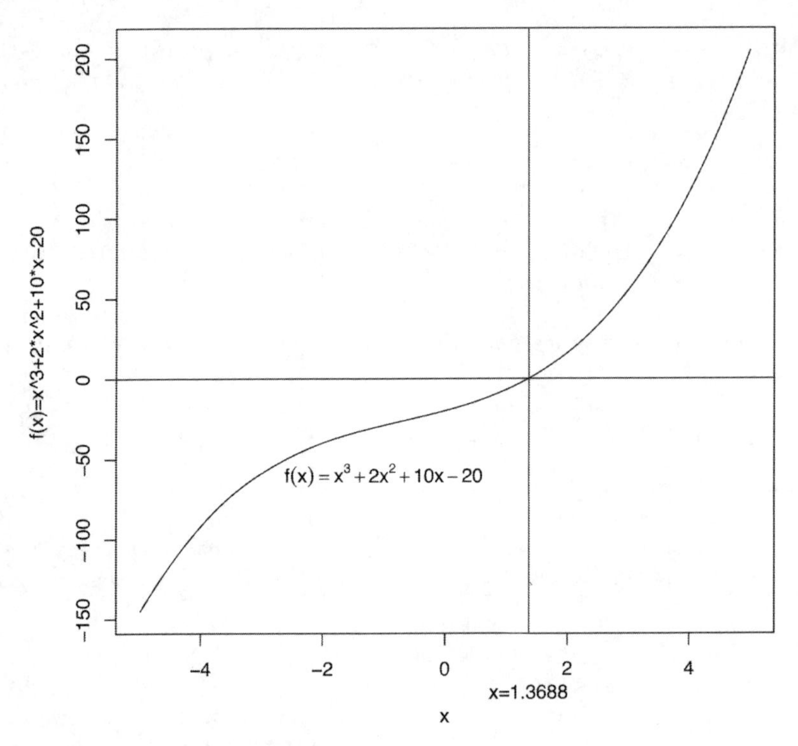

FIGURE 4.1
Newton iteration on $f(x) = x^3 + 2x^2 + 10x - 20$ to approximate its zero $x = 1.368808$.

If $x = p$ zeroes the function f, so that $f(p) = 0$, then $g'(p) = 0$. To evaluate the second-order derivative we compute

$$g'' = \left(\frac{f \cdot f''}{(f')^2}\right)' = \frac{(f' \cdot f'' + f \cdot f''')f'^2 + f \cdot f'' \cdot 2 \cdot f' \cdot f''}{(f')^4},$$

which evaluated at p simplifies to

$$g''(p) = \frac{f''(p)}{f'(p)}. \tag{4.4}$$

This is non-zero for almost all g that occur in practical settings. Hence, if the Newton algorithm converges, then its convergence is quadratic and a few steps often give sufficient precision for practical purposes.

Note that we do need $|g(x)| < 1$ in an interval around the solution p for the sufficient conditions of convergence to hold. The interval within which we obtain convergence to the true solution is also known as the "trust region."

Often, chemical processes can be expressed by a function for which we need to find its maximum or minimum value. A necessary condition for such

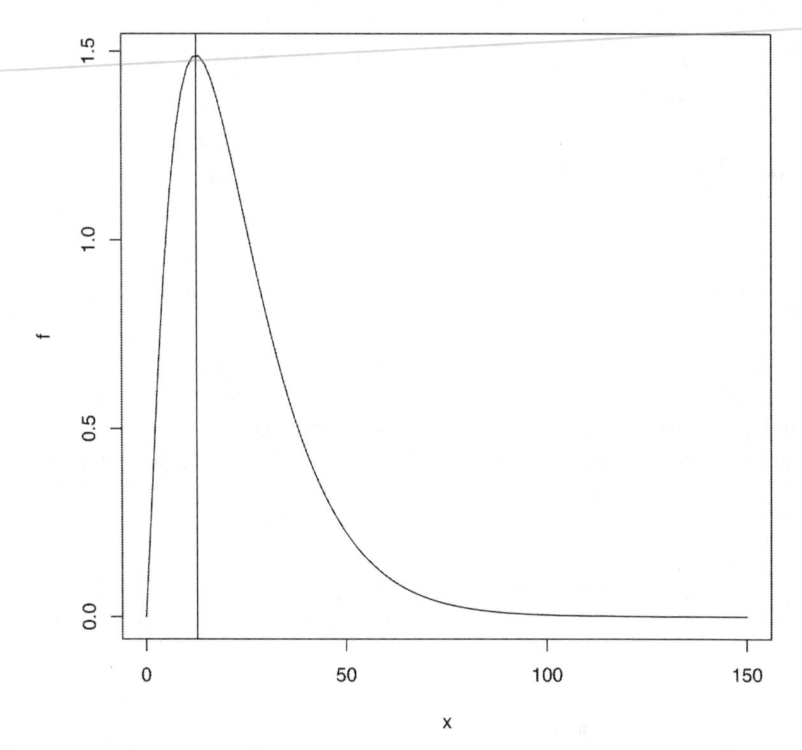

FIGURE 4.2
Plot of the concentration curve $f(t) = 0.2214 \cdot t^{1.2295} \cdot e^{-0.0959 \cdot t}$ attaining its maximum 1.489617 at $t = 12.81448$.

an extremum is that the first order derivative $f'(x) = 0$. Applying this to Equation (4.2), we find

$$x_{n+1} = x_n - \frac{f'(x_n)}{f''(x_n)}. \tag{4.5}$$

The importance of Newton's method is for the purpose of optimizing a function that depends on several variables for which plotting becomes difficult or even impossible.

Example 4.7 *The concentration function $f(t) = a \cdot t^b \cdot e^{-K \cdot t}$ mentioned in Chapter 1 with parameter values $a = 0.2214$, $b = 1.2295$, $K = 0.0959$ attains a unique maximum. The reader is asked in Exercise 1 to use first-order differentiation in order to show that this maximum occurs at $t = b/K$. An application of Newton's method gives, after a few iterative steps, the value $t = 12.81448$ for which t attains its maximum function value 1.489617, see Figure 4.2.*

4.2 Numerical Methods for Solving Differential Equations

Many chemical and physical laws are formulated in terms of an **ordinary differential equation** (ODE). For domain $[a, b]$ of y an ODE is given by

$$y'(x) = f(x, y), \tag{4.6}$$

with initial value condition

$$y(x_0) = y_0. \tag{4.7}$$

The problem is to find a function y of $x \in [a, b]$ with derivative f that starts at y_0, where f and initial value y_0 are given. The latter is necessary for the solution $y(x)$ to be unique. From calculus we know that a solution exists if f is continuous and bounded on an interval $[a, b]$ (Apostol, 2007a; Kreyszig et al., 2011; Stewart, 2007). This generality does cover almost all ODE problems arising in chemistry. However, even a simple differential equation[1] may not have an analytical solution, or it may happen that the solution of an ODE is analytically so complex that it hardly adds to our knowledge. For such problems it is useful to have an approximate solution given by an iterative method.

4.2.1 Euler's iterative method

For the interval $[a, b]$, we may choose a grid of N ordered equidistant points

$$a = x_0 < x_1 < x_2 < \cdots < x_n < \cdots < x_N = b.$$

To construct these we take $x_n = x_0 + nh$, with step size $h = (b - a)/N$ and $n = 0, 1, 2, \cdots, N$. Note that when N is fixed, then we obtain h. Usually, however, a and b are given and we choose step-size h which fixes N. Now the idea is to use x_n recursively to generate y_n in order to approximate the solution $y(x_n)$ with sufficient precision.

To define Euler's method we need a Taylor approximation of $y(x)$ around x_n, that is

$$y(x_{n+1}) = y(x_n) + \frac{h}{1!}\frac{dy}{dx}(x_n) + \frac{h^2}{2!}\frac{d^2y}{dx^2}(x_n) + \cdots$$

If we now take the first two terms on the right-hand side and use, from Equation (4.6), the substitution

$$\frac{dy}{dx}(x_n) = y'(x_n) = f(x_n, y(x_n)),$$

[1] $y' = y^2 + x$ is an example of a ODE without an analytic solution.

TABLE 4.2
Approximation of the solution $y(x) = e^x$ by Euler's method taking $h = 0.2, 0.1, 0.01$, respectively. The error ϵ_n is the difference between the true solution and Euler's approximation at location x_n.

		$h = 0.2$	ε_n	$h = 0.1$	ε_n	$h = 0.01$	ε_n
n	x_n	y_n	$y_n - y(x_n)$	y_n	$y_n - y(x_n)$	y_n	$y_n - y(x_n)$
1	0.2	1.200000	−0.021403	1.210000	−0.011403	1.220190	−0.001213
2	0.4	1.440000	−0.051825	1.464100	−0.027725	1.488864	−0.002961
3	0.6	1.728000	−0.094119	1.771561	−0.050558	1.816697	−0.005422
4	0.8	2.073600	−0.151941	2.143589	−0.081952	2.216715	−0.008826
5	1.0	2.488320	−0.229962	2.593742	−0.124539	2.704814	−0.013468
6	1.2	2.985984	−0.334133	3.138428	−0.181689	3.300387	−0.019730
7	1.4	3.583181	−0.472019	3.797498	−0.257702	4.027099	−0.028101
8	1.6	4.299817	−0.653215	4.594973	−0.358059	4.913826	−0.039206
9	1.8	5.159780	−0.889867	5.559917	−0.489730	5.995802	−0.053845
10	2.0	6.191736	−1.197320	6.727500	−0.661556	7.316018	−0.073038

then

$$y(x_{n+1}) = y(x_n) + h \cdot f(x_n, y(x_n)) + hT_n,$$

where T_n is the error term. Neglecting the latter and writing y_n for $y(x_n)$, the iterative method of Euler becomes

$$y_{n+1} = y_n + h \cdot f(x_n, y_n). \tag{4.8}$$

Taking $n = 0$, x_0, and y_0 from the initial condition, equation 4.8 gives y_1. Next, taking $n = 1$, x_1, and y_1, into 4.8 gives y_2. Then, taking $n = 2$, x_2, and y_2, into 4.8 gives y_3, etc. Hence, Equation (4.8) generates an approximating sequence of y_n values for each of the x_n values fixed beforehand from the interval $[a, b]$.

Example 4.8 *Suppose we have ODE $y' = y$, with initial value condition $y(0) = 1$. Taking primitives the immediate general solution to the ODE is $y(x) = ae^x$. Now, the initial value condition implies $y(0) = ae^0 = a = 1$, so that the unique solution $y(x) = e^x$. Using the definition that $f(x_n, y_n) = y_n$ and starting at $y_0 = 1$, the method of Euler specifies into*

$$y_{n+1} = y_n + hy_n = (1 + h)y_n.$$

To generate approximations to the solution function for $x \in [0, 2]$, we take three decreasing values $h = 0.2$, $h = 0.1$, and $h = 0.01$ to study the sequences generated by Euler's algorithm. From Table 4.2, it can be observed that the absolute error $\varepsilon_n = y_n - y(x_n)$ decreases for smaller values of h, and that it increases with x_n.

4.2.2 Runge-Kutta iterative method

The Runge-Kutta order 4 method (RK4) is very popular in chemistry for solving ODEs. It is based upon a weighted average of four estimates $k_1, k_2, k_3,$

TABLE 4.3
Exact solution $y(x) = x^2 - 2x + 2 - e^{-x}$ and its RK4 approximation y_n to $y' = x^2 - y$, with initial condition $y(0) = 1$.

n	$y(x)$	y_n
0	1.000000	1.000000
1	0.821269	0.821273
2	0.689680	0.689688
3	0.611188	0.611200
4	0.590671	0.590686
5	0.632121	0.632138
6	0.738806	0.738826
7	0.913403	0.913425
8	1.158103	1.158128
9	1.474701	1.474727
10	1.864665	1.864692

and k_4, taken from Equation (4.6), one after the other

$$\begin{aligned}
k_1 &= hf(x_n, y_n) \\
k_2 &= hf(x_n + \frac{h}{2}, y_n + \frac{k_1}{2}) \\
k_3 &= hf(x_n + \frac{h}{2}, y_n + \frac{k_2}{2}) \\
k_4 &= hf(x_{n+1}, y_n + k_3). \\
y_{n+1} &= y_n + \frac{1}{6}(k_1 + 2k_2 + 2k_3 + k_4)
\end{aligned}$$

Each step takes about four times as much computation time compared to Euler's iteration steps in Equation (4.8). Its error, however, is smaller and its stability larger compared to Euler's method.

Example 4.9 *Suppose we have on domain* $[0, 2]$ *the ODE*

$$y' = x^2 - y, \text{ with initial value condition } y(0) = 1.$$

The exact solution $y(x) = x^2 - 2x + 2 - e^{-x}$ *is derived in Exercise 10. Taking* $h = 0.2$, $N = 10$, $x_n = x_0 + nh$, $y_0 = y(0) = 1$, *and* $f(x_n, y_n) = x_n^2 - y_n$, *for* $n = 1, 2, \cdots, N$, *we obtain from the RK4 algorithm the sequence given in Table 4.3. The error seems negligible from a practical point of view.*

Example 4.10 *The reaction between nitrous oxide and oxygen to form nitrogen dioxide is given by the* **balanced equation**: $2NO + O_2 = 2NO_2$. *At* 25^oC *the rate at which* $2NO_2$ *is formed obeys the law of mass action, which is given by the rate equation*

$$y' = k(\alpha - y)^2(\beta - y/2), \text{ with initial value condition } y(0) = 0.$$

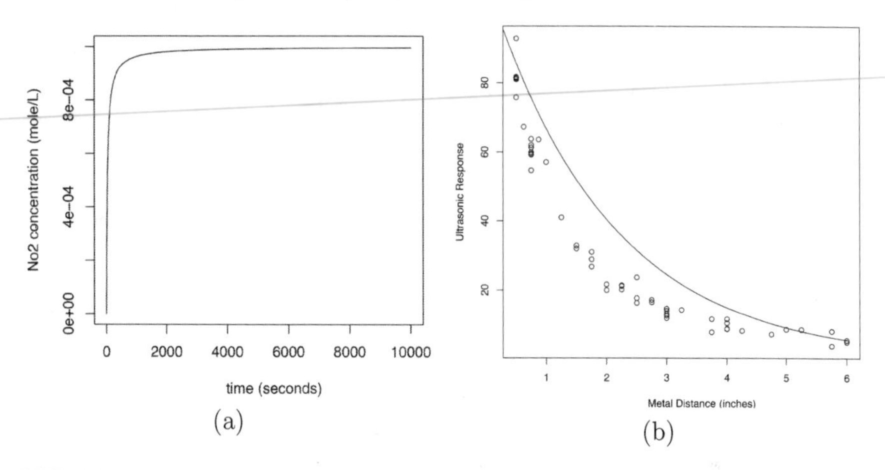

FIGURE 4.3
(a) RK4 approximation of the solution $y(t)$, describing the concentration $2NO_2$ as a function of time in seconds. (b) Ultrasonic response depending on metal distance measured in inches.

Here $y(t)$ is the concentration of $2NO_2$ in time t, the rate constant $k = 7.13 \cdot 10^3, \text{liter}^2/\text{mole}^2 sec.$, $\alpha = 0.0010$ mole/L, and $\beta = 0.0041$ mole/L. Note that the concentration starts at zero, where the slope $y'(0) = k\alpha^2\beta$ is largely positive. By continuity $y(t)$ increases until $y'(t)$ nears its smallest zero at $y = \alpha$, see Figure 4.3a.

Example 4.11 *In a NIST study (Chwirut, 1979) involving ultrasonic calibration, several measurements were obtained on ultrasonic response depending on metal distance measured in inches. From the data in Figure 4.3b, we observe that the ultrasonic response y is monotonically decreasing as the metal distance x increases. As a first educated guess we may conjecture that $y' = -by$, with $b = 0.5$, and initial value condition $y(0) = 110$. Taking domain $[0, 6]$ for x, $h = 0.1$, so that $y_0 = y(0) = 110$, and $f(x_n, y_n) = -0.5y_n$, for $n = 1, 2, \cdots, N$ we obtain an approximation from the RK4 algorithm as it is given by the smoothed curve in Figure 4.3b. It seems clear that the approximation can be improved by a change in parameters resulting in a steeper decrease of the solving function. Also note that some error in the observations is indicated by metal distance having different values measured for the same ultrasonic response.*

It seems clear from the previous example that the fit of the RK4 generated sequence to experimental data depends crucially upon the starting point $y(0)$ and the precision of the coefficients. The first seems not problematic for differential equations that have $y(0) = 0$ as its natural initial condition, but in general the coefficients and the initial values can be estimated from the data, as described in Chapter 9.

4.3 Differential Algebraic Equations

The class of differential algebraic equations matches quite closely with a system of chemical reactions in time. The concept can best be explained by an example.

Example 4.12 *Robertson (1966) gave a problem consisting of an autocatalic reaction between three chemical species A, B, C (Hairer and Wanner, 2010), as*

$$
\begin{aligned}
A &\overset{0.04}{\longrightarrow} B \\
B+B &\overset{3\cdot 10^7}{\longrightarrow} C+B \\
B+C &\overset{10^4}{\longrightarrow} A+C
\end{aligned}
$$

This leads to the equations

$$
\begin{aligned}
y_1' &= -0.04y_1 + 10^4 y_2 y_3 \\
y_2' &= 0.04y_1 - 10^4 y_2 y_3 - 3\cdot 10^7 y_2^2 \\
y_3' &= 3\cdot 10^7 y_2^2
\end{aligned}
\tag{4.9}
$$

with initial conditions $y_1(0) = 1$, $y_2(0) = y_3(0) = 0$. The problem can be reformulated as an initial value differential algebraic equation (DAE) reading Soetaert et al. (2010)

$$
\begin{aligned}
y_1' &= -0.04y_1 + 10^4 y_2 y_3 \\
y_2' &= 0.04y_1 - 10^4 y_2 y_3 - 3\cdot 10^7 y_2^2 \\
1 &= y_1 + y_2 + y_3
\end{aligned}
$$

The first two equations specify the dynamics of chemical species y_1 and y_2, and the third is that the summed concentrations of the three species remains 1. The DAE can be re-written, by bringing terms to the right-hand side, as residual functions to become as small as possible

$$
\begin{aligned}
r_1 &= -y_1' - 0.04y_1 + 10^4 y_2 y_3 \\
r_2 &= -y_2' + 0.04y_1 - 10^4 y_2 y_3 - 3\cdot 10^7 y_2^2 \\
r_3 &= -1 + y_1 + y_2 + y_3
\end{aligned}
$$

In the accompanying R code, the function `daefun`*, the current time t, and the residual equations are specified. These are used as input into the* `daspk` *solver for DAEs dealing using so-called implicit methods to solve this famous stiff ODE problem (Brenan, 1996). It returns the values of time t, that of the state variables $y_1(t)$, $y_2(t)$, $y_3(t)$, and the error term of the algebraic equation at time point t. By the latter, the accuracy of the results can be checked. The*

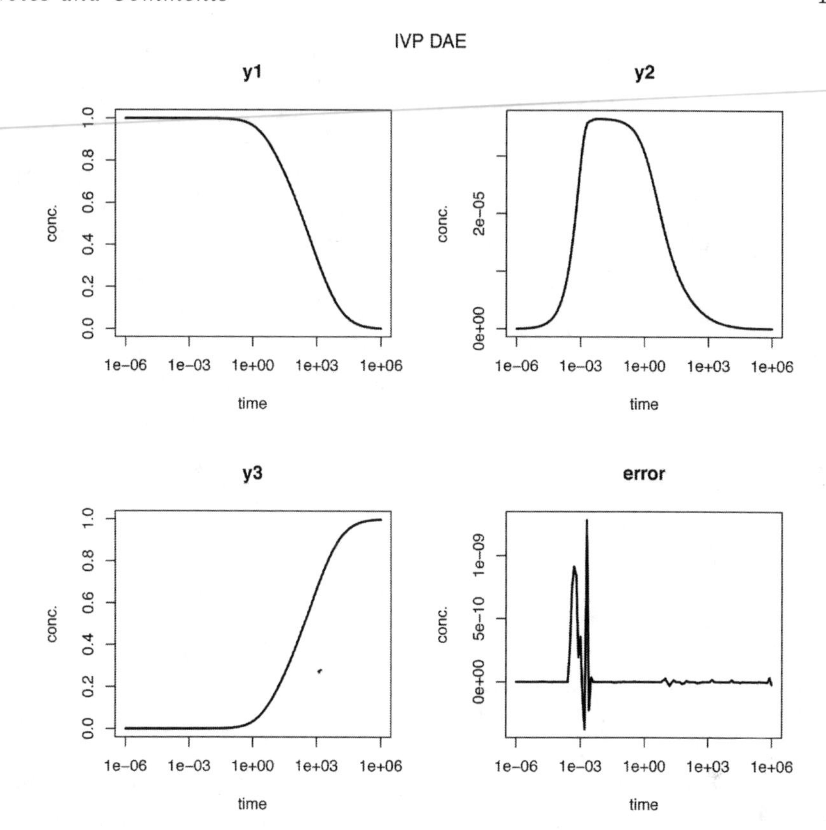

FIGURE 4.4
Solution of DAE problem for substances y_1, y_2, y_3, and the mass balance error; the deviation of total sum of substances to 1.

resulting approximations of the solution functions of the concentrations are plotted in Figure 4.4. It can be observed that the reaction between y_1 and y_2 takes place rather fast causing a monotonic decrease in y_1 concentration to zero and an initial increase of y_2 concentration immediately followed by a decrease to zero. The substance y_3 monotonically increases to 1. This apparently slow reaction speed suddenly changing into large differences in concentrations of e.g., y_2 is typical for stiff problems.

4.4 Notes and Comments

Newton's iterative method can be generalized to optimization for functions that depend on several variables even in Banach spaces (Luenberger, 1997). Additionally, a numerical approximation to derivatives can be used in developing algorithms (Quarteroni et al., 2010). Curve fitting based on multivariate Newton iteration is the basis of non-linear least squares estimation in order to find the best parameter values given measurements e.g., time x and chemical outcome y.

Simple differential equations can also be solved by symbolic systems such as Wolfram Alpha. The latter may be useful in distinguishing problems that have an analytic solution from those that only have a numerical solution. Differential equations are discussed in detail in books on Calculus by Stewart (2007) and Apostol (2007a), as well as by Kreyszig et al. (2011), and in books taking a numerical perspective by Hairer and Wanner (1992, 2010).

4.5 Notes on Using R

The above brief R code for the Newton algorithm illustrates high similarity in basic programming among many languages. Apart from directly programming the above algorithms (which is useful!), one may use R its build-in function `optim` to find extreme value(s) of a function, `uniroot` to find zeros of a function, `integrate` to numerically compute the integral of a function, and `rk` from the `deSolve` library to find Runge-Kutta type of approximations to solutions of ODEs. The latter contains several procedures to approximate solutions to ODEs by several algorithms and to plot these.

Explanations and examples on how to solve a wide scope of differential equations are publicly available (Soetaert et al., 2010).

A line can be added to a plot with data points by the `abline` and a curve by the `curve` R function.

4.6 Exercises

1. **Class of concentration curves.** A class of curves giving the concentration in time are given by

$$y(t) = at^b e^{-Kt}.$$

   (a) Verify that the curve y is the solution to the ODE $y' = (b/t - K)y$ with initial condition $y(0) = 0$.
   (b) Find the maximum of y by differentiation, that is setting the first-order derivative to zero.
   (c) Does the concentration tend to zero as time tends to infinity?
   (d) In a $^{13}CO_2$ excretion study it was found that $a = 0.07$, $b = 1.15$ and $K = 0.08$ with time in minutes. Find its time of maximum concentration. Plot the concentration curve.

2. **Two equations.** Solve the equations by one of the numerical methods. Hint: Use `uniroot` on the difference of functions.
   (a) $e^{-x} = \ln(x)$
   (b) $e^{x} + x^4 + x = 2$
3. **Heating equals cooling.** At what time will the heating process $f_1(t) = 100(1 - e^{-0.02t})$ and the cooling process $f_2(t) = 40e^{-0.01t}$ reach equal the same temperature? How large is this temperature? Hint: Use `uniroot` on the difference of functions.
4. **Maxwell–Boltzmann.** The Maxwell–Boltzmann probability distribution is the function

$$f(v) = \sqrt{\left(\frac{m}{2\pi kT}\right)^3} 4\pi v^2 e^{-\frac{mv^2}{2kT}},$$

   where m is the particle mass and kT is the product of Boltzmann's constant and thermodynamic temperature.
   (a) Simplify the formula by taking $a = \sqrt{kT/m}$.
   (b) Take $a = 5$ and find the v for which f is maximal.
5. **Lennard-Jones potential.** The Lennard-Jones potential for the interaction of two molecules separated by distance R is

$$U(R) = \frac{A}{R^{12}} - \frac{B}{R^6},$$

   where A and B are constants of the Buckingham potential. Suppose that $A = 0.01$ and $B = 0.5$. Does the potential have a minimum or maximum for $R \in (0, 2)$? What is its value?
6. **Bateman equation.** The concentration of a species B in a rate process is given by $A \overset{k_1}{\to} B \overset{k_2}{\to} C$, consisting of two consecutive irreversible first-order reactions, is given by

$$B(t) = \frac{k_1}{k_2 - k_1}(e^{-k_1 t} - e^{-k_2 t})$$

   Take $k_1 = 0.1$ and $k_2 = 0.2$. This is a form of the Bateman Equation.
   (a) Find the time t at which the concentration attains its maximum value.
   (b) What is the maximum concentration?
   (c) During which period is the concentration larger than 0.05? Find the two limit points of the time interval for which this holds.
7. **Michaelis-Menten.** Biochemical reactions involving a single substrate are often assumed to follow Michaelis–Menten kinetics. The reaction rate v is a function of the concentration of a substrate S according to

$$v(S) = \frac{V_{max} S}{K_m + S}.$$

(a) Plot the function with $K_m = 17.07$ and $V_{max} = 126.03$

(b) Show that V_{max} is the maximum rate achieved by the system, at saturating substrate concentration.

(c) Show that the constant K_m is the substrate concentration at which the reaction rate is half of V_{max}.
Hint: If may want to compare your analytical answer with that from numerical methods.

8. **Second-order reactions.** Second-order chemical reactions result in a differential equation of the form $y' = -ky^2$. Suppose that $k = 0.75$ and $y(0) = 1/100$.

   (a) Solve the differential equation analytically.

   (b) Approximate the solution with the RK4 algorithm and compare with the exact solution.

9. **ODE 1.** Consider the differential equation $y' = (y - x - 1)^2 + 2$, with initial value condition $y(0) = 1$, over domain $x \in [0, 1.5]$.

   (a) Approximate the solution with the RK4 algorithm. Use the programmes algorithm from Nestor or the `rk4` function from the R library `deSolve`.

   (b) Repeat (a) with the Euler algorithm.

10. **ODE 2.** Consider on domain $[0, 2]$ the ODE $y' = x^2 - y$ with initial value condition $y(0) = 1$.

    (a) Show that it has the exact solution $y(x) = x^2 - 2x + 2 - e^{-x}$.

    (b) Compare the numerical approximation to this solution by the methods "euler," second and fourth-order Runge Kutta ("rk2," "rk4"), and Runge-Kutta-Fehlberg ("rk34f"). What is your conclusion? Hint: Conveniently use the functions from the R library `deSolve`.

11. **Minimal value.** It may happen that we want to minimize a function and that this turns out to be impossible. Consider the function

$$f(x, y) = x^2 + (xy - 1)^2,$$

over $(x, y) \in \mathbb{R}^2$.

   (a) Show that zero is a lower bound of f.

   (b) Find the partial derivative of f with respect to y, set it to zero and solve.

   (c) Substitute the solution to the (b) question into the function and minimize the new function of x.

5

Laws on Propagation of Error

This chapter gives an introduction to the subject of error propagation. It is important to realize that there is always some error in the measurements of a chemical quantity. The size of the error becomes feasible if we obtain replicated measurements of the same quantity under equal circumstances. The main question of concern in this chapter is: How does the error in the measurements for x transfer into error for a known function g of x? The basic concepts of mean, variance, covariance, standard deviation are introduced from the point of view of the measurements of a chemical quantity.

In this chapter you learn

- In Section 5.1 how error propagation affects the absolute and relative error of the measurement of a chemical quantity.
- In Section 5.2 how the error of measurement propagates for a linear function in terms of the mean and the variability of a series of chemical measurements.
- In Sections 5.3 and 5.4 how the error of measurement propagates for functions of one variable or two variables.

5.1 Absolute and Relative Error of Measurement

Suppose we are able to measure a chemical quantity x, although we are not primarily interested in x itself. Rather, we are interested in a functional dependency of x, namely $y = g(x)$. Suppose, however, that it is very difficult or even impossible to measure $g(x)$ directly by a chemical experiment. The question that concerns us is: What is the implication of a measurement error in x for the error in y?

Suppose that the actual measurement x_a differs from true value x_t. Then the **primary error** of measurement $\triangle x = x_t - x_a$. The latter leads to the **secondary error** of measurement $\triangle y = g(x_t) - g(x_a)$. The $\triangle x$ and $\triangle y$ are also called **absolute errors** of measurement. In case $\triangle x$ is sufficiently small we have the first-order Taylor approximation

$$g(x_t) \approx g(x_a) + g'(x_a)(x_t - x_a)$$

DOI: 10.1201/9781003178194-5

or

$$g(x_t) - g(x_a) \approx g'(x_a)(x_t - x_a).$$

In terms of absolute error of measurement this reads as

$$\triangle y \approx g'(x_a)\triangle x$$

Example 5.1 *Suppose we wish to measure the volume V of certain ball-shaped protein molecules and we are only able to measure its radius r. The function g now becomes $V(r) = \frac{4}{3}\pi r^3$. Suppose furthermore that we measure $r_a = 10\mu m$, although it is known from previous research that the primary measurement error is 0.1 μm. Then $V(10) = 4188.79\,\mu\text{m}^3$. Since $V'(r) = 4\pi r^2$, we find $V'(10) = 4\pi 10^2 = 1256.637\,\mu\text{m}^2$, so that $\triangle V = V'(10) \cdot \triangle r = 125.6637\,\mu\text{m}^3$.*

In practical applications we are often much more interested in the **relative errors** $\triangle x/x$ and $\triangle y/y$. Assuming that x_a is close to x and that y_a is close to y, we find from $\triangle y \approx g'(x_a)\triangle x$ that[1]

$$\begin{aligned} \frac{\triangle y}{y} &\approx \frac{g'(x_a) \cdot \triangle x}{y_a} = \frac{g'(x_a) \cdot \triangle x}{g(x_a)} \\ &\approx \frac{x_a \cdot g'(x_a)}{g(x_a)} \cdot \frac{\triangle x}{x} \end{aligned} \tag{5.1}$$

We observe that $x_a \cdot g'(x_a)/g(x_a)$ serves as a conversion factor from the relative error in x to that of y.

Example 5.2 *In Example 5.1 we found a relative error $\triangle r/r = 0.1/10 = 0.01 = 1\%$ for the radius and a relative error $\triangle V/V = 125.6637/4188.79 = 0.03 = 3\%$ for volume. Observe that the relative error for the volume is three times larger than that of the radius. This also follows from*

$$\frac{r_a V'(r_a)}{V(r_a)} = \frac{r_a \cdot 4\pi r_a^2}{\frac{4}{3}\pi r_a^3} = \frac{1}{\frac{1}{3}} = 3.$$

This holds in fact more general for any power function of x. That is, if $y = g(x) = cx^r$, then $g'(x) = crx^{r-1}$. Using Equation (5.1), we find

$$\frac{g'(x_a) \cdot \triangle x}{g(x_a)} = \cdot \frac{x_a \cdot crx_a^{r-1}}{cx_a^r} = r \Rightarrow \frac{\triangle y}{y} \approx r \cdot \frac{\triangle x}{x}$$

So that we obtain a simple expression for the relative error of y.

Example 5.3 *Suppose the concentration c of a substance X is to be measured by a photometer through the absorption quotient $A = I_{in}/I_{out}$, where I_{in} is the intensity of incoming light and I_{out} that of outgoing. A version of the*

[1] For the sake of clarity note that we simply substitute y_a for y, and multiply by $x_a/x \approx 1$.

Lambert-Beer law *states that* $I_{out} = I_{in} \cdot e^{-kc}$*, where* k *is the absorption coefficient of the substance* X *and* c *the unknown concentration of* X*. By substitution it follows that*

$$A = \frac{I_{in}}{I_{out}} = e^{kc} \Rightarrow c = \frac{\ln(A)}{k} = g(A),$$

where the last step defines the function g *of* A*. If* A_a *is the actual absorption measured, then the absolute error of the concentration*

$$\triangle c = g'(A_a)\triangle A = \frac{1}{kA_a} \cdot \triangle A = \frac{1}{k} \cdot \frac{\triangle A}{A_a} \approx \frac{1}{k} \cdot \frac{\triangle A}{A}.$$

In the last step we used $\triangle A/A_a$ *as an approximation for* $\triangle A/A$*. So the absolute error of concentration is proportional to the relative error of absorption. If, for example,* $k = 0.05$*, then a relative absorption error of 1%, e.g.,* $\triangle A/A = 0.01$*, implies an absolute concentration error* $\triangle c \approx \frac{1}{0.05} \cdot 0.01 = 0.2$*. From Equation (5.1) we find the relative error of concentration*

$$\frac{\triangle c}{c} \approx \frac{A_a g'(A_a)}{g(A_a)} \cdot \frac{\triangle A}{A} = \frac{A_a \cdot 1/(kA_a)}{\ln(A_a)/k} \cdot \frac{\triangle A}{A} = \frac{1}{\ln(A_a)} \cdot \frac{\triangle A}{A}.$$

Hence, the relative concentration error $\triangle c/c$ *is mathematically independent from the absorption coefficient* k*. Note that, due to the logarithm, an actual absorption* A_a *measured close to 1 causes a large increase on the relative concentration error* $\triangle c/c$*. That is, if* $A_a = 1.01$*, then the relative concentration error becomes* $1/\ln(A_a) = 100.5$ *times as large.*

5.2 Mean and Variance

A series of independently replicated measurements of a single unknown quantity gives a sequence $(x_1, \cdots, x_n)$ of numbers. Its **mean** or average

$$\overline{x} = \frac{x_1 + \cdots + x_n}{n} = \frac{1}{n}\sum_{i=1}^{n} x_i, \tag{5.2}$$

is the best approximation of the unknown quantity. The **variance** of the measurements is defined as

$$s_x^2 = \frac{(x_1 - \overline{x})^2 + \cdots + (x_n - \overline{x})^2}{n} = \frac{1}{n}\sum_{i=1}^{n}(x_i - \overline{x})^2. \tag{5.3}$$

The variance gives the mean of the squared differences between the measurements and their average. It is small when all measurements are close.[2] The **standard deviation** is the square root of the variance, $s_x = \sqrt{s_x^2}$.

[2]Note throughout that precision in inversely related to variance in the sense that small variance implies high precision.

Example 5.4 *Suppose that g is a linear function; $y_i = g(x_i) = ax_i + b$, for $i = 1, \cdots, n$, then*

$$\overline{y} = \frac{1}{n}\sum_{i=1}^{n}(ax_i + b) = a\frac{1}{n}\sum_{i=1}^{n} x_i + \frac{1}{n}\sum_{i=1}^{n} b = a\overline{x} + b.$$

That is, $\overline{y} = g(\overline{x})$, and we see that the "operator" of taking the average $\frac{1}{n}\sum_{i=1}^{n}(\cdot)$ acts linear.

For the variance, we have

$$\begin{aligned} s_y^2 &= \frac{1}{n}\sum_{i=1}^{n}(y_i - \overline{y})^2 \\ &= \frac{1}{n}\sum_{i=1}^{n}(ax_i + b - a\overline{x} - b)^2 \\ &= \frac{1}{n}\sum_{i=1}^{n} a^2 \cdot (x_i - \overline{x})^2 \\ &= a^2 \cdot s_x^2. \end{aligned}$$

Below we will use the obvious

$$\frac{1}{n}\sum_{i=1}^{n}(x_i - \overline{x}) = \frac{1}{n}\sum_{i=1}^{n} x_i - \frac{1}{n}\sum_{i=1}^{n} \overline{x} = \overline{x} - \overline{x} = 0. \tag{5.4}$$

5.3 Functions that Depend on One Variable

5.3.1 First-order approximation

If all $(x_1, \cdots, x_n)$ values are sufficiently close to $\overline{x}$ and g has a second-order derivative, then a first-order approximation of g around $\overline{x}$ gives

$$y_i = g(x_i) \approx g(\overline{x}) + g'(\overline{x})(x_i - \overline{x}), \text{ for } i = 1, \cdots, n. \tag{5.5}$$

If we compute the mean $\overline{y}$ from this, we find

$$\frac{1}{n}\sum_{i=1}^{n} y_i \approx \frac{1}{n}\sum_{i=1}^{n} g(\overline{x}) + g'(\overline{x})\frac{1}{n}\sum_{i=1}^{n}(x_i - \overline{x}) \Rightarrow \tag{5.6}$$

$$\overline{y} \approx g(\overline{x}) \tag{5.7}$$

Example 5.5 *Suppose we obtained the following five measurements for the radius of molecules $r_1 = 10.09, r_2 = 10.39, r_3 = 9.32, r_4 = 10.99, r_5 = 10.62$ μm. Then $\overline{r} = 10.28339$, and for the volume function $V(r) = \frac{4}{3}\pi r^3$, we find $V(\overline{r}) = 4555.091\,\mu\text{m}^3$.*

In order to derive an approximation to the variance of y, we substitute $\overline{y}$ for $g(\overline{x})$ into Equation (5.5) for each i, then we have

$$y_i - \overline{y} \approx g'(\overline{x})(x_i - \overline{x})$$

If we now take the square and apply our average "operator" $\frac{1}{n}\sum_{i=1}^{n}(\cdot)$, then we find the important result

$$s_y^2 = \frac{1}{n}\sum_{i=1}^{n}(y_i - \overline{y})^2 \approx [g'(\overline{x})]^2 \cdot \frac{1}{n}\sum_{i=1}^{n}(x_i - \overline{x})^2 = [g'(\overline{x})]^2 \cdot s_x^2 \tag{5.8}$$

and, by taking square roots, that

$$s_y \approx |g'(\overline{x})| \cdot s_x. \tag{5.9}$$

Example 5.6 *For the volume function $V(r) = \frac{4}{3}\pi r^3$ in Example 5.5 we find $V'(r) = 4\pi r^2$. Hence, for $\overline{r} = 10.28$ and $s_r = 0.6295$, we find $|V'(\overline{r})| \cdot s_r = 836.5399$. This seems fairly close to the standard deviation 814.0202 of V.*

Example 5.7 *The absorbance of a solution is functionally related to the transmittance by $A(T) = -\log_{10}(T)$. Suppose a number of measurements resulted in $\overline{T} = 0.501$ and $s_T = 0.001$. Then $A(\overline{T}) = -\log_{10}(0.501) = 0.300$. Using that $\log_{10}(T) = \frac{\ln(T)}{\ln(10)}$ we find*

$$\frac{dA}{dT} = -\frac{1}{\ln(10)}\frac{1}{T} \Rightarrow s_A \approx \left|-\frac{1}{\ln(10)}\frac{1}{0.501}\right| \cdot 0.001 = 0.0008.$$

5.3.2 Second-order approximation

In some cases a first-order approximation can be improved by a second-order approximation. That is, if all values $(x_1, \cdots, x_n)$ are sufficiently close to $\overline{x}$, then a second-order approximation of x_i around $\overline{x}$, for $i = 1, \cdots, n$ is

$$y_i = g(x_i) \approx g(\overline{x}) + g'(\overline{x})(x_i - \overline{x}) + \left(\frac{1}{2}g''(\overline{x})\right)(x_i - \overline{x})^2, \tag{5.10}$$

If we now take the average to the left and the right, using Equation (5.4), then

$$\begin{aligned} \overline{y} = \frac{1}{n}\sum_{i=1}^{n} y_i &\approx g(\overline{x}) + g'(\overline{x})\frac{1}{n}\sum_{i=1}^{n}(x_i - \overline{x}) + \left(\frac{1}{2}g''(\overline{x})\right)\frac{1}{n}\sum_{i=1}^{n}(x_i - \overline{x})^2 \Rightarrow \\ \overline{y} &\approx g(\overline{x}) + \left(\frac{1}{2}g''(\overline{x})\right) \cdot s_x^2 \end{aligned} \tag{5.11}$$

This equation shows that the difference between the first and the second-order approximation is proportional to the variance of the x-measurements.

Example 5.8 *For the volume function $V(r) = \frac{4}{3}\pi r^3$ in Example 5.5, we find second-order derivative $V''(r) = 8\pi r$. From the measurement we had $\overline{r} = 10.28339$, and $s_r = 0.6295\mu m$, so that $V(\overline{r}) + \frac{1}{2}V''(\overline{r}) \cdot s_r^2 = 4606.301$, which is closer to $\overline{V} = 4595.642$ than the first-order approximation.*

Some remarks are in order:

1. The suppositions about second-order differentiability of g hold for almost all chemical applications.
2. Recall from Calculus Stewart (2007); Apostol (2007a) that a function g is convex if and only if $g''(x) > 0$ for all x. Since the variance $s_x^2 > 0$, the correction term $\left(\frac{1}{2}g''(\overline{x})\right) \cdot s_x^2$ is positive, so that for all convex functions g, we have that $g(\overline{x})$ is an under estimation of $\overline{y}$. Similarly, a function g is concave if and only if $g''(x) < 0$ for all x. From $s_x^2 > 0$, we observe that the correction term $\left(\frac{1}{2}g''(\overline{x})\right) \cdot s_x^2$ is negative, so that $g(\overline{x})$ is an over estimation of $\overline{y}$.
3. The two approximations (5.7) and (5.11) of $\overline{y}$ are equal if and only if $g''(\overline{x}) = 0$, that is, when g is e.g., linear $g(x) = ax + b$. In such a case there is no over or under estimation of $\overline{y}$.

5.4 Functions that Depend on Two Variables

Suppose that the outcome z depends on two variables x and y, that is $z = g(x, y)$. How can we generalize the previous approximations? A flavor of what happens is given by an example dealing with a linear function.

Suppose that $z_i = g(x_i, y_i) = a + bx_i + cy_i$, then

$$\overline{z} = \frac{1}{n}\sum_{i=1}^{n}(a + bx_i + cy_i) = a + b\overline{x} + c\overline{y}.$$

In order to compute the variance of z, we compute the deviation from the mean,

$$z_i - \overline{z} = b(x_i - \overline{x}) + c(y_i - \overline{y}).$$

From this we find

$$\begin{aligned}
s_z^2 &= \frac{1}{n}\sum_{i=1}^{n}(z_i - \overline{z})^2 = \frac{1}{n}\sum_{i=1}^{n}\Big(b(x_i - \overline{x}) + c(y_i - \overline{y})\Big)^2 \\
&= \frac{1}{n}\sum_{i=1}^{n}\Big(b^2(x_i - \overline{x})^2 + 2bc \cdot (x_i - \overline{x})(y_i - \overline{y}) + c^2 \cdot (y_i - \overline{y})^2\Big)^2 \\
&= b^2 \cdot \frac{1}{n}\sum_{i=1}^{n}(x_i - \overline{x})^2 + 2bc \cdot \frac{1}{n}\sum_{i=1}^{n}(x_i - \overline{x})(y_i - \overline{y}) + c^2 \cdot \frac{1}{n}\sum_{i=1}^{n}(y_i - \overline{y})^2 \\
&= b^2 \cdot s_x^2 + 2bc \cdot s_{xy} + c^2 \cdot s_y^2, \qquad (5.12)
\end{aligned}$$

where the sample **covariance** is defined as

$$s_{xy} = \frac{1}{n}\sum_{i=1}^{n}(x_i - \overline{x})(y_i - \overline{y}).$$

It is immediate that

$$s_z = \sqrt{b^2 \cdot s_x^2 + 2bc \cdot s_{xy} + c^2 \cdot s_y^2}$$

Example 5.9 *In a titration the initial reading of a burette is 4.51 ml and the final reading is 16.67 ml, both with standard deviation 0.03 ml. The two readings are conducted independently by two different analysts, so that it can safely be assumed that the covariance is zero. The volume of the titrant is the difference* $16.67 - 4.51 = 12.16$ ml. *The standard deviation of the difference is*

$$s_z = \sqrt{1^2 \cdot 0.03^2 + (-1)^2 \cdot 0.03^2} = 0.042 \text{ ml}.$$

Now suppose that the readings are conducted by a single analyst and that the covariance is 0.00072 ml. The standard deviation of the difference

$$s_z = \sqrt{1^2 \cdot 0.03^2 + 2 \cdot 1 \cdot -1 \cdot 0.00072 + (-1)^2 \cdot 0.03^2} = 0.019 \text{ ml}$$

is about half of the size of that in the previous. Note that the product bc is negative due to taking the difference and the standard deviation of the difference becomes smaller in case of a positive covariance and larger in case of negative covariance. An adequate interpretation of the formula's provides us with the tools to correctly unravel the situation.

5.4.1 Covariance and correlation

The Cauchy-Schwartz inequality gives an upper bound to the covariance

$$|s_{xy}| \leq s_x \cdot s_y.$$

Build upon this idea the **correlation coefficient** is defined as

$$r_{xy} = \frac{s_{xy}}{s_x \cdot s_y}.$$

The correlation coefficient is between -1 and 1. If $r_{xy} = 1$, then there is a linear relationship $y = ax + b$ between the measurements x and y. The correlation coefficient is among the most important coefficients to express the degree of linear relationship between measurements x and y.

Example 5.10 *A new flame atomic-absorption spectroscopic method of determining antimony (Castillo et al., 1982) in the atmosphere was compared with the recommended calorimetric method. For six samples from an urban atmosphere, the measurements were obtained as reported in Table 5.1. It can be*

TABLE 5.1
Antimony (mg m^{-3}) determined by a new method and a standard method.

Sample nr	**1**	**2**	**3**	**4**	**5**	**6**
New method	22.2	19.2	15.7	20.4	19.6	15.7
Standard method	25.0	19.5	16.6	21.3	20.7	16.8

expected that when both methods yield very similar measurements, then both are small or large implying a correlation close to 1. The means of the new and the standard method are 18.80 and 19.98, respectively. The variances are 6.828 and 9.846, respectively, and the covariance 7.980. From this we find the correlation equals $7.980/\sqrt{6.828 \cdot 9.846} = 0.973$.

5.4.2 First-order approximation

For a non-linear function g of both x and y we consider a first-order Taylor approximation around $(\overline{x}, \overline{y})$, given by

$$g(x_i, y_i) \approx g(\overline{x}, \overline{y}) + \frac{\partial g}{\partial x}(\overline{x}, \overline{y}) \cdot (x_i - \overline{x}) + \frac{\partial g}{\partial y}(\overline{x}, \overline{y}) \cdot (y_i - \overline{y}) \tag{5.13}$$

It is convenient to define

$$g_x = \frac{\partial g}{\partial x}(\overline{x}, \overline{y}), \; g_y = \frac{\partial g}{\partial y}(\overline{x}, \overline{y}),$$

where $\frac{\partial g}{\partial x}$ is the partial derivative of g with respect to x. Using Equations (5.12) and (5.13), it follows that

$$\begin{aligned} s_z^2 &= \frac{1}{n}\sum_{i=1}^{n}(z_i - \overline{z})^2 = \frac{1}{n}\sum_{i=1}^{n}\Big(g(x_i, y_i) - g(\overline{x}, \overline{y})\Big)^2 \\ &= \frac{1}{n}\sum_{i=1}^{n}\Big(g_x \cdot (x_i - \overline{x}) + g_y \cdot (y_i - \overline{y})\Big)^2 \\ &= g_x^2 \cdot s_x^2 + 2g_x \cdot g_y \cdot s_{xy} + g_y^2 \cdot s_y^2. \end{aligned} \tag{5.14}$$

In terms of standard deviations, this can be written as

$$s_z = \sqrt{g_x^2 \cdot s_x^2 + 2g_x \cdot g_y \cdot s_{xy} + g_y^2 \cdot s_y^2}.$$

5.4.3 Second-order approximation

The estimator may be improved by a second-order approximation. For this we need the definitions

$$g_{xx} = \frac{\partial^2 g}{\partial x \partial x}(\overline{x}, \overline{y}), \; g_{yy} = \frac{\partial^2 g}{\partial y \partial y}(\overline{x}, \overline{y}), \; g_{xy} = \frac{\partial^2 g}{\partial x \partial y}(\overline{x}, \overline{y})$$

where $g_{xy} = g_{yx}$ holds if the derivatives are continuous. A second-order Taylor approximation of $g(x_i, y_i)$ around $(\overline{x}, \overline{y})$ given by

$$\begin{aligned} g(x_i, y_i) \approx\ & g(\overline{x}, \overline{y}) + g_x \cdot (x_i - \overline{x}) + g_y \cdot (y_i - \overline{y}) \\ & + \frac{1}{2} g_{xx} \cdot (x_i - \overline{x})^2 + g_{xy} \cdot (x_i - \overline{x})(y_i - \overline{y}) + \frac{1}{2} g_{yy} \cdot (y_i - \overline{y})^2 \end{aligned}$$

If we compute the mean then we find

$$\overline{z} \approx g(\overline{x}, \overline{y}) + \frac{1}{2} g_{xx} \cdot s_x^2 + g_{xy} \cdot s_{xy} + \frac{1}{2} g_{yy} \cdot s_y^2. \tag{5.15}$$

5.5 Notes and Comments

Recall from Chapter 3 that if measurements X and Y are normally distributed, then they correlate zero if and only if they are stochastically independent. Furthermore, if they are stochastically independent, then they correlate zero, but this does not hold in general the other way around (Billingsley, 2012; Casella and Berger, 2021).

5.6 Notes on Using R

Basic functions such as `mean`, `sd`, `cov`, `var`, `cor` are available from R and various other statistical software applications.

5.7 Exercises

1. **Some functions**. Compute the approximations of the mean and standard deviation of z by using the Equations (5.14) and (5.15) for the following functions. Assume that the x and y have zero covariance.

   (a) $z = g(x, y) = a\frac{x}{y}$
   (b) $z = g(x, y) = \ln(x + y)$
   (c) $z = g(x, y) = e^{ax+by}$

2. **Michaelis-Menten kinetics.** Suppose that the reaction rate follows Michaelis–Menten kinetics, where the reaction rate v is a function of the concentration of a substrate S according to

$$v(S) = \frac{V_{max} S}{K_m + S},$$

where $V_{max} = 200$ and $K_m = 0.04$. For samples of substrate measurements the standard deviation $s_S = 0.1$.

   (a) Approximate the standard deviation of v for $\overline{S} = 0.05$.
   (b) Approximate the standard deviation of v for $\overline{S} = 0.80$.
   (c) How do you explain the difference in your findings?
   (d) Show that $1/v$ is linearly related to $1/S$.

3. **Bateman equation.** The concentration of species B in a rate process is given by $A \stackrel{k_1}{\to} B \stackrel{k_2}{\to} C$, consisting of two consecutive irreversible first-order reactions, is given by the Bateman equation.

$$B(t) = \frac{k_1}{k_2 - k_1}(e^{-k_1 t} - e^{-k_2 t}),$$

where $k_1 = 0.1$ and $k_2 = 0.2$. Suppose that time measurements of the duration of the reaction have $\overline{t}_1 = 2$ and $\overline{t}_2 = 9$ and that the standard deviation in both cases $s_t = 0.1$.

   (a) Approximate the standard deviation of B for $\overline{t}_1$.
   (b) Approximate the standard deviation of B for $\overline{t}_2$.
   (c) How do you explain the difference in your findings?

4. **Quantum yield of fluorescence.** A quantum yield of fluorescence, ϕ, of a material in a solution is calculated from the expression

$$\phi = \frac{I}{k \cdot c},$$

where k is a constant, I incident light intensity, and c concentration. Suppose that measurements gave $\overline{I} = 0.5$, $s_I = 0.05$, $\overline{c} = 0.3$, $s_c = 0.03$ with correlation $r_{I,c} = 0.8$.

   (a) Find an approximate for s_ϕ.

5. **Cement.** The heat evolved in 13 types of cement was measured together with four other variables. The variable y is heat in calories per gram of cement, and the remaining variables consist of percentages of weight of the constituents tricalcium aluminate x1, tricalcium silicate x2, tetracalcium alumumino ferrite x3 and dicalcium silicate x4 (Woods et al., 1932). The cement data are available from the MPV R library or from `cement.csv`.

(a) Compute the variance of y and that of x1 as well as the covariance between y and x1.

(b) Observe that the covariance is larger than one of the variances. Does this violate the Cauchy-Schwartz inequality?

(c) Compute the matrix containing all correlations between the variables. Report and briefly comment of the pairs with the largest (apart from sign) and the smallest correlation.
Hint: It is helpful to also plot the data.

6. **Correlation coefficient**. Here the reader will show that the correlation coefficient is invariant to linear transformations.

   (a) Use the definition of the correlation coefficient to shown that for all b_1, b_2 and all positive a_1 and a_2 that

$$\mathrm{COR}(a_1X + b_1, a_2Y + b_2) = \mathrm{COR}(X, Y).$$

Part III

Various Types of Models and Their Estimation

6

Measurement Models for a Chemical Quantity

By the formulation of a model the chemical engineer makes explicit how (s)he conceives the process in nature is that actually produces the measurements. Such guides our scientific thinking and provides ideas for hypotheses and their testing.

In almost any scientific measurement, a key role is played by some form of the Law of Large Numbers and the Central Limit Theorem. These are explained and discussed in the current chapter. Basic statistics such as the mean, variance, standard deviation, and coefficient of variation are introduced as well as two ways to construct a confidence interval. A basic testing procedure called the t-test is defined and illustrated by various examples.

It is quite useful for engineers and scientists to realize that there are three basic points of view on measurements. First, the actual measurements obtained from conducting an experiment; second, the pseudo measurements from a computer simulation; and third, the mathematical representation as a random variable. Each of these has various advantages to offer to the engineer.

In this chapter you learn:

- In Section 6.1 how a basic measurement model for a chemical quantity can be formulated what its basic properties are with respect to bias and precision.
- In Section 6.2 how the law of large numbers guarantees convergence of the mean to a chemical target.
- In Section 6.3 how confidence interval can be constructed that summarize the knowledge from an experiment about the value of a chemical target.
- In Section 6.4 how hypothesis about the equality of chemical targets can be tested.
- In Section 6.5 how any chemical parameter can be estimated by maximizing the likelihood of its underlying distribution.

DOI: 10.1201/9781003178194-6

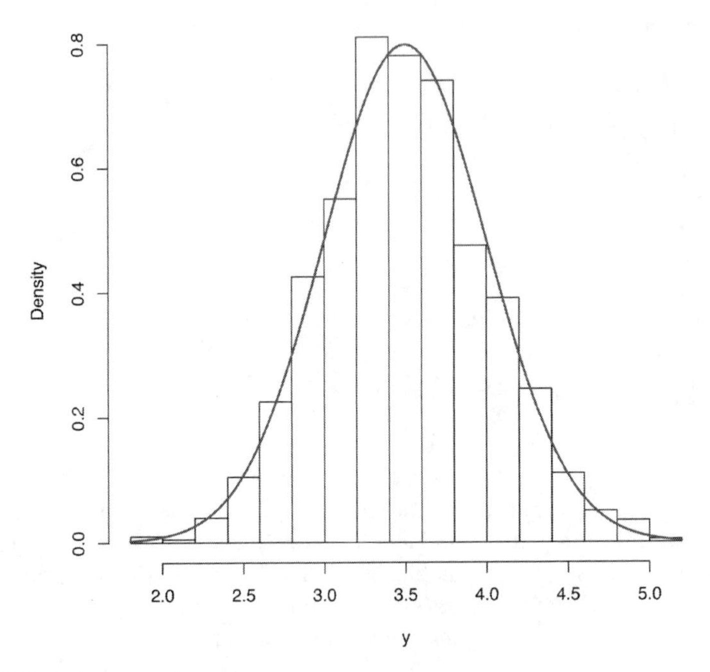

FIGURE 6.1
Histogram of 1000 determinations of in ($\mu g g^{-1}$) of copper in wholemeal flour with target $\mu = 3.5$, variance $\sigma^2 = 0.5^2$ and the density function from the normal distribution.

6.1 Measurement Model

Suppose the target of a number of measurements of the concentration of a substance in a liquid equals μ and that the error ε of the measurement procedure has mean 0 and variance $\sigma^2 > 0$.[1] Any chemical measurement procedure has some degree of error, the size of which is expressed in σ. In many applications the measurements error is symmetrically distributed around the target with measurements further away from the target to occur with smaller probability. If there are many small independently acting "causes" for the errors, then the measurements are normally distributed.

Suppose that n measurements $Y_1, \ldots, Y_n$ of the target μ have independent normally distributed errors. Then in terms of random variables our measurement model reads

$$Y_i = \mu + \varepsilon_i, \quad \text{where } \varepsilon_i \sim N(0, \sigma^2), \quad \text{for } i = 1, \ldots, n.$$

[1] Indeed, if $\sigma^2 = 0$, then $\varepsilon = 0$ and the measurements would be equal to μ without error.

The smaller the variance σ^2, the closer the measurements are distributed around the target μ, reason for which variance is inversely interpreted as the **precision of the measurements**. Due to the randomness of the error it is impossible to know on beforehand the exact value of a measurement.

Example 6.1 *Suppose that 1000 measurements in* $(\mu g g^{-1})$ *of copper in wholemeal flour are normally distributed around target* $\mu = 3.5$ *with variance* $\sigma^2 = 0.5^2$. *We summarize the measurements in a histogram, see Figure 6.1, with non-overlapping intervals* $(1.8, 2.0]$, $(2.0, 2.2]$, $\ldots$, $(5.0, 5.2]$ *horizontally and the height of the bars representing the proportion of measurements falling in each interval vertically. The proportion of measurements falling in* $(3.0, 4.0]$ *is 0.671 and the proportion in* $(2.5, 4.5]$ *is 0.956.*

In case the target μ and the variance σ^2 are known, then the probability that the measurements fall in a certain interval (a, b) can be computed exactly. For this we need the **density function** from the normal distribution

$$f(y) = \frac{1}{\sigma\sqrt{2\pi}} e^{-\frac{(y-\mu)^2}{2\sigma^2}}. \tag{6.1}$$

Its well-known bell-shaped curve completely characterizes all measurement properties of Y. When the number of measurements n increases without bound, then the histogram would converge exactly to this density curve taking the width of the bars smaller and smaller. The probability that a measurement is between a and b is

$$P(a \leq Y \leq b) = P(Y \leq b) - P(Y \leq a) = \int_a^b f(y)dy = F(b) - F(a).$$

The probability that the measurement is smaller or equal to y is

$$P(Y \leq y) = \int_{-\infty}^{y} f(x)dx = F(y),$$

where F is the cumulative **distribution function** of the random variable Y. Note that F is a monotonically increasing function of y over the interval where f is positive, so that the inverse F^{-1} is well-defined. The quantile y_α is the value such that $P(Y \leq y_\alpha) = \alpha$, which is in fact the value from the inverse distribution function $F^{-1}(\alpha)$. Quantiles play an important role in the derivation of confidence intervals, as we will see below.

The target μ and the variance σ^2 also follow as properties from the density function. That is, the expected value of the measurements is

$$\mu = E[Y] = \int_{-\infty}^{\infty} y f(y)dy = \int_{-\infty}^{\infty} y \cdot \frac{1}{\sigma\sqrt{2\pi}} e^{-\frac{(y-\mu)^2}{2\sigma^2}} dy.$$

The variance is the expected value of the squared difference between the measurement and the target

$$\mathrm{Var}[Y] = \sigma^2 = E(Y - \mu)^2 = \int_{-\infty}^{\infty} (y - \mu)^2 \cdot f(y)dy.$$

Clearly, the target μ and the inverse precision σ^2 can vary independently, so the setup is quite general.

Example 6.2 *Suppose that the measurements of copper in wholemeal flour have a normal distribution with mean* $\mu = 3.5\ \mu gg^{-1}$ *and standard deviation* $\sigma = 0.5\ \mu gg^{-1}$. *That is,* $Y \sim N(3.5, 0.5^2)$. *Note that it is impossible to say what exactly the measured value is, but it is possible to give the probability by which a measurement falls within a certain interval. That is, the probability that the measurements fall in the interval from 3.0 to 4.0 is*

$$P(3.0 \leq Y \leq 4.0) = F(3.0) - F(4.0) = 0.8413 - 0.1586 = 0.6829.$$

The probability that the measurements are between 2.5 and 4.5 is

$$P(2.5 \leq Y \leq 4.5) = F(2.5) - F(4.5) = 0.9772 - 0.0228 = 0.9545.$$

We are 95% certain that the measured value is within two standard deviations from the mean (target).

A **random sample** of n chemical measurements is often written as

$$\{Y_1, Y_2, \cdots, Y_n\} \overset{IID}{\sim} N(\mu, \sigma^2).$$

The measurements are independently and identically distributed according to the normal distribution with mean μ and standard deviation σ. All measurements are independent in the sense that the outcome of any Y_i does not have any influence on another outcome Y_j. Independence can often be guaranteed by the measurement procedure. For practical purposes, it is convenient that the unit of measurement for the mean as well as for the standard deviation are equal to that of the objects (sample).

The measurements are seen as independent replicates each providing information about the unknown target μ. The average or sample mean $\overline{Y}$ of the measurements is an unbiased *estimator* of the target μ. It is distributed as

$$\frac{1}{n}\sum_{i=1}^{n} Y_i = \overline{Y} \sim N(\mu, \sigma^2/n).$$

The precision of the estimator $\overline{Y}$ increases with the sample size n as the variance of the estimator decreases with n.

Example 6.3 *Suppose that 100 measurements in* (μgg^{-1}) *of copper in wholemeal flour are a random sample from* $Y \sim N(3.5, 0.5^2)$, *but that the target* $\mu = 3.5$ *and the variance* 0.5^2 *are unknown to us. Then the sample mean* $\overline{Y}$ *is distributed as*

$$\overline{Y} \sim N(3.5, 0.5^2/10^2) = N(3.5, 0.05^2).$$

The probability that the mean is between 3.40 and 3.60 is

$$P(3.4 \leq \overline{Y} \leq 3.6) = F(3.6) - F(3.4) = 0.9772 - 0.0228 = 0.9545.$$

We are 95% certain that the mean is within the interval $(3.4, 3.6)$.

The standard deviation $\sigma/\sqrt{n}$ of the estimator $\overline{Y}$ is called the **standard error** (SE). It is a key concept measuring the precision of a statistic and will appear throughout the book. It often seems the case that the variance σ^2 is determined by fixed properties of the measurement procedure, which are not under the control of the chemical engineer. However, the precision of the estimator increases monotonically with the sample size n.

6.2 Law of Large Numbers

Suppose that the measurements from a chemical experiment have target μ and variance $\sigma^2 > 0$, however, their distribution is unknown. A random sample of such measurements is written in terms of random variables as

$$\{Y_1, Y_2, \cdots, Y_n\} \sim IID(\mu, \sigma^2),$$

indicating that all measurements are independent and identically distributed with mean μ and variance σ^2. Generally speaking, the target μ and the variance σ^2 are unknown to us and these have to be estimated from the values of the measurements. A good question to ask is: What happens to a "statistic" such as the sample mean or the sample variance if the number of measurements increases without bound?

Theorem 6.1 Law of large numbers. *The sample mean tends to the population mean in probability as the sample size n increases to infinity,*

$$\overline{Y} \xrightarrow{P} \mu,$$

that is, for any $\epsilon > 0$ and $\delta > 0$ there exists an $n_{\epsilon,\delta}$ such that for all $n \geq n_{\epsilon,\delta}$,

$$P(|\overline{Y} - \mu| < \epsilon) \geq 1 - \delta,$$

where $\overline{Y} = \sum_{i=1}^{n} Y_i/n$ implicitly depends on n.

This property is called **law of large numbers** and can be seen to hold as follows. Let f_Y be the density function of Y and let g be a non-negative function. Then, the Markov inequality

$$P[g(Y) \geq k] \leq \frac{E[g(Y)]}{k},$$

follows from

$$\begin{aligned} E[g(Y)] &= \int_{-\infty}^{\infty} g(y) f_Y(y) dy \\ &= \int_{\{y:g(y)\geq k\}} g(y) f_Y(y) dy + \int_{\{y:g(y)<k\}} g(y) f_Y(y) dy \\ &\geq \int_{\{y:g(y)\geq k\}} g(y) f_Y(y) dy \\ &\geq k \int_{\{y:g(y)\geq k\}} f_Y(y) dy \\ &= k \cdot P[g(Y) \geq k]. \end{aligned}$$

Using for the probability of the event that the distance between $\overline{Y}$ and μ is smaller than ϵ, we find

$$\begin{aligned} P(|\overline{Y} - \mu| < \epsilon) &= P(|\overline{Y} - \mu|^2 < \epsilon^2) \\ &\geq 1 - \frac{E(\overline{Y} - \mu)^2}{\epsilon^2} = 1 - \frac{\sigma^2/n}{\epsilon^2} \\ &\geq 1 - \delta \end{aligned}$$

if $n > \sigma^2/\epsilon^2\delta$.

Hence, by increasing the sample size n, the probability of the event that $\overline{Y}$ and μ are within distance ϵ, can be made arbitrarily close to 1. To stress that the convergence is in probability, this property is conveniently written as

$$\overline{Y} = \frac{1}{n}\sum_{i=1}^{n} Y_i \xrightarrow{P} EY_1 = \mu.$$

The sample mean tends in probability to the target μ as the sample size increases without bound. This immediately generalizes to a continuous function g of the sample mean, that is

$$g(\overline{Y}) \xrightarrow{P} g(EY_1) = g(\mu),$$

such as the power of the random variables

$$\frac{1}{n}\sum_{i=1}^{n} Y_i^k \xrightarrow{P} EY_1^k.$$

It then follows that

$$\begin{aligned} S^2 &= \frac{1}{n-1}\sum_{i=1}^{n}(Y_i - \overline{Y})^2 \\ &= \frac{n}{n-1}\left(\frac{1}{n}\sum_{i=1}^{n} Y_i^2 - \overline{Y}^2\right) \\ &\xrightarrow{P} EY_1^2 - (EY_1)^2 = \mathrm{Var}(Y_1) = \sigma^2 \end{aligned} \tag{6.2}$$

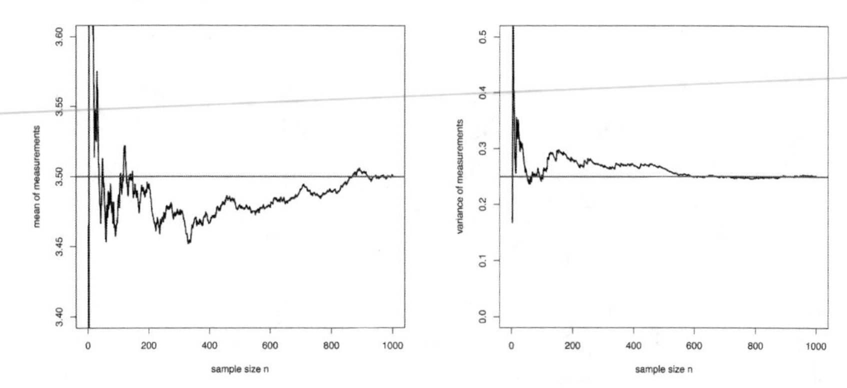

FIGURE 6.2
Sample mean as number of measurements increases from 1 to 1000.

FIGURE 6.3
Sample variance as number of measurements increases from 1 to 1000.

The sample variance S^2 tends in probability to the true variance σ^2, as the sample size n increases without bound. The sample mean and variance are examples of estimators of target values that are important to the chemical engineer.

Example 6.4 *A random sample of 1000 measurements of copper in wholemeal flour was taken with mean* $\mu = 3.5\ \mu gg^{-1}$ *and* $\sigma = 0.5\ \mu gg^{-1}$. *The sample mean* $\overline{y}_n = \frac{1}{n}\sum_{i=1}^{n} y_i$ *and the sample variance* s_n^2 *were computed for each value of* n *increasing from 1 to 1000. The graph of the 1000 means* $(\overline{y}_1, \cdots, \overline{y}_{1000})$ *in Figure 6.2 and that of the sample variances* $(s_1^2, \cdots, s_{1000}^2)$ *in Figure 6.3 illustrate that the sample statistics converge to the target* $\mu = 3.5$ *and the variance* $\sigma^2 = 0.5$, *as* n *increases. Note that* $\sigma = 0.5$ *is taken relatively large here compared to the scale of the y-axis. Obviously, the reader can easily construct such plots by a few lines of code.*

6.3 Constructing Confidence Intervals

Depending on the assumptions of the measurement model and their validity, there are several ways to construct a confidence interval for the values of a chemical quantity on the basis of a number of measurements.

6.3.1 Confidence interval from the central limit theorem

Under reasonable assumptions (Ferguson, 1996; Rao, 2009), the distribution of the sample mean will tend to the normal distribution as the sample size increases. This is stated in the **central limit theorem**, which is among the most important for empirical sciences.

Theorem 6.2 Central limit theorem. *If* $\{Y_1, Y_2, \cdots, Y_n\} \sim IID(\mu, \sigma^2)$, *then, as n tends to infinity, we have*

$$\frac{\sqrt{n}(\overline{Y} - \mu)}{\sigma} = \frac{\overline{Y} - \mu}{\sigma/\sqrt{n}} \xrightarrow{D} Z \sim N(0, 1). \tag{6.3}$$

That is, the distribution of the term on the left-hand side tends to the standard normal distribution as n increases without bound. An equivalent formulation of the conclusion is

$$P\left(\frac{\sqrt{n}(\overline{Y} - \mu)}{\sigma} \leq z\right) \longrightarrow P(Z \leq z), \tag{6.4}$$

for a standard normal distributed random variable Z and any real value z. If we standardize the mean $\overline{Y}$ by subtracting its target μ and divide it by its standard deviation $\sigma/\sqrt{n}$, then its distribution tends to the standard normal. The approximation holds generally for random samples from any unknown distribution.

The CLT follows from the convergence of the characteristic function to that of the normal. That is, using the independence of the random variables, that these are identically distributed, that $\varphi_{(Y_i-\mu)}(0) = 1$, $\dot{\varphi}_{(Y_i-\mu)}(0) = 0$, the second-order Taylor expansion

$$\varphi_Y(x_0 + t) = \varphi_Y(x_0) + \dot{\varphi}_Y(x_0) + t^2 \int_0^1 \int_0^1 v\ddot{\varphi}_Y(x_0 + uvt)dudv,$$

it follows by using the basic properties of the characteristic function together with independence

$$\begin{aligned}
\varphi_{\sqrt{n}(\overline{Y}-\mu)}(t) &= \varphi_{\sqrt{n}(\sum_{i=1}^n (Y_i-\mu)}(t) = \varphi_{\sum_{i=1}^n (Y_i-\mu)}(t/\sqrt{n}) \\
&= \prod_{i=1}^n \varphi_{(Y_i-\mu)}(t/\sqrt{n}) = \left(\varphi_{(Y_1-\mu)}(t/\sqrt{n})\right)^n \\
&= \left(\varphi_{(Y_i-\mu)}(0) + \dot{\varphi}_{(Y_i-\mu)}(0)\right. \\
&\quad \left. + \frac{t}{\sqrt{n}} \int_0^1 \int_0^1 v\ddot{\varphi}_{(Y_i-\mu)}(0 + uv\frac{t}{\sqrt{n}})dudv \frac{t}{\sqrt{n}}\right)^n \\
&= \left(1 + \frac{t^2}{n} \int_0^1 \int_0^1 v\ddot{\varphi}_{(Y_i-\mu)}(uv\frac{t}{\sqrt{n}})dudv\right)^n.
\end{aligned}$$

Now we use the general limit result $\lim_{n\to\infty}(1 + a_n)^n = \exp\{\lim_{n\to\infty} na_n\}$, the continuity of the second-order derivative of the characteristic function in a neighborhood of zero $\lim_{\epsilon\to 0} \ddot{\varphi}_{(Y_i-\mu)}(\epsilon) = \ddot{\varphi}_{(Y_i-\mu)}(0) = -\sigma^2$, to obtain for the last display that

$$
\begin{aligned}
&\to \exp\left(\lim_{n\to\infty} n\cdot\frac{t^2}{n}\int_0^1\int_0^1 v\ddot{\varphi}_{(Y_i-\mu)}(uv\frac{t}{\sqrt{n}})dudv\right)\\
&= \exp\left(t^2\int_0^1\int_0^1 v\lim_{n\to\infty}\ddot{\varphi}_{(Y_i-\mu)}(uv\frac{t}{\sqrt{n}})dudv\right)\\
&= \exp\left(t^2\int_0^1\int_0^1 v\ddot{\varphi}_{(Y_i-\mu)}(0)dudv\right)\\
&= \exp\left(t^2\int_0^1\int_0^1 -\sigma^2 vdudv\right) = \exp\left(-t^2\sigma^2/2\right)
\end{aligned}
$$

which is the characteristic function of the standard normal distribution Z. Hence, the characteristic function $\varphi_{\sqrt{n}(\overline{Y}-\mu)\sigma}(t)$ for any IID sequence of random variables tends to that of the standard normal $\exp(-t^2/2)$, which is equivalent to the claimed convergence in distribution.

The random variable Z has the so-called standard normal distribution $N(0,1)$, which is completely characterized by the density function

$$f(z) = \frac{1}{\sqrt{2\pi}}e^{-z^2/2}. \tag{6.5}$$

It has the well-known bell-shaped form, symmetric around zero, with tails tending to zero relatively fast. The probability that its values are smaller than z is given by the integral

$$P(Z \le z) = \int_{-\infty}^{z} f(x)dx = \int_{-\infty}^{z}\frac{1}{\sqrt{2\pi}}e^{-x^2/2}dx = F(z).$$

Although its distribution function F does not have an explicit mathematical expression, its values are easily computed by any programming language.

We may rewrite the convergence in the central limit theorem more directly for the mean as

$$\overline{Y} \overset{D}{\approx} N(\mu, \sigma^2/n).$$

That is, for a sufficiently large n the distribution of the $\overline{Y}$ is approximately normal. The standard deviation $\sigma/\sqrt{n}$ is also called the standard error (SE) of the statistic $\overline{Y}$. Note that the SE decreases with n, so that the distribution becomes more and more concentrated around the target μ.

Example 6.5 *Suppose that independent measurements of concentration come from the unit interval* $[0,1]$ *each of which is equally likely. That is, the data are a random sample from the uniform* $[0,1]$ *distribution, for which we have expected value* $\mu = 1/2$ *and variance* $\sigma^2 = 1/12$*; see also Chapter 3. Then the central limit theorem implies that*

$$\overline{Y} \overset{D}{\approx} N(\mu, \sigma^2/n) = N\left(\frac{1}{2}, \frac{1}{12n}\right).$$

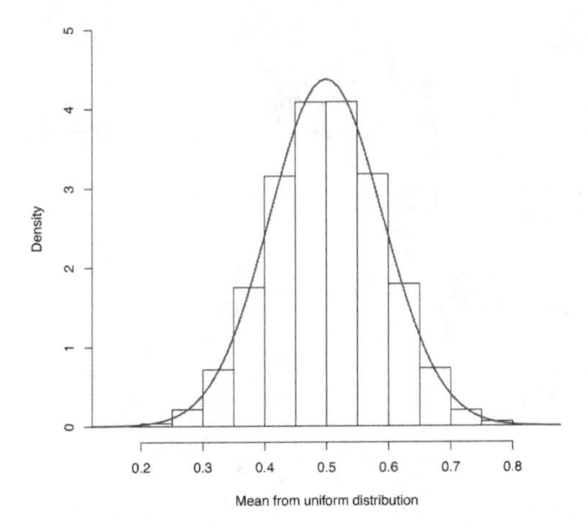

FIGURE 6.4
Distribution of mean from 10 observations from uniform distribution on $[0,1]$ and the corresponding density of the normal.

We may simulate the distribution of the mean $\overline{Y}$ by independently repeating the draw of a sample of only $n = 10$ measurements 10,000 times. Through the histogram of the 10,000 sample means in Figure 6.4 the corresponding normal density curve is drawn for comparison. It can be observed that even for small samples the distribution of the mean is close to normal. This illustrates that the convergence of the mean to the normal can be rather fast.

A symmetric $1-\alpha$ probability interval for the standard normal distribution is defined as

$$1 - \alpha = P(z_{\alpha/2} \leq Z \leq z_{1-\alpha/2}).$$

Using the approximation from the CLT, we find a $1 - \alpha$ confidence interval for the mean by

$$\begin{aligned} 1-\alpha &\approx P\left(z_{\alpha/2} \leq \frac{\sqrt{n}(\overline{Y}-\mu)}{\sigma} \leq z_{1-\alpha/2}\right) \\ &= P\left(\overline{Y} + z_{\alpha/2}\sigma/\sqrt{n} \leq \mu \leq \overline{Y} + z_{1-\alpha/2}\sigma/\sqrt{n}\right) \end{aligned}$$

So that the $1 - \alpha$ confidence interval for the target μ becomes

$$[\overline{Y} + z_{\alpha/2}\sigma/\sqrt{n}, \overline{Y} + z_{1-\alpha/2}\sigma/\sqrt{n}].$$

If we would apply this infinitely many times for a fixed sufficiently large n, then μ would be a member of the interval in 95 percent of the cases.

Example 6.6 *Suppose that 10 independent measurements from the uniform $[0,1]$ distribution give $\overline{y} = 0.5790$. Then, we obtain $\mu = 1/2$ and variance*

$\sigma^2 = 1/12$. *For a 95 percent confidence interval we have* $\alpha = 5/100$ *and* $z_{1-\alpha/2} = z_{0.975} = 1.9599$, *so that*

$$\begin{aligned} 0.95 &\approx P\left(\overline{Y} + z_{\alpha/2}\sigma/\sqrt{n} \le \mu \le \overline{Y} + z_{1-\alpha/2}\sigma/\sqrt{n}\right) \\ &= P\left(\overline{Y} - \frac{1.9599}{\sqrt{120}} \le \mu \le \overline{Y} + \frac{1.9599}{\sqrt{120}}\right) \\ &= P\left(\overline{Y} - 0.1789 \le \mu \le \overline{Y} + 0.1789\right) \end{aligned}$$

Based on the data $\overline{y} = 0.5790$ *and the known* $\sigma^2 = 1/12$, *we are 95% certain that* μ *is inside the interval* $[0.4001, 0.7579]$.

The importance of the central limit theorem is due to its generalizations. In case σ^2 is unknown, then we may substitute a consistent estimator to obtain

$$\frac{\sqrt{n}(\overline{Y} - \mu)}{s_y} = \frac{\overline{Y} - \mu}{s_y/\sqrt{n}} \xrightarrow{D} Z \sim N(0, 1). \tag{6.6}$$

This implies that the confidence intervals for the target μ becomes

$$[\overline{Y} + z_{\alpha/2} s_y/\sqrt{n}, \overline{Y} + z_{1-\alpha/2} s_y/\sqrt{n}].$$

Another generalization is multivariate. Suppose that we have p measurements collected in a vector and that we repeat these n times. That is, if $\{\mathbf{y}_1, \mathbf{y}_2, \ldots, \mathbf{y}_n\} \sim IID(\boldsymbol{\mu}, \boldsymbol{\Sigma})$, then, as n tends to infinity, we have

$$\sqrt{n}(\overline{\mathbf{y}} - \boldsymbol{\mu}) \xrightarrow{D} N(\mathbf{0}, \boldsymbol{\Sigma}).$$

Yet, another generalization of the CLT is known as the **delta theorem**. If $g : \mathbb{R}^p \to \mathbb{R}^k$ is a differentiable function for all $\mathbf{y}$ within a small distance to $\boldsymbol{\mu}$, then we have

$$\sqrt{n}\left(g(\overline{\mathbf{y}}) - \boldsymbol{g}(\boldsymbol{\mu})\right) \xrightarrow{D} N\left(\mathbf{0}, \dot{g}(\boldsymbol{\mu})^T \boldsymbol{\Sigma} \dot{g}(\boldsymbol{\mu})\right). \tag{6.7}$$

Example 6.7 *To illustrate the rate of convergence in the central limit theorem and the delta theorem, a small simulation was conducted by drawing 5 determinations in* ($\mu g g^{-1}$) *of copper in wholemeal flour 100,000 times. Each sample of size 5 was drawn from a normal distribution with* $\mu = 3.5$ *and* $\sigma = 0.5$, *and the* **coefficient of variation** *was computed 100,000 times in total. The latter* $S/\overline{Y}$ *is defined as the standard deviation divided by the mean expressing the relative error of measurement.*

If the $\{Y_1, Y_2, \ldots, Y_n\} \sim IID(\mu, \sigma^2)$, *and the fourth-order moment exists, then we find for the asymptotic distribution of the mean and the sample variance that*

$$\sqrt{n}\left[\begin{pmatrix} \overline{Y} \\ S_y^2 \end{pmatrix} - \begin{pmatrix} \mu \\ \sigma^2 \end{pmatrix}\right] \xrightarrow{D} N\left(\begin{pmatrix} 0 \\ 0 \end{pmatrix}, \begin{bmatrix} \sigma^2 & \mu_3 \\ \mu_3 & \mu_4 - \sigma^4 \end{bmatrix}\right)$$

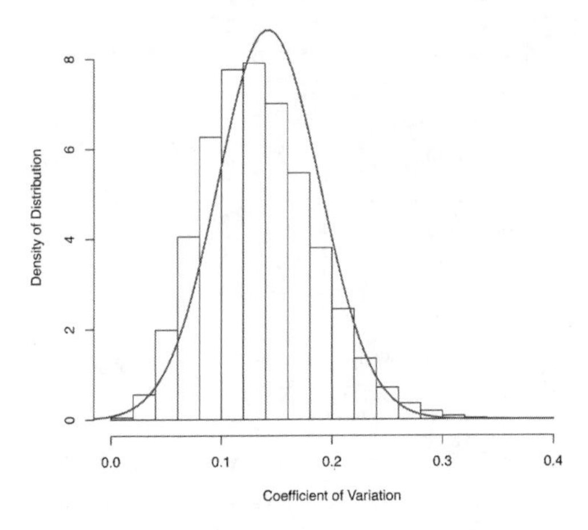

FIGURE 6.5
Distribution of coefficient of variation from an experiment with $n = 5$ measurements of the copper in wholemeal flour repeated 100,000 times in total.

If the random variables have a normal distribution, then the third-order moment $\mu_3 = E(Y-\mu)^3 = 0$ *and* $\mu_4 = E(Y-\mu)^4 = 3\sigma^4$. *Defining* $g(x,y) = x^{-1}y^{1/2}$, *we find* $\dot{g}(x,y) = (-\frac{\sqrt{y}}{x^2}, \frac{1}{2x\sqrt{y}})$, *so that* $\dot{g}(\mu,\sigma^2) = (-\frac{\sigma}{\mu^2}, \frac{1}{2\mu\sigma})$. *Hence, from the delta theorem, we obtain*

$$\sqrt{n}\left(\frac{s_y}{\bar{Y}} - \frac{\sigma}{\mu}\right) \xrightarrow{D} N\left(0, \frac{\sigma^4}{\mu^4} + \frac{\sigma^4}{2\mu^2\sigma^2}\right)$$

In order to plot its density we reformulate into

$$\frac{s_y}{\bar{Y}} \overset{D}{\approx} N\left(\frac{\sigma}{\mu}, \frac{1}{n}\sigma^2\left(\frac{1}{2} + \frac{\sigma^2}{\mu^2}\right)/\mu^2\right) = N(0.1429, 0.0106/n).$$

From the 100,000 values of the coefficient of variation, its distribution was represented by a histogram together with its approximating asymptotic normal distribution in Figure 6.5. It can be observed that the distribution of the statistic is fairly close to the normal, even with a sample size as small as five.

6.3.2 Confidence interval from the bootstrap

In case the measurements $\{Y_1, Y_2, \cdots, Y_n\} \sim IID(\mu, \sigma^2)$, we may also apply a data driven method, called the **bootstrap** , to construct a confidence interval. After having obtained the measurements

$$\{y_1, y_2, \cdots, y_n\}$$

the idea is to randomly re-sample, say $k = 1000$ times, n values from these with replacement and to compute the mean from each of the re-samples.

That is, a first re-sample would yield

$$\{y_{11}^*, y_{12}^*, \cdots, y_{1n}^*\}$$

from which we compute $\bar{y}_1^*$. A second re-sample would yield

$$\{y_{21}^*, y_{22}^*, \cdots, y_{2n}^*\}$$

from which we compute $\bar{y}_2^*$. Repeating this k-times, we obtain at the k-th step

$$\{y_{k1}^*, y_{k2}^*, \cdots, y_{kn}^*\}$$

from which we compute $\bar{y}_k^*$. This bootstrap procedure yields k values of the mean from the re-sampled data

$$\{\bar{y}_1^*, \bar{y}_2^*, \cdots, \bar{y}_k^*\}.$$

The distribution of this sequence is called empirical distribution, which can be estimated or inspected by a histogram. The percentile bootstrap confidence interval follows from taking the 2.5 and 97.5 percentile from the empirical distribution as left and right limit of a confidence interval. Inferences from e.g., its percentile method are valid if the measurements are a random sample from any unknown distribution with finite mean and variance (Efron and Tibshirani, 1993; Davison and Hinkley, 1997).

Example 6.8 *The table in Figure 6.6(a) gives the measurements of 5 determinations of copper in wholemeal flour in* $(\mu g g^{-1})$. $\bar{y} = 3.5554$, $s^2 = 0.2536/4 = 0.0634$, $s = 0.2518$ $(\mu g g^{-1})$. *Re-sampling 1000 times, with replacement, from the 5 measurements of copper determinations, each of the resulting means are computed. The histogram of the empirical distribution of the mean in Figure 6.6 appears to be fairly normal given the small sample size of 5. From the 2.5 and 97.5 percentile interval* $(3.37, 3.77)$, *we are 95% certain that it covers the true target.*

6.3.3 Confidence interval from the normal distribution

Suppose that our measurements are a random sample from a normal distribution. That is, our measurement model is

$$\{Y_1, Y_2, \cdots, Y_n\} \overset{IID}{\sim} N(\mu, \sigma^2).$$

We wish to construct a confidence interval for the target μ. The mean of the n measurements serves as an estimator for μ having a normal distribution

$$\bar{Y} = \frac{1}{n}\sum_{i=1}^{n} Y_i \sim N(\mu, \sigma^2/n).$$

nr	y	$(y_i - \overline{y})$	$(y_i - \overline{y})^2$
1	3.25	−0.31	0.09
2	3.57	0.01	0.00
3	3.46	−0.09	0.01
4	3.94	0.39	0.15
5	3.56	0.00	0.00
sum	17.777	0	0.2536

(a)

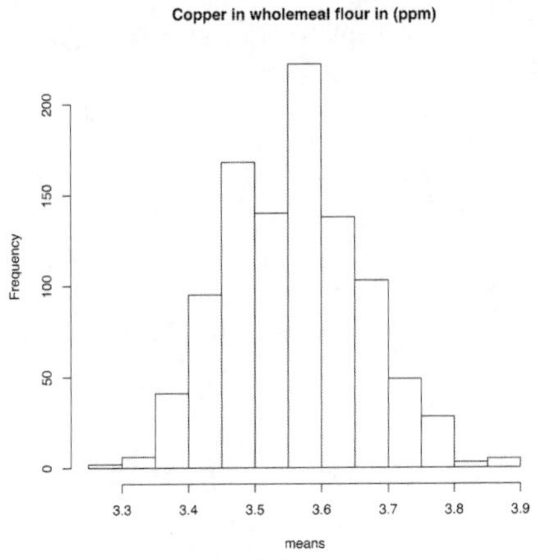

Copper in wholemeal flour in (ppm)

Frequency: 0, 50, 100, 150, 200

means: 3.3, 3.4, 3.5, 3.6, 3.7, 3.8, 3.9

(b)

FIGURE 6.6
Five determinations of copper ($\mu g g^{-1}$) in wholemeal flour (a) and their empirical distribution of the mean.

Obviously, in many situations σ^2 is unknown and has to be estimated by the sample variance

$$S^2 = \frac{1}{n-1}\sum_{i=1}^{n}(Y_i - \bar{Y})^2.$$

In order to construct a confidence interval we need the quantiles of the distribution of $\sqrt{n}(\bar{Y}-\mu)/S$. To derive this distribution we conveniently use vector notation for $\bar{Y} = n^{-1}\sum_{i=1}^{n} Y_i = (\mathbf{1}^T\mathbf{1})^{-1}\mathbf{1}^T\mathbf{y}$, where $\mathbf{1}$ is the n-vector with unit elements. Then, after defining the projection matrix $\mathbf{J} = \mathbf{1}(\mathbf{1}^T\mathbf{1})^{-1}\mathbf{1}^T$, we obtain the Chi-squared distribution for

$$\begin{aligned}
\frac{(n-1)S^2}{\sigma^2} &= \sum_{i=1}^{n}\left(\frac{Y_i-\mu}{\sigma} - \frac{\bar{Y}-\mu}{\sigma}\right)^2 = \sum_{i=1}^{n}(Z_i - \bar{Z})^2 \\
&= \|\mathbf{Z} - \mathbf{1}(\mathbf{1}^T\mathbf{1})^{-1}\mathbf{1}^T\mathbf{Z}\|^2 = \|(\mathbf{I}-\mathbf{J})\mathbf{Z}\|^2 \\
&= \mathbf{Z}^T(\mathbf{I}-\mathbf{J})\mathbf{Z} \\
&\sim \chi^2_{n-1},
\end{aligned}$$

since $(\mathbf{I}-\mathbf{J})^2 = (\mathbf{I}-\mathbf{J})$ with rank equal to $n-1$, see Equation (1). From $(\mathbf{I}-\mathbf{J})\mathbf{1} = \mathbf{0}$, it follows that the numerator and the denominator in the following definition of the t-statistic are independent, so that

$$t = \frac{\bar{Y}-\mu}{S/\sqrt{n}} = \frac{(\bar{Y}-\mu)/(\sigma/\sqrt{n})}{\sqrt{\frac{(n-1)S^2}{\sigma^2}/(n-1)}} \sim t_{n-1},$$

where the latter denotes the t-distribution with $n-1$ degrees of freedom. Now a $(1-\alpha)\cdot 100\%$ confidence interval for the target μ is given by

$$(\overline{Y} - t_{1-\alpha/2}\cdot S/\sqrt{n}, \overline{Y} + t_{1-\alpha/2}\cdot S/\sqrt{n}).$$

Here $t_{1-\alpha/2}$ is the $(1-\alpha/2)\cdot 100\%$ percentile of the t-distribution with $n-1$ degrees of freedom. A 95% confidence interval nicely summarizes the knowledge obtained from the measurements about the true value of the chemical parameter.

Example 6.9 *To compute a 95% confidence interval for the mean from the data in Example 6.8, we recall that* $\overline{y} = 3.5554$ *and* $s = 0.2518$ $(\mu g g^{-1})$. *The quantile* $t_{(1-0.05/2)}(\mathrm{df}=4) = 2.7764$. *We find*

$$[3.5554 - 2.7764\cdot 0.2518/\sqrt{5}, 3.5554 + 2.7764\cdot 0.2518/\sqrt{5}] = [3.2428, 3.8681],$$

which is fairly close the previous 95% bootstrap interval, given the small sample size of 5 measurements. From this we are 95% certain that the target value is between $[3.24, 3.87]$.

6.4 Testing Chemical Hypotheses related to Measurement Models

Chemical researchers formulate their ideas about reality in terms of an hypothesis, such as the absence of systematic bias in the measurements, or the equality of means or precision of the measurements.

6.4.1 Testing for the presence of bias

Suppose the target of measurement is μ, but there is some **bias** or **systematic error** δ in the measurement procedure. The measurement model now becomes

$$Y_i = \mu + \delta + \varepsilon_i, \quad \text{where } \{\varepsilon_1, \cdots, \varepsilon_n\} \overset{IID}{\sim} N(0, \sigma^2)$$

If $\delta < 0$, then the measurements are biased to the left as these underestimate the target μ. If $\delta > 0$, then the measurements are biased to the right as these overestimate the target μ. The smaller the variance σ^2, the more precise the measurements are. The notions bias and precision are independent concepts: Unbiased measurements may be highly precise or not and exactly the same holds for biased measurements. Obviously, highly precise, but biased estimates are invalid and can be very misleading.

Suppose that a chemical experiment yields fewer than 40 measurements ($n < 40$) that follow a normal distribution with unknown variance. In such a setting the **null hypothesis** $H_0 : \mu = \mu_0$ can be tested against the alternative $H_A : \mu \neq \mu_0$ by a **one-sample t-test**. The null-hypothesis may e.g., correspond to the absence of bias, that is $H_0 : \delta = 0$. The testing procedure is based on the fact that under the null hypothesis the test statistic t has a t-distribution with $n - 1$ degrees of freedom

$$t = \frac{\sqrt{n}(\overline{Y} - \mu_0)}{S} \sim t_{n-1}.$$

We take the decision to reject the null hypothesis if $|t| > t_{1-\alpha/2}(n-1)$, that is, if it is larger than the $1 - \alpha/2$ quantile of the t distribution with $n - 1$ degrees of freedom. The key idea is that if $\overline{Y}$ deviates largely from μ_0 relative to the standard error $S/\sqrt{n}$, then t becomes so large that the data from the experiment provide sufficient evidence to reject the null hypothesis.

From the actual measurements $\{y_1, y_2, \cdots, y_n\}$ of the experiment we compute $\overline{y}$, s, and $t = \sqrt{n}(\overline{y} - \mu_0)/s$. The p-value is defined as

$$2 \cdot P(t_{n-1} > |t|) = P(t_{n-1} < -|t|) + P(t_{n-1} > |t|),$$

the probability of more extreme outcomes than the actually obtained $|t|$, given that the null-hypothesis holds true. The null hypothesis is rejected if and only if the p-value is smaller than the **significance level** α, commonly taken equal to 0.05.

Example 6.10 *The quality of a new method for determining selenourea in water is tested for the presence of a systematic error. Five tap water samples where spiked with 50* ng ml^{-1} *selenoutea (Aller, 1998). We take* $H_0 : \mu = 50$ *and* $H_A : \mu \neq 50$. *This yielded the measurements*

$$50.4, 50.7, 49.1, 49.0, 51.1 \text{ ng ml}^{-1}.$$

We find $\overline{y} = 50.06$ *and* $s_y = 0.956$, *so that the test statistic*

$$t = \frac{\sqrt{5}(50.06 - 50)}{0.956} = 0.14041 < 2.78 = t_{0.975}(\text{df} = 5 - 1).$$

This suggests the decision not to reject the null hypothesis. The conclusion is that the data provide no evidence for a systematic bias. From the confidence interval, we are 95% certain that the target μ *is in* $[48.87, 51.25]$. *The p-value 0.8951, the probability that the statistic attains values larger than* 0.14 *or smaller than* -0.14, *is illustrated by the shaded surface in Figure 6.7.*

6.4.2 Testing for equality of two means

Suppose we wish to test the null hypothesis of equal population means in a setting of two independent sets of measurements. That is, we have a first

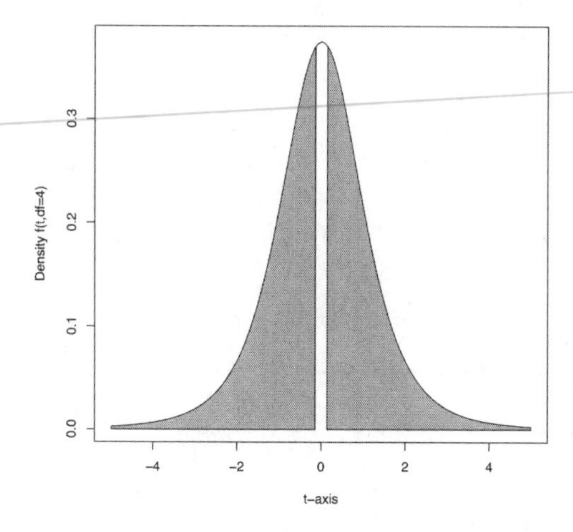

FIGURE 6.7
The p-value 0.8951 from the t-test as the area of the shaded surface enclosed by the t_4 density function and t-values to the left of $t = -0.14$ and the right of $t = 0.14$.

random sample $\{Y_{11}, \cdots, Y_{n_1 1}\} \overset{IID}{\sim} N(\mu_1, \sigma_1^2)$ and a second random sample $\{Y_{12}, \cdots, Y_{n_1,2}\} \overset{IID}{\sim} N(\mu_2, \sigma_2^2)$. We do not assume equality of variances ($\sigma_1^2 = \sigma_2^2$). We have the hypotheses

$$H_0 : \mu_1 - \mu_2 = 0, \text{ against } \quad H_A : \mu_1 - \mu_2 \neq 0$$

It can be shown Welch (1947) that under the null hypothesis, we have

$$T = \frac{\overline{Y}_1 - \overline{Y}_2 - (\mu_1 - \mu_2)}{\sqrt{\frac{S_1^2}{n_1} + \frac{S_2^2}{n_2}}} \sim t(\text{df}), \text{ where df} = \frac{\left(\frac{S_1^2}{n_1} + \frac{S_2^2}{n_2}\right)^2}{\frac{(S_1^2/n_1)^2}{n_1 - 1} + \frac{(S_2^2/n_2)^2}{n_2 - 1}}.$$

Similar to the above, we reject the null hypothesis if $|t| > t_{1-\alpha/2}(\text{df})$.

Example 6.11 *Arsenic concentration in public drinking water is a potential risk to health. A newspaper reported the following concentrations in ten Metropolitan Phoenix communities* $3, 7, 25, 10, 15, 6, 12, 25, 15, 7$ *parts per billion (ppb) and in ten rural Arizona communities 48,44,40,38,33,21,20,12,1,18 ppb. The research question is whether there is a difference in mean concentration between rural and metropolitan communities. That is, if the mean Arsenic concentration in Phoenix is* μ_1 *and that of Rural Arizona* μ_2*, we test the null-hypothesis* $H_0 : \mu_1 - \mu_2 = 0$ *against the alternative* $H_1 : \mu_1 - \mu_2 \neq 0$*. From the data we have* $n_1 = n_2 = 10$*,* $\overline{y}_1 = 12.5, \overline{y}_2 = 27.5$*,* $s_1 = 7.63, s_2 = 15.3$ *ppb. Observe that the assumption of equal variances seems not reasonable for* *the current data. We compute*

$$\text{df} = \frac{\left(\frac{s_1^2}{n_1} + \frac{s_2^2}{n_2}\right)^2}{\frac{(s_1^2/n_1)^2}{n_1-1} + \frac{(s_2^2/n_2)^2}{n_2-1}} = \frac{\left(\frac{7.63^2}{10} + \frac{15.3^2}{10}\right)^2}{\frac{(7.63^2/10)^2}{9} + \frac{(15.3^2/10)^2}{9}} = 13.196$$

$$t = \frac{\overline{x}_1 - \overline{x}_2}{\sqrt{\frac{s_1^2}{n_1} + \frac{s_2^2}{n_2}}} = \frac{12.5 - 27.5}{\sqrt{\frac{7.63^2}{10} + \frac{15.3^2}{10}}} = -2.77$$

The $t_{0.975,df} = t_{0.975,13.196} = 2.160$. *Since* $|t| = 2.77 > 2.160$ *the decision indicated is to reject the null hypothesis. The conclusion is that the data provided evidence for rural Arizona to have larger Arsenic concentrations in public drinking water compared to Metropolitan Phoenix.*

6.4.3 Testing for equality of variance

Suppose we wish to test the null hypothesis of equal population variance in a setting of two independent sets of measurements. That is, we have a first random sample $\{Y_{11}, \cdots, Y_{n_1}\} \overset{IID}{\sim} N(\mu_1, \sigma_1^2)$ being independent to a second random sample $\{Y_{12}, \cdots, Y_{n_2,2}\} \overset{IID}{\sim} N(\mu_2, \sigma_2^2)$. We have the hypotheses

$$H_0 : \sigma_1 = \sigma_2, \text{ against } \quad H_A : \sigma_1 \neq \sigma_2$$

Obviously, due to the strict positivity of variance, these hypotheses may also stated as

$$H_0 : \frac{\sigma_1}{\sigma_2} = 1, \text{ against } \quad H_A : \frac{\sigma_1}{\sigma_2} \neq 1,$$

and a test statistic may be based on the ration of sample variances. Since,

$$\frac{(n-1)S_1^2}{\sigma_1^2} \sim \chi^2_{n_1-1} \text{ is independent to } \frac{(n-1)S_2^2}{\sigma_2^2} \sim \chi^2_{n_2-1}.$$

Under the null hypothesis we find that the ratio

$$\frac{S_1^2/\sigma_1^2}{S_2^2/\sigma_2^2} = \frac{\frac{(n_1-1)S_1^2}{\sigma_1} \cdot \frac{1}{n_1-1}}{\frac{(n_2-1)S_2^2}{\sigma_2} \cdot \frac{1}{n_2-1}} = \frac{\chi^2_{n_1-1}/(n_1-1)}{\chi^2_{n_2-1}/(n_2-1)} \sim F_{n_1-1,n_2-1}$$

has an F-distribution with parameters $n_1 - 1, n_2 - 1$. From this the testing procedure for the null hypothesis on the precision of two sets of measurements $H_0 : \sigma_1 = \sigma_2$ is immediate.

Example 6.12 *For the Arsenic concentrations in public drinking water from Metropolitan Phoenix communities* $3, 7, 25, 10, 15, 6, 12, 25, 15, 7$ *parts per billion (ppb) and from rural Arizona communities* $48, 44, 40, 38, 33, 21, 20, 12, 1, 18$ *ppb we test the hypothesis of equal variance. The sample variance for the first is 58.27778 and for the second 235.6111, so that its ratio is 0.2473473. Under the null hypothesis this has an* $F(10 - 1, 10 - 1)$ *distribution resulting in p-value 0.04936. Hence, the null hypothesis of equal variances is rejected.*

6.5 General Inference Paradigm

In this chapter we have learned how the average $\overline{Y}$ of a random sample $Y_1, \ldots, Y_n \sim \mathcal{L}(\mu, \sigma^2)$ is a good way to estimate the unknown mean μ. First we saw that this was the case if the measurements were assumed normally distributed. Moreover, the law of large numbers showed that also for non-normal data, the average will approach the true mean. Furthermore, the central limit theorem showed that irrespective of the distribution of the data, the average always tends to be normally distributed. This was then used to construct confidence intervals for the unknown target μ.

This section will mimic the above approach for an arbitrary target θ of interest. For example, assume that we are interested in estimating the reaction rate of a particular chemical reaction. The three questions we will answer in this section are the core elements of a general estimation paradigm, that closely mimics the approach for the mean:

1. In Section 6.5.1 we derive a general method, called maximum likelihood estimation, for combining the data into an estimate of the target θ.
2. In Section 6.5.2 we show that the maximum likelihood approaches the target, just like the law of large numbers for the average.
3. In Section 6.5.3 we show that analogous to the central limit theorem for the mean, the maximum likelihood estimator tends to be normally distributed.
4. In Section 6.5.4 and Section 6.5.5. we derive confidence intervals and testing procedures, respectively, for the target of interest.
5. In Section 6.5.6. and Section 6.5.7. we develop model evaluation procedures for nested and non-nested models, respectively.

6.5.1 Maximum likelihood estimation (MLE)

Whereas the average $\bar{Y}$ was an intuitive way to combine the data to estimate the mean μ, this may in general not be so obvious. For example, if we have a set temporal measurements of interarrival times for a particular reaction, then how should we combine them to estimate the reaction rate of that reaction? There are a number of methods available to create estimators, such as the method-of-moments and Bayesian posterior estimation. Here we introduce maximum likelihood estimation, as it has superior performance over method-of-moment estimators and in general it is computationally less involved than Bayesian inference.

Definition 6.1 *Consider data Y distributed according to some parametric distribution $Y \sim \mathcal{L}(\theta)$ with parameter θ. The log-likelihood of θ is defined as,*

$$\ell(\theta) = \log f_\theta(Y),$$

and the **maximum likelihood estimator** *$\hat{\theta}$ of θ is defined as*

$$\hat{\theta}(Y) = \arg\max_\theta \ell(\theta).$$

So the maximum likelihood estimator is defined as that value of θ that makes the data as likely as possible.

Example 6.13 *Consider a sample $Y_1, \ldots, Y_n \overset{IID}{\sim} N(\mu, \sigma^2)$. Then the log-likelihood function is given as,*

$$\begin{aligned} \ell(\mu) &= \log \prod_{i=1}^n \frac{1}{\sqrt{2\pi\sigma^2}} e^{\frac{1}{2\sigma^2}(Y_i-\mu)^2} \\ &= -\frac{n}{2}\log(2\pi\sigma^2) - \frac{1}{2\sigma^2}\sum_{i=1}^n (Y_i - \mu)^2 \end{aligned}$$

Maximizing this function can be done by setting its derivative to zero,

$$\frac{d\ell}{d\mu} = \frac{1}{2\sigma^2}\sum_{i=1}^n (Y_i - \mu) \overset{set\ to}{=} 0,$$

Working out that $\sum_{i=1}^n Y_i - n\mu = 0$, we obtain the maximum likelihood estimator,

$$\hat{\mu}(Y) = \bar{Y}.$$

By plugging this into the second derivative, $\frac{d^2\ell}{d\mu^2}(\bar{Y}) = -\frac{n}{\sigma^2} < 0$, it is easy to check that this is indeed a maximum.

Therefore, the maximum likelihood estimator of for the mean of normals is indeed the average, as we have used in the rest of the chapter. The advantage is that maximum likelihood estimation can be applied in any situation.

Example 6.14 Reaction rate. *Consider the simple reaction of the falling apart of an oxygen molecule in two oxygen atoms $O_2 \xrightarrow{\lambda} O + O$. Consider the idealized case in which we observe a single O_2 molecule n times and record the amount of time T_i ($i = 1, \ldots, n$) it takes for each spontaneous reaction to occur. Assuming that the waiting times are exponentially distributed, $T_1, \ldots, T_n \overset{IID}{\sim} Exp(\lambda)$, we obtain the log-likelihood function*

$$\begin{aligned} \ell(\lambda) &= \prod_{i=1}^n \lambda e^{-\lambda T_i} \\ &= n\log(\lambda) - \lambda\sum_{i=1}^n T_i. \end{aligned}$$

Setting its derivative to zero, $\frac{d\ell}{d\lambda} = \frac{n}{\lambda} - \sum_{i=1}^{n} T_i \overset{\text{set to}}{=} 0$, *gives us* $\lambda = n/\sum_i T_i$ *and checking that its second derivative is negative,* $\frac{d^2\ell}{d\lambda^2}(\frac{n}{\sum_i T_i}) = -\frac{n}{(n/\sum_i T_i)^2} < 0$, *we obtain that the maximum likelihood estimator for the reaction rate is given as*

$$\hat{\lambda}(T) = \frac{n}{\sum_{i=1}^{n} T_i} = 1/\bar{T}.$$

In this section we have seen how finding an estimator for any parameter of interest does not have to be a matter of insight or creativity, but simply of optimizing the log-likelihood function. In the next two sections, we will find out how good maximum likelihood estimators actually are.

6.5.2 Consistency of the MLE

With the use of the law of large numbers we have seen that the average converges to the mean, as long as the variance of individual random variables is finite. Similarly, for general maximum likelihood estimators it holds that the estimator converges to the true value of the parameter, under some conditions. First, we give this property a formal name, namely **consistency**.

Definition 6.2 Consistency. *Consider data* $Y_1, \ldots, Y_n \sim f_{\theta_0}$ *is distributed according to some density* f_{θ_0} *with true value* θ_0. *An estimator* $\hat{\theta}_n(Y)$ *is called* **consistent**, *if*

$$\hat{\theta}_n(Y) \xrightarrow{P} \theta_0 \text{ as } n \longrightarrow \infty.$$

As we have already seen in this chapter, the average is a consistent estimator of the mean.

Example 6.15 The mean again. *Consider a random sample* $Y_1, \ldots, Y_n \overset{IID}{\sim} \mathcal{L}(\mu, \sigma^2)$, *then*

$$\hat{\mu}_n(Y) = \frac{1}{n}\sum_{i=1}^{n} Y_i \text{ is consistent,}$$

because with the law of large numbers we have that $\hat{\mu}_n(Y) \xrightarrow{P} \mu$.

The wonderful thing is that consistency holds more generally for maximum likelihood estimators under some **regularity conditions**. As these conditions are used a number of times in this section, we will define these regularity conditions explicitly.

Definition 6.3 Regularity conditions. *A distribution* f_{θ_0} *satisfies the so-called* regularity conditions, *if*

- *the parameter* θ_0 *belongs to the interior of the parameter space,*

- *and the log-likelihood function is trice differentiable w.r.t. θ with a bounded third derivative.*

Theorem 6.3 Consistency of the maximum likelihood estimator. *Consider a random sample $Y_1, \ldots, Y_n \stackrel{IID}{\sim} f_{\theta_0}$ that satisfies the regularity conditions. In that case the maximum likelihood estimator $\hat{\mu}_n(Y)$ is consistent,*

$$\hat{\theta}_n(Y) \xrightarrow{P} \theta_0$$

Example 6.16 Reaction rate. *Consider the simple reaction we considered above of the disintegration of an oxygen molecule in two oxygen atoms $O_2 \xrightarrow{\lambda} O + O$. Again, we consider the idealized case in which we observe a single O_2 molecule n times and record the amount of time T_i $(i = 1, \ldots, n)$ it takes for each spontaneous reaction to occur. Assuming that the waiting times are exponentially distributed, $T_1, \ldots, T_n \stackrel{IID}{\sim} Exp(\lambda_0)$ with $\lambda_0 \in (\epsilon, \infty)$ for some $\epsilon > 0$, then we note that*

- *the true value λ_0 is in the interior of $[\epsilon, \infty)$,*
- *and the log-likelihood is three times (actually infinitely) differentiable with a bounded third derivative,*

$$\begin{aligned} \frac{d\ell}{d\lambda} &= \frac{n}{\lambda} - \sum_{i=1}^{n} T_i \\ \frac{d^2\ell}{d\lambda^2} &= -\frac{n}{\lambda^2} \\ \frac{d^3\ell}{d\lambda^3} &= \frac{2n}{\lambda^3}. \end{aligned}$$

Note that the third derivative is bounded as $\lambda > \epsilon > 0$.

So, in that case the maximum likelihood estimator is consistent,

$$\hat{\lambda}_n(T) = \frac{n}{\sum_{i=1}^{n} T_i} \xrightarrow{P} \lambda_0.$$

We had to make the slightly uncomfortable assumption that $\lambda_0 > \epsilon$ is bounded away from zero, but in practice this is not a problem because the results holds for an arbitrarily small ϵ. So, only if $\lambda_0 = 0$, then there is an issue: this is clear, because if the rate of a certain reaction is zero, then the reaction will not happen and the reaction times T_i would not be observed.

In the rest of the section, we will give a feeling for why the maximum likelihood estimator is consistent. We consider the following Taylor approximation of the derivative of the log-likelihood function,

$$\dot{\ell}_n(\hat{\theta}_n) \approx \dot{\ell}(\theta_0) + (\theta_0 - \hat{\theta}_n)\ddot{\ell}(\theta_0),$$

where $\hat{\theta}_n$ is the maximum likelihood estimate, θ_0 the true value, $\dot{\ell}_n(\theta) = \frac{d\ell}{d\theta}$ and $\ddot{\ell}_n(\theta) = \frac{d^2\ell}{d\theta^2}$ the first and second derivative of the log-likelihood, respectively, based on n observations. This Taylor expansion is quite good, because we know that the third derivative is bounded. By using the fact that $\hat{\theta}_n$ is a maximum and therefore $\dot{\ell}_n(\hat{\theta}_n) = 0$, we can re-arrange the above equation as,

$$\hat{\theta}_n - \theta_0 \approx \frac{\frac{1}{n}\dot{\ell}_n(\theta_0)}{\frac{1}{n}\ddot{\ell}_n(\theta_0)}. \tag{6.8}$$

We take numerator and denominator in turn.

- **Numerator.** First we note that the log-likelihood function of a set of independent observations is the sum of the individual log-likelihood functions, i.e.,

$$\ell(\theta) = \log f_Y(\theta) = \log \prod_{i=1}^{n} f_{Y_i}(\theta) = \sum_{i=1}^{n} \log f_{Y_i}(\theta) = \sum_{i=1}^{n} \ell_{Y_i}(\theta).$$

  Taking the derivative on both sides, we get that

$$\frac{1}{n}\dot{\ell}_n(\theta_0) = \frac{1}{n}\sum_{i=1}^{n} \dot{\ell}_{Y_i}(\theta_0).$$

  As the Y_i are independent, also the $\dot{\ell}_{Y_i}(\theta_0)$ are independent, and so we can use the law of large numbers,

$$\frac{1}{n}\dot{\ell}_n(\theta_0) = \frac{1}{n}\sum_{i=1}^{n} \dot{\ell}_{Y_i}(\theta_0) \xrightarrow{P} E\dot{\ell}_{Y_1}(\theta_0),$$

  where, using the interchange of integration and differentiation (see Section 2.6.4),

$$E\dot{\ell}_{Y_1}(\theta_0) = \int \dot{\ell}_y(\theta_0) f_{\theta_0}(y)\, dy = \int \frac{\frac{df_{\theta_0}(y)}{d\theta}}{f_{\theta_0}(y)} f_{\theta_0}(y)\, dy = \frac{d}{d\theta}\int f_{\theta_0}(y)\, dy = 0,$$

  as long as variance of $\dot{\ell}_{Y_1}(\theta_0)$ is finite. For the moment we assume that this is the case.

- **Denominator.** Similarly, the second derivative of the log-likelihood can be written as a sum and therefore, using again the law of large numbers, we obtain

$$\frac{1}{n}\ddot{\ell}_n(\theta_0) = \frac{1}{n}\sum_{i=1}^{n} \ddot{\ell}_{Y_i}(\theta_0) \xrightarrow{P} E\ddot{\ell}_{Y_1}(\theta_0),$$

  as long as the variance is again finite. This quantity $E\ddot{\ell}_{Y_1}(\theta_0)$ is called the Fisher Information,

$$I(\theta_0) = E\ddot{\ell}_{Y_1}(\theta_0),$$

  and it will return next section as a important precision bound.

Putting the results for the numerator and denominator together, we can now conclude that

$$\hat{\theta}_n - \theta_0 \approx \frac{\frac{1}{n}\dot{\ell}_n(\theta_0)}{\frac{1}{n}\ddot{\ell}_n(\theta_0)} \xrightarrow{P} \frac{0}{I(\theta_0)} = 0.$$

The result requires us to take the limit separately in the numerator and denominator, which is allowed by a Slutsky's theorem. Although the above derivation is for univariate parameters, it can straightforwardly be extended to a multivariate setting. We do not provide the extension here, but we do provide an example.

Example 6.17 Simultaneous consistency of mean and variance. *Consider $Y_1, Y_2, \ldots \overset{IID}{\sim} N(\theta_1, \theta_2)$, so the parameter $\theta = (\theta_1, \theta_2)$ is bivariate. Assuming that the parameter space is $\Theta = \mathbb{R} \times (\epsilon, \infty)$ for some arbitrarily small ϵ, then*

- *the true value θ_0 lies in the interior of Θ,*
- *the log-likelihood $\ell(\theta) = -\frac{n}{2}\log(2\pi) - \frac{n}{2}\log(\theta_2) - \frac{1}{2\theta_2}\sum_{i=1}^n (Y_i - \theta_1)^2$ is three times differentiable with a bounded third derivative,*

$$\begin{aligned}
\dot{\ell}(\theta) &= \begin{bmatrix} \frac{1}{\theta_2}\sum_{i=1}^n (Y_i - \theta_1) \\ -\frac{n}{2\theta_2} + \frac{1}{2\theta_2^2}\sum_{i=1}^n (Y_i - \theta_1)^2 \end{bmatrix} \\
\ddot{\ell}(\theta) &= \begin{bmatrix} -\frac{n}{\theta_2} & -\frac{1}{\theta_2^2}\sum_{i=1}^n (Y_i - \theta_1) \\ -\frac{1}{\theta_2^2}\sum_{i=1}^n (Y_i - \theta_1) & \frac{n}{2\theta_2^2} - \frac{1}{\theta_2^3}\sum_{i=1}^n (Y_i - \theta_1)^2 \end{bmatrix} \\
\frac{d}{d\theta_1}\ddot{\ell}(\theta) &= \begin{bmatrix} 0 & \frac{n}{\theta_2^2} \\ \frac{n}{\theta_2^2} & \frac{2}{\theta_2^3}\sum_{i=1}^n (Y_i - \theta_1) \end{bmatrix} \\
\frac{d}{d\theta_2}\ddot{\ell}(\theta) &= \begin{bmatrix} \frac{n}{\theta_2^2} & \frac{2}{\theta_2^3}\sum_{i=1}^n (Y_i - \theta_1) \\ \frac{2}{\theta_2^3}\sum_{i=1}^n (Y_i - \theta_1) & -\frac{n}{\theta_2^3} + \frac{3}{\theta_2^4}\sum_{i=1}^n (Y_i - \theta_1)^2 \end{bmatrix}
\end{aligned}$$

where $\dot{\ell}(\theta) = \frac{\partial \ell}{\partial \theta}$ and $\ddot{\ell}(\theta) = \frac{\partial^2 \ell}{\partial \theta \partial \theta}$ are the Jacobian ∇_ℓ and Hessian H_ℓ, respectively. Because the third derivative would be an array, we represent it instead by the two partial derivatives of the Hessian.

6.5.3 Efficiency of the MLE

In this section we show that, similar to central limit theorem for the sample mean, also the maximum likelihood estimator is asymptotically normally distributed. But it even gets better than that. The asymptotic variance of the maximum likelihood estimator is as good as it can get. We refer to the asymptotic normality and minimum variance property as *efficiency* . Although the efficiency of the maximum likelihood estimator is quite general, certain conditions need to be fulfilled first. They are the same regularity conditions that we have seen above.

There is a fundamental "no free lunch" theorem in statistics that states that if you want to estimate a quantity accurately on average, then under the same regularity conditions as before it will have at least a minimum amount of variance attached to it. This theorem is known as the Cramér-Rao Lowerbound or the Cramér-Rao inequality.

Theorem 6.4 Cramér-Rao inequality
Consider a random sample $Y_1, \ldots, Y_n \overset{IID}{\sim} f_{\theta_0}$ *for some continuous differentiable density* f_{θ_0} *whose support does not depend on* θ, *and an* unbiased *estimator* $\hat{\theta}(Y)$, *then*

$$V(\hat{\theta}(Y)) \geq [nI(\theta_0)]^{-1}$$

where the Fisher Information $I(\theta)$ *can be calculated in two equivalent ways,*

$$\begin{aligned} I(\theta) &= E\dot{\ell}_{Y_1}(\theta)\dot{\ell}_{Y_1}(\theta_0)^t &\quad (6.9) \\ &= -E\ddot{\ell}_{Y_1}(\theta_0), &\quad (6.10) \end{aligned}$$

if the density is also twice continuously differentiable.

Example 6.18 *Consider a sample of 3 normally distributed variables,* $Y_1, Y_2, Y_3 \overset{IID}{\sim} N(\mu, 1)$. *We are interested in estimating* μ *by means of a weighted sum of the individual values, i.e.,*

$$\hat{\mu}(Y) = w_1 Y_1 + w_2 Y_2 + w_3 Y_3.$$

First we ask for the estimator to be unbiased. Using the linearity of the expected value, we obtain the following constraint on the weights:

$$\begin{aligned} \mu &= E\hat{\mu} \\ &= w_1 EY_1 + w_2 EY_2 + w_3 EY_3 \\ &= \mu(w_1 + w_2 + w_3). \end{aligned}$$

In other words, $w_1 + w_2 + w_3$ *have to add up to 1. In that case, the variance of the estimator is given as,*

$$V(\hat{\mu}) = w_1^2 V(Y_1) + w_2^2 V(Y_2) + w_3^2 V(Y_3) = w_1^2 + w_2^2 + w_3^2$$

On the other hand, the Cramér-Rao inequality states that for any unbiased estimator $\hat{\mu}^*$ *the variance is at least,*

$$V(\hat{\mu}^*) \geq \frac{1}{3E\ddot{\ell}_{Y_1}(\mu)} = \frac{1}{3}.$$

If we select $w_1 = w_2 = w_3 = 1/3$, *then* $V(\hat{\mu}) = 1/3$ *achieving the Cramér-Rao lowerbound, and therefore we know that this is the best possible unbiased estimate. We call such estimate* **efficient**.

Under the usual regularity conditions, the maximum likelihood estimator has the property that it is asymptotically normally distributed with the true value as mean and the smallest possible variance. We refer to this property as the efficiency of the maximum likelihood.

Theorem 6.5 Efficiency of the maximum likelihood estimator *Consider a random sample* $Y_1, \ldots, Y_n \overset{IID}{\sim} f_{\theta_0}$ *for some density* f_{θ_0}*, for which the regularity conditions in definition 6.3 hold. The maximum likelihood estimator* $\hat{\mu}_n(Y)$ *is asymptotically efficient, i.e.,*

$$\sqrt{n}(\hat{\theta}_n(Y) - \theta_0) \xrightarrow{D} N(0, I(\theta_0)^{-1}),$$

where $I(\theta_0) = E\ddot{\ell}_{Y_1}(\theta_0)$ *is called the Fisher information.*

To get an idea where this comes from we take expression (6.8) and multiply both sides by $\sqrt{n}$,

$$\sqrt{n}(\hat{\theta}_n - \theta_0) \approx \frac{\frac{1}{\sqrt{n}}\dot{\ell}_n(\theta_0)}{\frac{1}{n}\ddot{\ell}_n(\theta_0)}.$$

Just as before in the proof of consistency, the denominator converges to the Fisher information $I(\theta_0)$. For the nominator, we can apply directly the central limit theorem by noting that the $\dot{\ell}_n(\theta_0)$ is the sum of independent random functions $\dot{\ell}_{Y_i}(\theta_0)$, each with mean 0 and variance $I(\theta_0)$,

$$\frac{1}{\sqrt{n}}\dot{\ell}_n(\theta_0) = \sqrt{n}\overline{\dot{\ell}_n(\theta_0)} \longrightarrow N(0, I(\theta_0)).$$

Putting numerator and denominator together, we get the result stated in the theorem.

Example 6.19 Reaction rate. *Consider n independent reaction times,* $T_1, \ldots, T_n \overset{I.I.D.}{\sim} Exp(\lambda_0)$*, described in example 6.16. We saw that the maximum likelihood estimator is given as,*

$$\hat{\lambda} = 1/\bar{T}.$$

The distribution of the maximum likelihood is not a standard distribution. In fact, it is a rescaled inverse Gamma distribution, which not in most books. Nevertheless, if n is moderately large, then the distribution of the estimator is given as,

$$\hat{\lambda} \sim N(\lambda_0, \lambda_0^2/n),$$

as $I(\lambda) = -E\ddot{\ell}_{Y_1}(\lambda) = 1/\lambda^2$*, as was derived in example 6.16.*

6.5.4 Confidence intervals using the MLE

The asymptotically efficient distribution of the maximum likelihood estimator now allows us to define approximate interval estimates for a univariate parameter θ. We assume in this and next section that we are in the situation where the regularity conditions are satisfied. In that case, we use the finite approximation of the distribution of the maximum likelihood estimator as a result of theorem 6.5,

$$\hat{\theta}_{\text{MLE}} \overset{\text{approx}}{\sim} N\left(\theta_0, \frac{1}{nI(\theta_0)}\right),$$

we obtain the following $1-\alpha$ coverage probibility,

$$1-\alpha = P\left(\hat{\theta}_{\text{MLE}} - \frac{z_{\alpha/2}}{\sqrt{nI(\theta_0)}} \leq \theta_0 \leq \hat{\theta}_{\text{MLE}} + \frac{z_{\alpha/2}}{\sqrt{nI(\theta_0)}}\right),$$

where z_p is defined as the upper p quantile of a standard normal distribution.

This results in a $(\mathbf{1}-\alpha) \times \mathbf{100\%}$ approximate **confidence interval** for θ

$$\left[\hat{\theta}_{\text{MLE}} - \frac{z_{\alpha/2}}{\sqrt{nI(\hat{\theta}_{\text{MLE}})}}, \hat{\theta}_{\text{MLE}} + \frac{z_{\alpha/2}}{\sqrt{nI(\hat{\theta}_{\text{MLE}})}}\right].$$

There are two approximations in this interval. Firstly, the distribution is under most distributions not precisely normal. For most practical purposes, an n of at least 30 sample will result in a good approximation. Secondly, the variance of the estimator distribution should be calculate at the, unknown, true value of the parameter. Instead we plug in the estimate of the parameter in the Fisher information. Under certain, quite special, circumstances the lower and upper bound of the parameters can be solved explicitly, but in most cases we use the previous approximation.

Example 6.20 Reaction rate confidence interval. *Consider n independent reaction times, $T_1, \ldots, T_n \overset{I.I.D.}{\sim} Exp(\lambda_0)$, described in example 6.16. Our aim is to construct a 95% confidence interval. As $z_{0.025} = 1.96$ and the MLE is $\hat{\lambda} = 1/\bar{T}$, we obtain a confidence interval*

$$\left[\frac{1}{\bar{T}} - \frac{1.96}{\bar{T}\sqrt{n}}, \frac{1}{\bar{T}} + \frac{1.96}{\bar{T}\sqrt{n}}\right].$$

Assume that we observe $n = 10$ samples with average reaction time $\bar{T} = 0.4$, a 95% confidence interval for the reaction rate λ is given as,

$$[0.95, 4.05].$$

6.5.5 Testing hypotheses with the MLE

Hypothesis testing is one of the basic tools in the statistician's toolbox. Like any tool it can be misused, but with careful attention to its limits and with the correct interpretation of an hypothesis test it can be a powerful way not to be led astray by apparent patterns in small datasets.

According to Austrian-British philosopher Karl Popper (1902–1994) science progresses through falsifying current theories and ideas, rather than by finding evidence in favor of these theories. This can be formalized in a statistical framework, whereby the current theory framework is presented as the *null hypothesis* H_0. The idea is to frame null hypothesis as a statement about a parameter in a model, such as for instance the reaction rate of particular reaction,

$$H_0 : \theta = \theta_0.$$

Unlike before the value θ_0 does not refer to the true parameter, but about an hypothesized value of the parameter. Then we perform experiments in order to learn about the true value of θ. Here we consider a random sample $Y_1, \ldots, Y_n \overset{IID}{\sim} f_\theta$, whereby f_θ is assumed to satisfy the regularity conditions from definition 6.3. As the maximum likelihood is known to be a particularly informative estimate of θ, we propose to use it as the *test-statistic* ,

$$\text{test-statistic} = \hat{\theta}_{\text{MLE}}(Y).$$

Given the actual data $y_1, \ldots, y_n$, we can now calculate the actual estimate $\hat{\theta}_{\text{MLE}}(y)$. The idea is to compare $\hat{\theta}_{\text{MLE}}(y)$ with θ_0 and if they are "sufficiently different," then we reject the null hypothesis. For this we employ a *modus tollens*-like form of reasoning from propositional logic: if the statement P implies the statement Q, then if it can be shown that Q is false, then P must be false.

The idea is to find a probabilistic analogy to the *modus tollens.* This is done by defining the concept of a *p-value* as the probability of obtaining something as extreme as $\hat{\theta}_{\text{MLE}}(y)$ if the null hypothesis is true,

$$\text{p-value} = P_{\theta_0}(|\hat{\theta}_{\text{MLE}}(Y)| \geq |\hat{\theta}_{\text{MLE}}(y)|).$$

The formal argument now goes as follows:

1. Assume that the null hypothesis is true, then calculate the p-value expressing the probability of obtaining $\hat{\theta}_{\text{MLE}}(y)$ as estimate if θ is really θ_0.
2. If the p-value is *small*, then it is very unlikely that one would have obtained $\hat{\theta}_{\text{MLE}}(y)$ if the null hypothesis was true.
3. Reject the null hypothesis.

The concept of the *significance level* α formalizes the concept of a *small p-value.* Ever since 1925 when Ronald Aylmer Fisher (1890–1962) suggested that $\alpha = 0.05$ is a good trade-off, this has been used ubiquitously. However, it is worthwhile remembering that the significance level can be set more or less stringently, depending on the circumstances. The value $\alpha = 0.05$ means that for every 20 true null hypotheses, one will be rejected by mistake on average. We now summarize all the steps into a single practical testing procedure.

Generic testing procedure

1. Given the (univariate) parameter of interest θ, encode the current knowledge about θ as the null-hypothesis

$$H_0 : \theta = \theta_0$$

2. Set the appropriate significance level for your situation, often $\alpha = 0.05$.
3. Collect a random sample $Y_1 = y_1, \ldots, Y_n = y_n \overset{IID}{\sim} f_\theta$, obtain the maximum likelihood estimate $\hat{\theta}_{\text{MLE}}(y)$, and calculate the p-value

$$\text{p-value} = 2\left(1 - \Phi\left(\frac{|\hat{\theta}_{\text{MLE}}(y) - \theta_0|}{\sqrt{nI(\theta_0)}}\right)\right),$$

where Φ is the cumulative distribution function of the standard normal distribution.

4. Take a decision,
   - if p-value $\geq \alpha$, then do not reject the null hypothesis.
   - if p-value $< \alpha$, then reject the null hypothesis.

Example 6.21 Testing reaction rate. *A certain reaction observed under laboratory settings for a total of 10 cases. Standard textbook information describe the reaction rate as $\lambda = 2$ per minute. However, the given the average reaction time $\bar{t} = 0.4$ minute across the 10 observations, we want to test at the $\alpha = 0.05$ level whether we can reject the null hypothesis,*

$$H_0 : \lambda = 2.$$

We have seen in example 6.16 that if the null hypothesis is true, the maximum likelihood estimator $\hat{\lambda} = 1/\bar{T}$ is roughly distributed as

$$\hat{\lambda} \mid H_0 \sim N(2, 0.4).$$

The p-value is then given as,

$$\begin{aligned} p\text{-}value &= 2P(\hat{\lambda} \geq 1/0.4 \mid H_0) \\ &= 2(1 - \Phi(\frac{2.5 - 2)}{\sqrt{0.4}})) \\ &= 0.429 \end{aligned}$$

So, we cannot reject the null hypothesis at the $\alpha = 0.05$ level. The data are consistent with the textbook value $\lambda = 2$.

Power and sample size calculations.

If a null hypothesis does not get rejected, then there is in general very little that can be said about the truth of the null hypothesis: it could be that insufficient data were collected to find sufficient evidence to reject the null hypothesis or that indeed the null hypothesis, for all practical purposes, is true. In order to solve this problem, we introduce the concept of *power*. Formally, the power of a test is the probability of rejecting the null hypothesis, when it is in fact false. When considering this definition in some detail, it immediately exhibits some problematic aspects. If the true parameter is really really close to the hypothesized parameter θ_0, say $\theta = \theta_0 + 10^{-100}$, then clearly the null hypothesis is false, but no amount of data will be able to detect it in practice.

To make the concept of power practical, we have to accompany it in practice by the minimal difference with respect to the value in the null hypothesis we would want to be able to detect, the so-called *minimum detectable effect size* (MDES) $\Delta\theta$. This quantity has to be set by the experimenter and depends on the relevant background knowledge about what constitutes a relevant effect size. There are two things we can do with this quantity: (1) for a fixed number of observations n, we can calculate the power of the test to detect a difference of at least that amount $|\theta_0 - \theta_{\text{true}}| \geq \Delta\theta$; (2) for a desired amount of power to detect such difference, we can calculate the necessary amount amount of samples n.

Definition 6.4 *Given data $Y \sim f_{\theta_{true}}$, the aim is to test the null hypothesis*

$$H_0 : \theta = \theta_0$$

using a test-statistic T. Given a **minimal detectable effect size** *$\Delta\theta$ the* **power** *$1-\beta$ of the test is defined as the probability of rejecting the null hypothesis when the true parameter θ_{true} is at least $\Delta\theta$ away from the hypothesized value θ_0, i.e., $|\theta_0 - \theta_{true}| \geq \Delta\theta$.*

Using the generic testing framework based on the maximum likelihood estimator $\hat{\theta}_{\text{MLE}}$ as test-statistic, it is now possible to determine the (approximate) power of a test, the number of samples needed in order to achieve a certain power, or the maximum MDES for a certain power and sample size.

Generic power and sample size calculations Consider a random sample $Y_1, \ldots, Y_n \overset{IID}{\sim} f_\theta$, whereby f_θ is assumed to satisfy the regularity conditions from definition 6.3. We use the generic testing framework described above to test the null-hypothesis

$$H_0 : \theta = \theta_0$$

using the maximum likelihood estimate $\hat{\theta}_{\mathrm{MLE}}(y)$.

1. For a fixed number of samples n and minimum detectable effect size $\Delta\theta$, the **power** $1 - \beta$ of the test is approximately given as,

$$\text{Power} = \Phi(\Delta\theta\sqrt{nI(\theta_0)} - z_{\alpha/2}),$$

   where $I(\theta)$ is the Fisher information and $z_{\alpha/2}$ is the $(1 - \alpha/2)$ quantile of a standard normal distribution.

2. Given a desired level of power $1 - \beta$ and minimum detectable effect size $\Delta\theta$, the required sample size is given as,

$$n = \left(\frac{z_\beta + z_{\alpha/2}}{\Delta\theta}\right)^2 I(\theta_0)^{-1}.$$

3. Given a sample size n and a desired power of $1-\beta$, the minimum dectectable effect size $\Delta\theta$ is given as,

$$\Delta\theta = \frac{z_\beta + z_{\alpha/2}}{\sqrt{nI(\theta_0)}}.$$

In all of these calculations it is assumed that $I(\theta_0 \pm \Delta\theta) \approx I(\theta_0)$. This is often quite reasonable. However, especially when considering values of θ_0 close to the boundary of the parameter space, it may be worthwhile checking.

Example 6.22 Reaction rate power calculations. *A laboratory compares the reaction rates of a particular reaction with the standard textbook value of* $\lambda = 2$. *With a sample of 10 observations, they use the maximum likelihood to estimate that* $\hat{\lambda}_{10} = 2.5$. *This was described in example 6.21. However, the sample is too small at* $\alpha = 0.05$ *to conclude that the textbook value is incorrect. Still, also for other reasons, some doubt lingers in the minds of the scientists and they prepare to do an experiment to test the hypothesis again. However, they would like to be able to detect at least a* minimal detectable effect size *of* $\Delta\lambda = 0.1$. *What is the power of the current experimental set-up?*

$$\textit{Power} = \Phi(0.1\sqrt{0.4} - 1.96) = 0.036.$$

This means that this study has only a measly power of approximately 4%. This means that if the null hypothesis is false with a $\Delta\lambda = 0.1$, *then with 10 samples only 4% of the times the null hypothesis will be rejected. The laboratory decides to increase the power to 80%, i.e.,* $1-\beta = 0.8$*. How many observations does the laboratory need to collect in order to achieve this? Using the formula derived above, we find*

$$n = \left(\frac{0.842 + 1.96}{0.1}\right)^2 4 = 3139.6,$$

so at least 3140 samples are needed.

Multiple testing with the Bonferroni correction

Often experiments are performed not to test a single hypothesis, but for testing multiple hypotheses. The danger exists that the more hypotheses are formulated, the likelier it becomes that by chance the data turns out to support one of these hypotheses. In order to avoid this so-called *multiple testing problem*, the amount of evidence required to reject an individual hypothesis on a long list of hypothesis, should also depend on the length of this list.

Consider a list of m null hypotheses, $H_0^1, \ldots, H_0^m$, which we want to test with the data X from a single experiment. Associated with each of the null hypotheses, we calculate a p-value,

$$p_1(X), \ldots, p_m(X).$$

Given that a p-value is distributed as a Uniform$(0, 1)$ if the null hypothesis is true and, therefore, $P(p_i(X) \leq \alpha \mid H_0^i \text{ true}) = \alpha$. Then for any number m_0 of true null hypotheses $H_0^{i_1}, \ldots, H_0^{i_{m_0}}$, the **familywise error rate** is the total probability of rejecting at least one of these true hypotheses, i.e., the probability that at least one of their associated p-values is less than the significance level α,

$$\text{FWER} = P(\cup_{k=1}^{m_0} \{p_{i_k}(X) \leq \alpha\}) \leq \sum_{k=1}^{m_0} P(p_{i_k}(X) \leq \alpha) \leq m\alpha.$$

The inequality relies on a result discovered in 1936 by the Italian mathematician Carlo Emilio Bonferroni (1892 – 1960).

Bonferroni correction. When testing m hypotheses simultaneously at an overall significance level α,

- either one can reject each individual test at significance level α/m,
- or, equivalently, one can multiply each of the individual p-values by m and reject the individual tests at significance level α.

This is called the **Bonferroni correction** for controlling the familywise error rate.

6.5.6 Testing multiple parameters with likelihood ratio test

There are a number of circumstances where one wants to test a number of parameters simultanously. For example, when a chemical lab acquires a new measurement machine, it might want to test that it is both that its measurements are both *accurate* and *precise.* Accuracy means that the measurements are centred around the true value, whereas precision means that the measurements are highly reproducible. Statistically speaking, one may want to test the null hypothesis

$$H_0 : \mu = \mu_0 \text{ (accuracy) and } \sigma^2 = \sigma_0^2 \text{ (precision).}$$

where μ_0 is the true value of the quantity of interest and where σ_0^2 is the precision of the old measurement machine – we want to make sure that the new one is at least as good as the old one! In subsequent chapters, we will come across multiple testing scenarios in more generic situations, such as checking whether a set of effects can be igored or not, by testing whether or not they are zero,

$$H_0 : \beta_1 = \beta_2 = \ldots = \beta_q = 0.$$

Whatever the setting, the answer can be obtained via the generic *likelihood ratio test.*

Consider a probabilistic model for some data $Y = y$ with q parameters $\theta \in \Theta \subset \mathbb{R}^q$, i.e., probability density f_θ. Then the ratio

$$\frac{f_{\theta_0}(y)}{f_{\theta_1}(y)}, \tag{6.11}$$

is a scalar that expresses how much more (or less) likely θ_0 is with respect to θ_1 for the *same data* $Y = y$. Clearly we would like to accept θ_0 only if there are no other θ that are much more likely. Consider the following hypotheses,

$$\begin{aligned} H_0 &: \ \theta = \theta_0 \\ H_1 &: \ \theta \in \Theta \backslash \theta_0 \end{aligned}$$

then the idea is to compare the ratio (6.11) for all values in the denominator and look a the largest value. We refer to this value as the *likelihood ratio*,

$$\Lambda(y) = \frac{f_{\theta_0}(y)}{\sup_{\theta \in \Theta} f_\theta(y)}. \tag{6.12}$$

The idea is to reject the null hypothesis if $\Lambda(y)$ is close to zero and not to reject it when it is sufficiently close to one. By determining the distribution of the test-statistic $\Lambda(Y)$, or, more precisely, $-2 \log \Lambda(Y)$, we can make precise where the decision boundary lies.

First, note that the supremum in the denominator of (6.12) is obtained right at the maximum likelihood estimate $\hat{\theta}$. Second, the first derivative of the likelihood is zero at $\hat{\theta}$. Then combining these facts with a second-order Taylor expansion, we get

$$\begin{aligned} -2\log\Lambda(Y) &= -2(\ell(\theta_0) - \ell(\hat{\theta})) \\ &\approx -2\left(\ell(\hat{\theta}) + \frac{1}{2}(\hat{\theta}-\theta_0)^t\ddot{\ell}(\hat{\theta})(\hat{\theta}-\theta_0) - \ell(\hat{\theta})\right) \\ &= -(\hat{\theta}-\theta_0)^t\ddot{\ell}(\hat{\theta})(\hat{\theta}-\theta_0) \end{aligned}$$

Given the asymptotic normality of the maximum likelihood estimator $\hat{\theta} \sim N(\theta_0, I(\theta_0)^{-1})$, we see that under the null hypothesis $H_0 : \theta = \theta_0$, we get that the log likelihood ratio statistic is a sum of q squares of standard normal distributions and therefore,

$$-2\log\Lambda(Y) \mid H_0 \sim \chi^2_q.$$

Remember that one would reject the null hypothesis for small values of $\Lambda(y)$, which are large values of $-\log\Lambda(y)$. In fact, for significance level α, the null hypothesis is rejected if

$$-2\log\Lambda(y) \geq \chi^2_{q,1-\alpha},$$

where $\chi^2_{q,1-\alpha}$ is the $1-\alpha$ quantile of a χ^2_q distribution. As a nice rule of thumb, one can use the fact that

$$\chi^2_{q,0.95} \approx q + 2.5\sqrt{q}.$$

See Exercise 18 for more information on this derivation.

Example 6.23 Testing accuracy and precision simultaneously. *After purchasing a new surface tension meter, a lab wants to check first whether the device is both accurate, i.e., having zero bias, and at least as precise as its current device that measure surface tension up to a precision of* $\sigma = 0.01$ *millinewton per square meter. It performs 10 bias measurements* $y_1, \ldots, y_{10}$ *and finds an average of* $\bar{y} = 0.003$ *and and standard deviation* $s_y = 0.008$. *Assuming that the bias measurements are independently normally distributed with mean* μ *and variance* σ^2, *test the hypothesis,*

$$H_0 : \mu = 0, \quad \sigma = 0.01.$$

Note that the log likelihood is given as $\ell(\mu, \sigma^2) = -5\log\sigma^2 - \frac{5}{\sigma^2}(s_y^2 + (\bar{y}-\mu)^2)$. *Then the log likelihood ratio can be calculated as,*

$$\begin{aligned} -2\log\Lambda(y) &= -2(\ell(0, 0.01^2) - \ell(0.003, 0.008^2)) \\ &= 2\left(5\log\frac{0.01^2}{0.008^2} + \frac{5}{0.01^2}(0.008^2 + 0.003^2) - \frac{5}{0.008^2}0.008^2\right) \\ &= 1.762 \end{aligned}$$

As $-2\log\Lambda(y) < 5.99 = \chi^2_{2,0.95}$, *we cannot reject the null hypothesis at the 5% significance level. Therefore, there is no evidence that the new machine is inaccurate or less precise than the previous instrument.*

6.5.7 Model comparison

Statistical modeling aims to determine what stochastic model best describes the available data. It therefore important to be able to compare models. In this section we will describe ways to compare models with each other. There are a number of approaches to perform model comparison depending on what criteria are used to tell how a model best fits the data. First, we define the concept of a statistical model.

Definition 6.5 Statistical model. *A* statistical model $\mathcal{M}$ *is a sequence of sets of probability measures index by the sample size* n,

$$\mathcal{M} = \{\mathcal{P}_n\}_{n \in \mathbb{N}},$$

whereby for a parametric *statistical model each set* $\mathcal{P}$ *has an associated parametric probability density* f_θ,

$$\mathcal{P}_n = \{f_\theta : \mathbb{R}^n \longrightarrow \mathbb{R}_0^+ \mid \theta \in \Theta\},$$

where $\Theta \in \mathbb{R}^p$ *is some fixed parameter space. In the special case of* random sampling, *the joint density can be simplified* $f_\theta(y_1, \ldots, y_n) = \prod_{i=1}^n f_\theta(y_i)$.[2]

A statistical model is a sequence of stochastic models. The reason for this somewhat complex definition is that it should, in principle, always be possible to collect more data. We do not necessarily assume that the data are always independent. In fact, when we consider mixed effect models in the later part of this book, we explicitly want to take into account that data from the same physical sample or data that was collected in close physical or temporal proximity might be correlated.

Example 6.24 Accuracy/Precision model. *A chemical wants to test the accuracy, i.e., bias, and precision, i.e., variance, of a novel measurement instrument. In principle, they can take any number* $n \in \mathbb{N}$ *of samples to test the instrument. Assuming that the underlying measurements are normally distributed and that the samples are independent, they consider the following parametric statistical model,*

$$\mathcal{M} = \left\{ \prod_{i=1}^n \varphi_{\mu,\sigma^2} \mid \mu \in \mathbb{R}, \sigma^2 \geq 0 \right\}_{n \in \mathbb{N}},$$

where $\varphi_{\mu,\sigma^2}(y) = \frac{1}{\sqrt{2\pi\sigma^2}} e^{-(y-\mu)^2/2\sigma^2}$ *is the density of a* $N(\mu, \sigma^2)$ *distribution.*

[2]In the definition, as well as throughout this book, we slightly abuse notation to let the domain of a probability density f_θ implicitly depend on the context.

Likelihood ratio test for comparing nested models.

Formally, a model $\mathcal{M}_1$ is *nested* in another model $\mathcal{M}_2$, if $\mathcal{M}_1 \subset \mathcal{M}_2$. Most often, we will consider nested *parametric* models, with consists of sets of the same distributions, such that the parameter space of one model is a subset of the other, $\Theta_1 \subset \Theta_2$. The likelihood ratio test, that we encountered in the previous section, gives us a way to compare two nested models.

Consider two nested parametric models $\mathcal{M}_1 \subset \mathcal{M}_2$ with parameter $\theta = (\theta_1, \ldots, \theta_q, \theta_{q+1}, \ldots, \theta_p) \in \Theta \subset \mathbb{R}^p$. Whereas the parameter in the larger model can take any value in Θ, the value for the smaller model is fixed for the first q elements. We obtain data $y = (y_1, \ldots, y_n)$ and we want to test the hypotheses,

$$\begin{aligned} H_0 : &\quad y \sim \mathcal{M}_{1,n} \\ H_1 : &\quad y \sim \mathcal{M}_{2,n} \end{aligned}$$

These hypotheses correspond to testing the value of the first q parameter elements to some fixed specified value, i.e.,

$$\begin{aligned} H_0 : &\quad \theta_1 = \theta_{01}, \ldots, \theta_q = \theta_{0q} \\ H_1 : &\quad \theta \in \Theta \end{aligned}$$

Let $\hat{\theta}^{(1)}$ and $\hat{\theta}^{(2)}$ be the maximum likelihood estimates of θ in models $\mathcal{M}_1$ and $\mathcal{M}_2$, respectively. The likelihood ratio test statistic,

$$-2 \log \Lambda(Y) = -2(\ell(\hat{\theta}^{(2)}) - \ell(\hat{\theta}^{(1)})),$$

measures how much more likely the larger model is with respect to the smaller one. Under the null hypothesis, it is asymptotically χ^2_q distributed. Therefore, we reject the null hypothesis at the α significance level, if

$$-2 \log \Lambda(y) > \chi^2_{q,1-\alpha}.$$

Example 6.23 can also be interpreted as testing two nested models. Consider, the smaller model with parameter space

$$\Theta_1 = (\mu = 0, \sigma = 0.01)$$

and the larger model with

$$\Theta_2 = \mathbb{R} \times \mathbb{R}^+,$$

then clearly $\Theta_1 \subset \Theta_2$, and given that the underlying distributions in both models are assumed normal, the models are clearly nested. As we have seen above, the null hypothesis cannot be rejected. Therefore, there is no evidence based on the ten available samples to doubt the validity of $\mathcal{M}_1$.

TABLE 6.1
Model comparison methods based on data $y = (y_1, \ldots, y_n)$ involving the model log-likelihood $\ell^{\mathcal{M}}$ for a model $\mathcal{M}$ with p parameters $\theta = (\theta_1, \ldots, \theta_p)$ and the maximum likelihood estimate $\hat{\theta}(y)$.

Criterion	Objective	Selection method
True model	$\max_{\mathcal{M}} P(\mathcal{M}\|y)$ $\approx -2\ell_y^{\mathcal{M}}(\hat{\theta}) + p \log n$	Bayesian information criterion (BIC)
Best prediction	$E_{y,y_{n+1}} \ell_{y_{n+1}}^{\mathcal{M}}(\hat{\theta}(y))$ $\approx -2\ell_y^{\mathcal{M}}(\hat{\theta}) + 2p$	Akaike information criterion (AIC), cross-validation

Akaike and other information criteria

The likelihood ratio test can only be used if models are nested, which is rather restrictive. When comparing multiple models, it would be much more attractive to attach a score to each model and select the model with the best score. In this section we describe the idea behind two information criteria than can be calculated easily on the data and that allow direct comparison between models.

Before introducing the criteria, we have to discuss a more philosophical question. What is a good model? There are at least two possible answers. It would seem sensible to call the true model the best model. But similarly, the model that results in the best predictions might also lay claim to that title. In turns out that both criteria are not the same and typically lead to a different choice of model. But irrespective of the model comparison procedure, the best model is not necessarily the best fitting model on the data at hand. As our model In Table 6.1 we summarize the most widely used moder comparison methods.

AIC: best predictive model. Despite our best assumptions, in reality the data that we observe are most likely not generated according to any of the parametric models $\mathcal{M}$ that we propose, but instead according to some unknown true process $\mathcal{T}$. Let's consider a particular parametric model,

$$\mathcal{M} \text{ with parameter space } \Theta \subset \mathbb{R}^p.$$

Given some data $y = (y_1, \ldots, y_n) \sim \mathcal{T}$, we can select one element from $\mathcal{M}$ by calculating the maximum likelihood estimate $\hat{\theta}_n$. In order to know how well $f_{\hat{\theta}_n}$ predicts new data Y_{n+1}, we define a kind of distance between the true data generating process and the model,

$$KL(\mathcal{T}, \mathcal{M}_{\hat{\theta}_n}) = E_{Y_{n+1} \sim \mathcal{T}} \log \frac{f_{\mathcal{T}}(Y_{n+1})}{f_{\hat{\theta}_n}(Y_{n+1})},$$

which is called the *Kullback-Leibler divergence.* Note that the divergence is zero if the estimated model happens to be the true model, $\mathcal{M}_{\hat{\theta}_n} = \mathcal{T}$, and

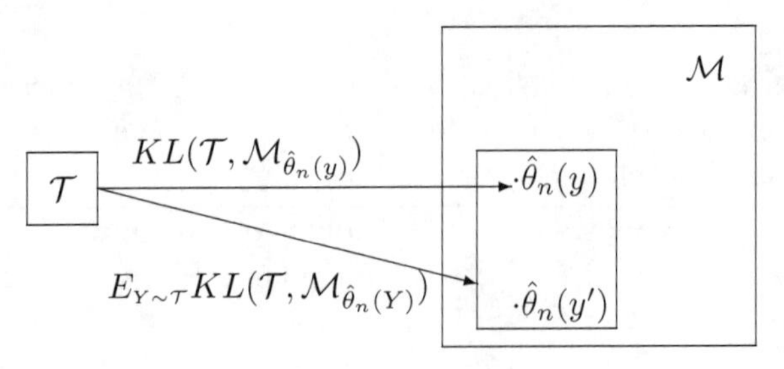

FIGURE 6.8
Schematic representation of the AIC criterion as an average Kullback-Leibler divergence from the parametric model $\mathcal{M}$ to the true data generating process $\mathcal{T}$.

greater than zero otherwise. The Kullback-Leibler divergence can also be seen as an expected likelihood ratio and is sometimes called the *relative entropy.*

Given a different sample y', we would find a different estimate $\hat{\theta}_n(y')$, which clearly should not affect the divergence between $\mathcal{T}$ and $\mathcal{M}$. We therefore define the "distance" to the true model as the expected Kullback-Leibler divergence,

$$\begin{aligned} D_{\mathcal{T}}(\mathcal{M}) &= E_{Y\sim\mathcal{T}}KL(\mathcal{T}, \mathcal{M}_{\hat{\theta}_n(Y)}) \\ &= C - E_{Y,Y_{n+1}}\ell^{\mathcal{M}}_{Y_{n+1}}(\hat{\theta}_n(Y)) \end{aligned}$$

where $C = E_{Y,Y_{n+1}}\ell^{\mathcal{T}}_{Y_{n+1}}(\hat{\theta}_n(Y))$ is a constant, in the sense that it does not depend on the choice of model $\mathcal{M}$. This idea is expressed in Figure 6.8. Let $\bar{\ell}(\hat{\theta}) = E_{Y_{n+1}}\ell^{\mathcal{M}}_{Y_{n+1}}(\hat{\theta}_n(Y))$ be the population log-likelihood that has a maximum at θ_0, then using the second-order Taylor expansion around the maximum we get

$$\bar{\ell}(\hat{\theta}) \approx \bar{\ell}(\theta_0) - \frac{1}{2}(\hat{\theta} - \theta_0)^t I(\theta_0)(\hat{\theta} - \theta_0),$$

as the first derivative at θ_0 is zero and minus the second derivative is the Fisher information $I(\theta_0)$. Taking the expectation with respect to Y, we then get,

$$\begin{aligned} E_{Y,Y_{n+1}}\ell^{\mathcal{M}}_{Y_{n+1}}(\hat{\theta}_n(Y)) &= E_{Y\sim\mathcal{T}}\bar{\ell}(\hat{\theta}(Y)) \\ &\approx \bar{\ell}(\theta_0) - \frac{1}{2}E_{Y_{\mathcal{T}}}(\hat{\theta}(Y) - \theta_0)^t I(\theta_0)(\hat{\theta}(Y) - \theta_0) \\ &\approx \bar{\ell}(\theta_0) - \frac{p}{2n} \qquad (6.13) \end{aligned}$$

as $n(\hat{\theta}(Y) - \theta_0)^t I(\theta_0)(\hat{\theta}(Y) - \theta_0) \sim \chi^2_p$ distributed, as we saw in derivation of the distribution of the log-likelihood ratio statistic in the previous section. The main issue now is that we do not know what the first term is. Fortunately, we can estimate it via the empirical likelihood,

$$
\begin{aligned}
\ell_y^{\mathcal{M}}(\hat{\theta}(y)) &= \sum_{i=1}^{n} \ell_{y_i}^{\mathcal{M}}(\hat{\theta}(y)) \\
&\approx \sum_{i=1}^{n} \ell_{y_i}^{\mathcal{M}}(\theta_0) + (\hat{\theta}(y) - \theta_0)^t \dot{\ell}_{y_i}^{\mathcal{M}}(\theta_0) + \frac{1}{2}(\hat{\theta}(y) - \theta_0)^t \ddot{\ell}_{y_i}^{\mathcal{M}}(\theta_0)(\hat{\theta}(y) - \theta_0)
\end{aligned}
$$

by noticing that $\frac{1}{n}\sum_{i=1}^{n} \ell_{y_i}^{\mathcal{M}}(\theta_0)$ is an unbiased estimate of $\bar{\ell}(\theta_0)$. Rearranging the terms, we get

$$
\begin{aligned}
\widehat{\bar{\ell}(\theta_0)} &= \frac{1}{n}\sum_{i=1}^{n} \ell_{y_i}^{\mathcal{M}}(\theta_0) \\
&= \frac{\ell_y^{\mathcal{M}}(\hat{\theta}(y))}{n} - (\hat{\theta}(y) - \theta_0)^t \frac{\dot{\ell}_y^{\mathcal{M}}(\theta_0)}{n} - \frac{1}{2}(\hat{\theta}(y) - \theta_0)^t \frac{\ddot{\ell}_y^{\mathcal{M}}(\theta_0)}{n}(\hat{\theta}(y) - \theta_0).
\end{aligned}
$$

By noticing that the second term can be approximated as

$$
\begin{aligned}
(\hat{\theta}(y) - \theta_0)^t \frac{\dot{\ell}_y^{\mathcal{M}}(\theta_0)}{n} &= (\hat{\theta}(y) - \theta_0)^t \frac{\dot{\ell}_y^{\mathcal{M}}(\theta_0) - \dot{\ell}_y^{\mathcal{M}}(\hat{\theta})}{n} \\
&= -(\hat{\theta}(y) - \theta_0)^t \frac{\ddot{\ell}_y^{\mathcal{M}}(\theta_0)}{n}(\hat{\theta}(y) - \theta_0)
\end{aligned}
$$

we obtain

$$
\begin{aligned}
\widehat{\bar{\ell}(\theta_0)} &\approx \frac{\ell_y^{\mathcal{M}}(\hat{\theta}(y))}{n} + \frac{1}{2}(\hat{\theta}(y) - \theta_0)^t \frac{\ddot{\ell}_y^{\mathcal{M}}(\theta_0)}{n}(\hat{\theta}(y) - \theta_0) \\
&\approx \frac{\ell_y^{\mathcal{M}}(\hat{\theta}(y))}{n} - \frac{p}{2n}, \qquad (6.14)
\end{aligned}
$$

where the final approximation comes from noticing that $\ddot{\ell}_y^{\mathcal{M}}(\theta_0)/n \to -I(\theta_0)$ and then using the average of the second term as an estimate. Putting now (6.13) and (6.14) together and setting C arbitrarily to zero, we see that

$$
\widehat{D_{\mathcal{T}}(\mathcal{M})} = -\frac{\ell_y^{\mathcal{M}}(\hat{\theta}(y))}{n} + \frac{p}{n}.
$$

Definition 6.6 Akaike information criterion. *If we are interested in selecting a model $\mathcal{M}$ with good predictive properties on the basis of data $y_1, \ldots, y_n \sim \mathcal{T}$, we minimize the estimated Kullback-Leibler divergence $\widehat{D_{\mathcal{T}}(\mathcal{M})}$. Multiplying it by an arbitrary constant $2n$ gives us the final expression of the* Akaike Information Criterion,

$$
\mathcal{M}_{aic} = \underset{\mathcal{M}}{\operatorname{argmin}}\, AIC(\mathcal{M})
$$

where

$$
AIC(\mathcal{M}) = -2\ell_y^{\mathcal{M}}(\hat{\theta}(y)) + 2p.
$$

BIC: most likely true model. Although often used alongside the AIC, the Bayesian Information Criterion has a very different origin story. Instead of considering predictive properties on future data, another model comparison criterion involves the probability that a parametric model $\mathcal{M}$ is the true model for generating the current data $y = (y_1, \ldots, y_n)$,

$$f(\mathcal{M} \mid y) \propto f_{\mathcal{M}}(y) f(\mathcal{M}).$$

Assuming no prior information about the true model, one can use a flat prior $f(\mathcal{M}) \propto 1$. Then integrating out the parametric space $\Theta \subset \mathbb{R}^p$ of $\mathcal{M}$, we obtain

$$\begin{aligned} f(\mathcal{M} \mid y) &= \int_\Theta e^{\ell_y^{\mathcal{M}}(\theta)} \, \partial\theta \\ &\approx e^{\ell_y^{\mathcal{M}}(\hat\theta)} \int_\theta e^{\frac{n}{2}(\hat\theta-\theta)^t \frac{\ddot\ell_y^{\mathcal{M}}(\hat\theta)}{n}(\hat\theta-\theta)} \, \partial\theta \end{aligned}$$

where the final expression is based on a Taylor expansion around the maximum likelihood estimate $\hat\theta$. Then using that $\ddot\ell_y^{\mathcal{M}}(\hat\theta)/n \approx -I(\theta_0)$ and the Laplace approximation of the integral, we get

$$\log f(\mathcal{M} \mid y) \approx \ell_y^{\mathcal{M}}(\hat\theta) - \frac{p}{2}\log 2\pi n - \frac{1}{2}\log|\ddot\ell^{\mathcal{M}}(\hat\theta)/n|$$

where the second term depends explicitly on the complexity p of the model.

Definition 6.7 Bayesian information criterion. *If we are interested in selecting a model $\mathcal{M}$ that has strong support given the current data $y_1, \ldots, y_n$, we can maximize an estimate of the logarithm of the posterior model probability,*

$$\mathcal{M}_{max} = \operatorname*{argmax}_{\mathcal{M}} \ell_y^{\mathcal{M}}(\hat\theta) - \frac{p}{2}\log 2\pi n.$$

In the literature, this has been somewhat curiously further simplified by ignoring the 2π term, resulting in the Bayesian Information Criterion,

$$\mathcal{M}_{bic} = \operatorname*{argmin}_{\mathcal{M}} BIC(\mathcal{M})$$

where

$$BIC(\mathcal{M}) = -2\ell_y^{\mathcal{M}}(\hat\theta(y)) + p\log n.$$

Example 6.25 *Reaction times model selection. A common model for reaction times t is the Exponential distribution model*

$$\mathcal{M}_{exp} = \left\{ f_T \mid f_T(t) = \begin{cases} \lambda e^{-\lambda t} & t \geq 0 \\ 0 & otherwise \end{cases} \right\}$$

In this example we also consider a Uniform distribution model, Unif(α, β),

$$\mathcal{M}_{unif} = \left\{ f_T \mid f_T(t) = \begin{cases} \frac{1}{\beta-\alpha} & \alpha \leq t \leq \beta \\ 0 & otherwise \end{cases} \right\}$$

TABLE 6.2

Twenty independently sampled reaction times $t_1, \ldots, t_{20}$ for a particular chemical reaction.

Reaction times										
0.04	0.39	0.53	0.58	0.66	0.78	1.04	1.43	1.57	1.68	average
1.89	2.11	2.13	2.14	2.25	2.45	3.14	3.62	4.38	4.84	1.88

We observe 20 independent reactions times $t_1, \ldots, t_{20}$, given in Table 6.2. Clearly, the two models are not nested, so we cannot use the likelihood ratio test to compare the two models. Instead, in this example we will apply the AIC and BIC criteria.

Exponential model. *As we have seen in example 6.16 the maximum likelihood estimate for the rate is $\hat{\lambda} = 1/\bar{t} = 1/1.88 = 0.53$. Plugging this into the likelihood, we get*

$$\ell_{exp}(\hat{\lambda}) = 20 \log 0.53 - 0.53 \times 20 \times 1.88 = -32.7,$$

so this results in an Akaike and Bayesian information criteria of

$$\begin{aligned} AIC(\mathcal{M}_{exp}) &= -2\ell_{exp}(\hat{\lambda}) + 2 \times 1 = 67.4 \\ BIC(\mathcal{M}_{exp}) &= -2\ell_{exp}(\hat{\lambda}) + \log(20) \times 1 = 68.4 \end{aligned}$$

Uniform model. *As will be derived in exercise 15, the maximum likelihood estimates for the parameters of the uniform distribution are given by the minimum and maximum reaction time respectively, $\hat{\alpha} = 0.04$ and $\hat{\beta} = 4.84$. Plugging this into the likelihood, we get*

$$\ell_{unif}(\hat{\alpha}, \hat{\beta}) = -20 \log(4.84 - 0.04) = -31.4,$$

so this results in an Akaike and Bayesian information criteria of

$$\begin{aligned} AIC(\mathcal{M}_{unif}) &= -2\ell_{unif}(\hat{\alpha}, \hat{\beta}) + 2 \times 2 = 66.7 \\ BIC(\mathcal{M}_{unif}) &= -2\ell_{unif}(\hat{\alpha}, \hat{\beta}) + \log(20) \times 2 = 68.7 \end{aligned}$$

Comparing the two Akaike information values with each other, we see that the larger Uniform model is the better model for when it comes to prediction. On the other hand, comparing the Bayesian information criteria with each other we notice that the smaller Exponential model has the edge in terms of posterior model probability given the current data. It is not uncommon that the AIC and BIC result in different choices of model, whereby the BIC typically selects a more parsimonious model.

In the end, it is worthwhile remembering in George Box's words (Wit et al., 2012) that "all models are wrong, but some are useful." What this means for model comparison methods is that if one is interested in good prediction then the AIC is recommended; on the other hand, if the main focus lies on parsimonious explanation of the observed data, then the BIC is often more useful.

6.6 Notes and Comments

We may now summarize the steps of a statistical test for a chemical hypothesis as follows (Fisher, 1925):

1. Specify a research question by translating it into a quantitative statement about a particular chemical target μ_0.
2. Specify the null hypothesis with respect to the target parameter, representing the current best available knowledge.
3. Define a test statistic, i.e., a function of the data that is sensitive to the target parameter, such as a parameter estimate $\hat{\mu}$.
4. Specify the significance level as the largest allowable type I error you are willing to accept and calculate the rejection region.
5. Obtain the data from the chemical experiment and check whether the test statistic falls in the rejection region; alternatively, calculate the p-value.
6. Formulate the conclusion in general non-statistical terms with respect to the aim of the research.

When the measurements are a random sample from the normal distribution, then the sample mean is the best estimator of the target (Ferguson, 1996; Rao, 2009). The *t*-test is optimal for testing the null hypothesis $H_0 : \mu = \mu_0$ if the data are normally distributed (Lehmann and Romano, 2005).

We have seen that the usual calculus type of limit relations do not hold for random variables (Lehmann, 2014). Therefore, we had to specify a mode of convergence, such as "in probability" or "in distribution." Other convergence definitions for random variables exist. For example, "convergence with probability one" or "almost sure convergence" is a stronger mode of convergence, which holds if and only if the expected values of the random variable exist (Rao, 2009).

The non-parametric bootstrap re-samples values from the actual data obtained from the experiment with replacement and provides an empirical distribution of the test statistic. This approximates the theoretical distribution of the test statistic, if the underlying model assumptions are valid. The bootstrap mimics a simulation study and it possesses favorable asymptotic properties (Hall, 2013). Davison and Hinkley (1997) is a seminal book that provides both a theoretical background and a practical implementation of the bootstrap.

As can be expected from the central limit theorem, the t-distribution nicely tends to that of the standard normal as the sample size increases (Ferguson, 1996; Lehmann, 2014). The slightly stronger result that the density of the t-distribution tends to the bell-shaped of the standard normal can be illustrated by a *Geogebra* demonstration. Such animations illustrate that the approximation is sufficiently precise for all practical purposes when $n \geq 30$; see also Chapter 3.

The proof for the Delta Theorem follows quite directly from the mean value theorem and the CLT (Billingsley, 2012; Ferguson, 1996).

In the above we refrained from so-called one-sided tests because these are typically defined post-hoc or, if they are not, are unable to handle unexpected results from the data. Obviously, these may certainly be useful in case the direction of the alternative hypothesis is clearly indicated by the experimental context.

Note that when a series of determinations are conducted on the same units by two different methods or at two different time points, then the data are considered to be paired (see Exercise 6). Testing for differences would involve a paired test, such as a paired t-test, if the underlying measurements are normal, or a Wilcoxon signed-rank test, if the data are not normally distributed. An approach to robust testing is undertaking in Chapter 12.

There are good books with a more detailed treatment of statistical testing for engineers (Bain and Engelhardt, 1992) as well as for more statistically interested readers (Casella and Berger, 2021; Lehmann and Romano, 2005).

6.7 Notes on Using R

The online NIST Engineering Statistics Handbook systematically summarizes a gallery of distributions and provides R-code for the statistical analysis of chemical data.

Every programming language provides four basic functions related to a distribution of outcomes of a chemical experiment. These are the density function $f = F'$, the distribution function $F(y) = P(Y \leq y)$, the quantile function F^{-1}, and, for simulation purposes, a function to draw pseudo random numbers from the distribution. The latter is crucial in developing a simulation study for an experiment in order e.g., to determine appropriate sample size, precision of prediction, etc.

For the normal distribution the density `dnorm`, cumulative probability `pnorm`, quantile `qnorm` and random deviate `qnorm` are available. The function `t.test` has various forms and design applications. The library `TeachingDemos` has a nice demonstration of the approximation of the t-distribution to the

TABLE 6.3
Ten determinations in ($\mu g g^{-1}$) of copper in wholemeal flour by four different laboratoria.

	Lab1	**Lab2**	**Lab3**	**Lab4**
1	3.277	1.535	5.835	4.077
2	2.897	2.391	3.886	5.899
3	3.521	3.743	4.917	3.087
4	3.820	3.258	4.766	3.570
5	3.107	0.828	4.177	0.968
6	3.307	3.840	4.802	4.879
7	3.262	3.810	3.608	3.768
8	3.860	3.550	4.667	6.615
9	3.491	0.593	4.780	3.017
10	2.813	3.926	5.110	1.803

standard normal as the degrees of freedom increase. The library `boot` has the function `boot` for bootstrapping a statistic such as the mean or the coefficient of variation. Build upon this, R has many wrapper functions that make the application of the bootstrap fairly easy.

6.8 Exercises

1. **Copper in wholemeal flour**. Suppose the following four data sets from measurements copper in wholemeal flour obtained from 4 different laboratories on exactly the sample sample. The measurements are given in Table 6.3. The data are available from the file `CopperInWholemealFlour.txt`.
   (a) Compute the mean, variance, and coefficient of variation for each laboratorium.
   (b) Characterize the measurement procedure by each laboratorium in terms of bias and precision if the true (target) mean equals $3.5(\mu g g^{-1})$.
   (c) Construct confidence intervals for the mean for each laboratorium based upon the t distribution.
   Hint: Use the R function `t.test`.
2. **Arsenic concentration.** In the above example on Arsenic concentration in public drinking water the null-hypothesis was tested that the mean concentrations of the metropolitan and the rural communities was equal, under the assumption that the concentrations are normally distributed. Note that there are many statistical developed

useful for testing many different types of chemical hypotheses, too many to detail them all in this book. It does, however, make sense to be able to use a test for normality and a test for the equality of variances. The data are available as `ArsenicConc.csv`.

(a) Test the null hypothesis that the data have a normal distribution.
Hint: Use the Shapiro-Wilk normality test available from the R function `shapiro.test`.

(b) Test the null hypothesis that the data have equal variances $H_0 : \sigma_1^2 = \sigma_2^2$.

(c) Compare the t-test with and without the assumption of equal variances.

3. **Nitrate ion concentration**. Suppose that nitrate ion concentration in μg ml^{-1} are normally distributed with mean 0.50 and variance 0.04.

   (a) What is the probability of smaller values than 0.51?

   (b) What is the probability that the measurements are between 0.48 and 0.52?

   (c) For what value (quantile) are the measurements smaller with probability than 0.52.

   (d) What is the probability that the mean of a sample of 100 measurements is between 0.48 and 0.52?

4. **Antibody concentration**. If the antibody concentration in human blood sera is measured for different persons, then the data are often not normal but log normal. That is, if log with basis ten is taken, then the distribution is normal. The data are contained in the file `ConcSerumImmunoglobulin.txt`.

   (a) Construct a histogram to verify that the concentrations are not normally distributed.

   (b) Compute the log10 transformed concentrations and verify whether these have a normal distribution.

   (c) Compute the mean of the transformed concentrations and compare this with the geometric mean of the concentrations. How are they related?

5. **Mass of tablets**. The mass of 30 headache tables was measured (unfortunately in unknown units). Suppose the target value is 0.55. The tablets data are available from the library `quantchem` and from `tablets.txt`.

   (a) Construct a histogram and apply a normality test to evaluate the distribution of the data.
   Hint: Conveniently use the R function `shapiro.test`.

TABLE 6.4
Albumin concentrations in blood sera for 8 males and 8 females.

	1	2	3	4	5	6	7	8
Male	37	39	37	42	39	45	42	39
Female	44	40	39	45	47	47	43	41

(b) Test the null hypothesis $\mu = 0.55$.

(c) Construct the 95 % CI for the mean.

6. **Shoes**. Ten boys wear shoes with two materials, A and B, randomly assigned to the left or right shoes of the boys. The data consist of the measured thickness of the wear (Box et al., 1978). The data are available from the library MASS or as `shoes.csv`.

   (a) Test the null hypothesis that the variances for material A and B are equal.
   Hint: Use the R function `var.test`.

   (b) Use the independent t-test for the null hypothesis that the means for material A and B are equal.

   (c) Test the null hypothesis that the correlation between material A and B is zero.
   Hint: Use the R function `cor.test`.

   (d) Use the paired t-test for the null hypothesis that the means for material A and B are equal.

7. **Cement**. The heat evolved in 13 types of cement was measured together with four other variables. The variable y is heat in calories per gram of cement, and the remaining variables consist of percentages of weight of the constituents tricalcium aluminate x1, tricalcium silicate x2, tetracalcium alumumino ferrite x3 and dicalcium silicate x4 Woods et al. (1932). The cement data are available from the MPV library or from `cement.csv`.

   (a) Test the hypothesis that the correlation between y and x3 is zero.
   Hint: Use the R function `cor.test`.

   (b) Compare the 95% CI with that from the adjusted bootstrap percentile (BCa) method.

8. **Albumin concentrations**. Albumin concentrations, in g l^{-1}, from blood sera for 8 males and 8 females were measured in order to investigate differences between these (Foote and Delves, 1983). The data presented in Table 6.4 and to be loaded from the file `Albumin.txt`.

   (a) Is there a difference in gender with respect to Albumin concentrations?

TABLE 6.5
Antimony found (mg m^{-3}) by a new method and a standard method.

Sample nr	1	2	3	4	5	6
New method	22.2	19.2	15.7	20.4	19.6	15.7
Standard method	25.0	19.5	16.6	21.3	20.7	16.8

9. **Antimony**. A new flame atomic-absorption spectroscopic method of determining antimony (Castillo et al., 1982) in the atmosphere was compared with the recommended calorimetric method. For six samples from an urban atmosphere the results were obtained as reported in Table 6.5 and to be loaded from the file `Antimony.txt`.

   (a) Do the results obtained differ?
   Hint: Take into consideration that each sample is measured by both methods.

10. **Coefficient of variation**. In the above the rate of convergence in the central limit theorem was illustrated by a small simulation using $n = 5$ determinations in ($\mu g g^{-1}$) of copper in wholemeal flour 100,000 times. The measurements were drawn from a normal distribution with $\mu = 3.5$ and $\sigma = 0.5$, and next the coefficient of variation $S/\overline{Y}$ was computed 100,000 times in total. The latter $S/\overline{Y}$ is defined as the standard deviation divided by the mean expressing the relative error of measurement. The delta theorem implies the convergence in distribution

$$\frac{S}{\overline{Y}} \xrightarrow{D} Z \sim N\left(\frac{\sigma}{\mu}, \frac{1}{n}\sigma^2\left(\frac{1}{2} + \frac{\sigma^2}{\mu^2}\right)/\mu^2\right) = N(0.1429, 0.0106/n).$$

   (a) Repeat the above simulation with $n = 50$ measurements from and use a histogram with 100 bars and the density function from the normal approximation.
   (b) Comment on your findings.

11. **Statistical significance**. Statistical significance expressed as the p-value is unrelated to chemical relevance in the sense that the false null-hypotheses is correctly rejected if the true effect is very small.

   (a) Sample $n = 20$ measurements from the $N(\mu + \delta, \sigma^2)$ distribution, where e.g., $\mu = 3.5$, $\delta = 10^{-6}$ and $\sigma = 10^{-7}$. Conduct the implied t-test.
   (b) Sample a data set of size $n = 10^7$ from a bivariate normal distribution with true correlation coefficient $\rho = 10^{-3}$ and test the hypothesis $H_0 : \rho = 0$.

12. **Some applications of the central limit theorem**

    (a) Suppose Y_i is the number of molecules in substrate $i = 1, \cdots, 10$. Let Y_i have a binomial distribution with $n = 1000$ and $p = 0.03$, for $i = 1, \cdots, 10$ for each i; the substrates are equally and independently distributed. Simulate 10 substrates, compute the mean and repeat this $N = 100,000$ times, so that you obtain N mean values. Construct a histogram with 100 bars and add the normal density curve. Are these equal?

    (b) Suppose the reaction time between two molecules is exponentially distributed with mean 0.025 seconds. Simulate 10 reactions, compute the mean reaction time, and repeat this 100,000 times. Construct a histogram with 100 bars and add the normal density curve. Are these equal?

    (c) Increase the number of reactions to be simulated to 50 and repeat forgoing question.

13. **Application of delta theorem**. The attentive reader noticed that the delta theorem can be applied to the main example of volume of e.g., a protein. Simulations may be used in modern chemometrics to answer research questions on the speed of convergence of to the normal distribution in Equation (6.7). We may simulate a measurement situation where the radius $R \sim N(\mu_R, \sigma_R^2)$, taking $\mu_R = 10$ and $\sigma_R^2 = 0.1^2$.

    (a) Show for the volume function $V(r) = \frac{4}{3}\pi r^3$, that

    $$V(\overline{R}) \xrightarrow{D} N\left(V(\mu_R), \frac{(4\pi\mu_R^2)^2}{n} \cdot \sigma_R^2\right). \tag{6.15}$$

    (b) By R function `rnorm` we easily simulate n measurements of the radius, and repeat this 10,000 times. Taking $n = 5$ we may compute 10,000 Volume measurements from the mean radius and study its distribution in a histogram. Next, we may add the density function from the indicated normal to better evaluate the precision of the approximation by the delta theorem. Conduct the suggested simulation to become surprised about the precision. Only a few lines code suffice to program the simulation.

14. **Simulating distributions.** We may verify theoretical properties by simulation. (See Chapter 3, exercise 18, for a similar exercise.)

    (a) Suppose that n independent random variables X_i have the same normal distribution $N(\mu, \sigma^2)$, then

    $$\frac{\sqrt{n}(\overline{Y} - \mu)}{S} = \frac{\overline{Y} - \mu}{S/\sqrt{n}} \sim t(n-1),$$

where $t(n-1)$ is the t-distribution with $n-1$ degrees of freedom. To verify this we take take $\mu = 10$, $\sigma = 2$, $n = 11$, and generate 10^5 times these n pseudo random variables and their average. Use these 10^5 averages to compute a histogram with 100 bars and add to this the proper density curve. Are the two almost equal?

(b) Suppose that n independent random variables X_i have the same normal distribution $N(\mu, \sigma^2)$, then

$$(n-1)S^2/\sigma^2 = \sum_{i=1}^{n} \left(\frac{Y_i - \bar{Y}}{\sigma}\right)^2 \sim \chi^2_{n-1},$$

where the latter denotes the Chi-squared distribution with $n-1$ degrees of freedom. To verify this we take take $\mu = 10$, $\sigma = 2$, $n = 11$, and generate 10^5 times these n pseudo random variables, their sample variance s^2 and the ratio. Use these 10^5 ratios to compute a histogram with 100 bars and add to this the proper density curve. Are the two almost equal?

15. **Maximum likelihood estimation.** Consider n independent draws, $T_1, \ldots, T_n$ from a Uniform(α, β) distribution.

(a) Show that the MLE for α is given by the minimum, $\hat{\alpha} = \min\{T_1, \ldots, T_n\}$.

(b) Show that the MLE for β is given by the maximum, $\hat{\beta} = \min\{T_1, \ldots, T_n\}$.

(c) Why are the regularity conditions in definition 6.3 not satisfied in this case?

(d) Calculate the mean of $\hat{\alpha}$ and $\hat{\beta}$ and show that they converge to the true value when n goes to infinity.

(e) Calculate the variance of $\hat{\alpha}$ and $\hat{\beta}$ and show that the rate of convergence is $1/n^2$, which is faster than the rate $1/n$ within Cramèr-Rao lower bound. Again, this is again a result from the fact that the uniform distribution does not satisfy the regularity conditions.

16. **Inference with the maximum likelihood estimate.** A particular chemical production process is subject to failures from time to time. A company is keeping track of the number of errors that occur each day. They obtain the data from an entire year $Y_1, \ldots, Y_{365}$ and they are interested in estimation the failure rate λ, assuming the underlying failure process is independent and distributed as Poisson(λ).

(a) Maximize the log-likelihood,

$$\ell_Y(\lambda) = \log \prod_{i=1}^{n} e^{-\lambda} \frac{\lambda^{Y_i}}{Y_i!},$$

to derive the maximum likelihood estimator for λ.

(b) Show that the MLE $\hat{\lambda}$ is unbiased.

(c) Derive the Cramér-Rao lowerbound,

$$\text{CRLB} = \frac{1}{nE[\dot{\ell}_Y(\lambda)^2]} = \frac{-1}{nE[\ddot{\ell}_Y(\lambda)]},$$

for $\hat{\lambda}$.

(d) By using the efficiency of the maximum likelihood estimator, derive the approximate distribution for $\hat{\lambda}$.

(e) Assume that during the year in total 73 failures were detected.

i. Estimate the failure rate λ.

ii. Determine a 95% confidence interval for λ.

iii. For legal reasons, the failure rate should be less than 0.25. Test the null-hypothesis

$$H_0 : \lambda = 0.25$$

with significance level $\alpha = 0.05$ using the maximum likelihood estimator as test-statistic.

iv. If we want to detect a minimum detectable effect size of $\Delta\lambda = 0.05$ with a power of $1 - \beta = 0.8$, then for how many days should the process be observed?

17. **Power calculations.** In example 6.22 it was derived that 3140 samples were needed to obtain a power of 80% to detect at least a $\Delta\lambda = 0.1$ difference with respect to the hypothesized $\lambda = 2$. In this exercise we see what this really means by calculating backwards. Consider performing 3140 independent time measurements of a certain type of reaction, $T_1, \ldots, T_{3140} \overset{IID}{\sim} \text{Exp}(\lambda)$.

(a) Verify that the maximum likelihood estimator is $\hat{\lambda} = 1/\bar{T}$ and that given the true value λ_0 it has the approximate distribution

$$\hat{\lambda} \sim N(\lambda_0, \lambda_0^2/3140).$$

(b) If we want to test the null hypothesis $H_0 : \lambda = 2$ with the maximum likelihood estimator $\hat{\lambda}$. What is the *rejection region* at the $\alpha = 0.05$ significance level, i.e., what are all values for which the null hypothesis would be rejected,

$$\text{Rejection region}(\alpha) = \left\{x \mid 2 \times \min\{P(\hat{\lambda} < x), P(\hat{\lambda} > x)\} \leq \alpha\right\},$$

if the null hypothesis is true.

(c) Assume that the unknown true value of λ is 2.1 reactions per minute. What is the probability that the null hypothesis will be rejected?

(d) Note how the answer to the last question should be related to the power of 80%. In fact, this is the true meaning of the concept of power. Give two reasons why the answer to (c) is not exactly 0.8?

18. **Chi square significance values.** We have seen that under the null hypothesis the log likelihood ratio statistic $-2\log\Lambda(Y)$ for q-dimensional parameter θ is distributed as a chi square distribution.

    (a) Use the CLT to show that for large q,

    $$-2\log\Lambda(Y) \mid H_0 \sim N(q, 2q).$$

    (b) Use the previous result to show that the critical value for a likelihood ratio test with $\alpha = 0.05$ is approximately,

    $$\chi^2_{q,0.95} \approx q + 2.5\sqrt{q}.$$

19. **Model comparison.** Two labs are asked to estimate the rate of the same reaction. Each lab takes 10 independent samples of the reaction times t_{lr}, where $l = 1, 2$ indicates the lab and $r = 1, \ldots, 10$ the sample. We want to check whether we can put the samples of the two labs together to estimate the true reaction rate λ or whether there are lab specific factors that influence the estimation of the rates. We therefore consider two models. The first model assumes that there are no lab specific effects,

    $$\mathcal{M}_1 = \{f_Y \mid Y_{lr} \sim \text{Exp}(\lambda)\},$$

    whereas the second model assumes that there are such effects,

    $$\mathcal{M}_2 = \{f_Y \mid Y_{lr} \sim \text{Exp}(\lambda_l)\}.$$

	Reaction times (min)										**Average**
Lab 1	1.18	0.32	0.15	0.28	0.05	0.03	0.29	1.98	0.59	0.50	0.54
Lab 2	0.62	0.02	0.14	0.57	0.09	0.44	0.13	0.32	0.33	0.10	0.28

    (a) Use the likelihood ratio test for testing the null hypothesis $H_0 : Y_{lr} \sim \mathcal{M}_1$.

    (b) Calculate the AIC and BIC to compare models $\mathcal{M}_1$ and $\mathcal{M}_2$.

7

Linear Models

Suppose that the yield of some chemical production process depends on the pressure, the temperature, and several other variables in a linear way. If a chemical experiment generates a sufficient amount of values for the yield and the explanatory variables, then the linear coefficients can be estimated quite precisely. The estimation procedures for the linear model also provide the basis of the analysis of data from designed experiments in Chapter 11.

We start with the general formulation of the linear model, its estimation, and the statistical testing of its parameters. Since unusual observations decrease our confidence in the validity of a model, a brief overview is given of the detection of so-called influential observations. Various sub-models are given and illustrated by applications in chemistry. Two important ideas for the selection problem of coefficients are given and illustrated. This is particularly relevant when the number of explanatory variables to be investigated is relatively large. Finally, it is explained how to incorporate random effects for potential nuisance factors into the linear model.

In this chapter you learn:

- In Section 7.1 how the linear model is defined.
- In Section 7.2 how the model can best be estimated and how its can be used to compute predictions of future chemical outcomes.
- In Section 7.3 about ways to diagnose the validity of linear model estimation.
- In Section 7.4 about ways to select a linear sub-model among many alternatives.
- In Section 7.5 about two important special definitions and applications of the linear model.

DOI: 10.1201/9781003178194-7

7.1 Linear Model

The linear model for some chemical yield Y is defined as

$$Y = \beta_0 + \beta_1 x_1 + \cdots + \beta_p x_p + \varepsilon, \quad \varepsilon \sim N(0, \sigma^2). \tag{7.1}$$

Note that $\frac{dY}{dx_j} = \beta_j$. Therefore, the yield Y changes by β_j, if x_j increases by one unit. If we let the error standard deviation σ tend to zero, then we obtain a completely deterministic model as a special case. The **outcome** Y is a linear combination of the **predictive variables** $\{x_1, x_2, \cdots, x_p\}$ with **intercept** β_0 and weights or slopes $\{\beta_1, \beta_2, \cdots, \beta_p\}$. If Y is the yield of a chemical process, then the predictive variables can include, for example, x_1 temperature, and x_2 catalyst concentration. Several special cases of model (7.1) will be considered below. For a random sample $(Y_1, \cdots, Y_n)$ of n outcomes, we write

$$\begin{aligned} Y_i &= \beta_0 + \beta_1 x_{i1} + \beta_2 x_{i2} + \cdots + \beta_p x_{ip} + \varepsilon_i, \text{ for } i = 1, \cdots, n, \\ & \{\varepsilon_1, \cdots, \varepsilon_n\} \overset{IID}{\sim} N(0, \sigma^2). \end{aligned}$$

It is handy to write $\mathbf{x}_i^T = (1, x_{i1}, x_{i2}, \cdots, x_{ip})$ for row i of the n by $p+1$ **design** or **model matrix X**. Very similar to the measurement model in the previous chapter, the variation in the yield is decomposed in a structural part $E[Y_i] = \mathbf{x}_i^T \boldsymbol{\beta}$ and an error part ε_i for each value of the index i. Our observed values of the yield contain some amount of error, the average size of which is given by σ. We now formulate the model in a more tractable way, writing the equations one beneath the other in matrix notation

$$\mathbf{y} = \begin{bmatrix} Y_1 \\ \vdots \\ Y_n \end{bmatrix} = \begin{bmatrix} \mathbf{x}_1^T \boldsymbol{\beta} \\ \vdots \\ \mathbf{x}_n^T \boldsymbol{\beta} \end{bmatrix} + \begin{bmatrix} \varepsilon_1 \\ \vdots \\ \varepsilon_n \end{bmatrix} = \begin{bmatrix} \mathbf{x}_1^T \\ \vdots \\ \mathbf{x}_n^T \end{bmatrix} \boldsymbol{\beta} + \varepsilon = \mathbf{X}\boldsymbol{\beta} + \varepsilon,$$

where $\varepsilon \sim N(\mathbf{0}, \sigma^2 \mathbf{I})$. That is, the error comes from a multivariate normal distribution centered around zero with $\sigma^2 \mathbf{I}$ as variance matrix. The fact that the latter is diagonal reflects the assumption that all error terms are stochastically independent. A succinct way the write the linear model is by $\mathbf{y} \sim N(\mathbf{X}\boldsymbol{\beta}, \sigma^2 \mathbf{I})$.

7.2 Estimation and Prediction

The unknown parameters are the weights $\boldsymbol{\beta}$ and the error variance σ^2 are to be estimated from the data collected in the vector $\mathbf{y}$ and the model matrix $\mathbf{X}$.

Given a new observation $\boldsymbol{x}_0^T$ is a new observation we often wish to predict the outcome Y_0. We are also often interested in the structural part of the model $\mathbf{X}\boldsymbol{\beta}$ not containing any error. The approximation of the latter is called prediction.

7.2.1 Parameter estimation

The coefficients in the vector $\boldsymbol{\beta}$ and the error variance σ^2 are the unknown parameters of the model. The maximum likelihood estimator has many attractive properties such as the smallest possible standard errors. Note that we assume for linear models that the measurements can be represented as n independent random variables for which the i-th observation Y_i is distributed as $N(\mathbf{x}_i^T\boldsymbol{\beta}, \sigma^2)$. Hence, under normality and independence of the error terms, we obtain the maximum likelihood estimator by maximizing

$$\begin{aligned} L(\boldsymbol{\beta}, \sigma) &= \prod_{i=1}^{n}(2\pi\sigma^2)^{-1/2}\exp\left(-\frac{1}{2\sigma^2}(Y_i - \mathbf{x}_i^T\boldsymbol{\beta})^2\right) \\ &= (2\pi\sigma^2)^{-n/2}\exp\left(-\frac{1}{2\sigma^2}(\mathbf{y} - \mathbf{X}\boldsymbol{\beta})^T(\mathbf{y} - \mathbf{X}\boldsymbol{\beta})\right) \end{aligned}$$

over the parameters $\boldsymbol{\beta}$ and σ^2. This is equivalent to maximizing the log-likelihood

$$\ell(\boldsymbol{\beta}, \sigma^2) = \log L(\boldsymbol{\beta}, \sigma) = -\frac{n}{2}\log(2\pi\sigma^2) - \frac{1}{2\sigma^2}(\mathbf{y} - \mathbf{X}\boldsymbol{\beta})^T(\mathbf{y} - \mathbf{X}\boldsymbol{\beta}). \quad (7.2)$$

Setting the first order derivatives to zero gives

$$\frac{\partial\ell(\boldsymbol{\beta}, \sigma)}{\partial\beta} = -\frac{1}{2\sigma^2}\left(-2\mathbf{X}^T\mathbf{y} + 2\mathbf{X}^T\mathbf{X}\boldsymbol{\beta}\right) = 0 \quad (7.3)$$

$$\frac{\partial\ell(\boldsymbol{\beta}, \sigma)}{\partial\sigma^2} = -\frac{n}{2\sigma^2} + \frac{1}{2\sigma^2}(\mathbf{y} - \mathbf{X}\boldsymbol{\beta})^T(\mathbf{y} - \mathbf{X}\boldsymbol{\beta}) = 0 \quad (7.4)$$

The information matrix defined as minus the expected value of the Hessian matrix becomes

$$\begin{aligned} I(\boldsymbol{\beta}, \sigma) &= -E\begin{bmatrix} -\frac{1}{\sigma^2}\mathbf{X}^T\mathbf{X} & -\frac{1}{\sigma^4}(\mathbf{X}^T\mathbf{y} - \mathbf{X}^T\mathbf{X}\boldsymbol{\beta}) \\ -\frac{1}{\sigma^4}(\mathbf{X}^T\mathbf{y} - \mathbf{X}^T\mathbf{X}\boldsymbol{\beta})^T & \frac{n}{2\sigma^4} - \frac{1}{\sigma^6}(\mathbf{y} - \mathbf{X}\boldsymbol{\beta})^T(\mathbf{y} - \mathbf{X}\boldsymbol{\beta}) \end{bmatrix} \\ &= \begin{bmatrix} \frac{1}{\sigma^2}\mathbf{X}^T\mathbf{X} & \mathbf{0} \\ \mathbf{0}^T & \frac{n}{2\sigma^4} \end{bmatrix} \end{aligned} \quad (7.5)$$

so that its inverse becomes

$$I(\boldsymbol{\beta}, \sigma)^{-1} = \begin{bmatrix} \sigma^2(\mathbf{X}^T\mathbf{X})^{-1} & \mathbf{0} \\ \mathbf{0}^T & \frac{2\sigma^4}{n} \end{bmatrix}.$$

Assuming that the matrix $\mathbf{X}^T\mathbf{X}$ is invertible, the solution to Equation (7.3) is given by

$$\hat{\boldsymbol{\beta}} = (\mathbf{X}^T\mathbf{X})^{-1}\mathbf{X}^T\mathbf{y}, \quad (7.6)$$

From Equation (7.2) it is clear that maximizing the log-likelihood is equivalent to minimizing the residual sum of squares

$$\begin{aligned} \text{RSS}(\boldsymbol{\beta}) &= \sum_{i=1}^{n}(Y_i - \mathbf{x}_i^T\boldsymbol{\beta})^2 = (\mathbf{y} - \mathbf{X}\boldsymbol{\beta})^T(\mathbf{y} - \mathbf{X}\boldsymbol{\beta}) \\ &= \mathbf{y}^T\mathbf{y} - 2\boldsymbol{\beta}^T\mathbf{X}^T\mathbf{y} + \boldsymbol{\beta}^T\mathbf{X}^T\mathbf{X}\boldsymbol{\beta}. \end{aligned}$$

Setting the first order derivative of RSS to zero, we obtain

$$\begin{aligned}\left.\frac{\partial RSS}{\partial \beta}\right|_{\beta=\hat{\beta}} &= -2\mathbf{X}^T\mathbf{y} + 2\mathbf{X}^T\mathbf{X}\hat{\beta} = 0 \Rightarrow \mathbf{X}^T\mathbf{X}\hat{\beta} = \mathbf{X}^T\mathbf{y} \Rightarrow \\ \hat{\boldsymbol{\beta}} &= (\mathbf{X}^T\mathbf{X})^{-1}\mathbf{X}^T\mathbf{y}, \end{aligned} \tag{7.7}$$

so that the maximum likelihood estimator equals the least squares estimator $\hat{\boldsymbol{\beta}}$. Using vector calculus it is fairly straightforward to see that for the RSS function the second-order derivative or Hessian matrix

$$\frac{\partial^2 RSS}{\partial \boldsymbol{\beta} \partial \boldsymbol{\beta}^T} = 2 \cdot \mathbf{X}^T\mathbf{X} \tag{7.8}$$

is positive definite. From this the convexity of the RSS function follows. Hence, the least squares estimator $\hat{\boldsymbol{\beta}}$ globally minimizes the residual sums of squares. It gives the best linear combination $\mathbf{X}\hat{\boldsymbol{\beta}}$ of the columns of the predictor matrix $\mathbf{X}$ to approximate the observed outcomes $\mathbf{y}$. Furthermore, note that the estimator $\hat{\boldsymbol{\beta}}$ is unbiased, since

$$E[\hat{\boldsymbol{\beta}}] = (\mathbf{X}^T\mathbf{X})^{-1}\mathbf{X}^T E[\mathbf{y}] = (\mathbf{X}^T\mathbf{X})^{-1}\mathbf{X}^T\mathbf{X}\boldsymbol{\beta} = \boldsymbol{\beta},$$

so that it attains the Cramer-Rao lower bound. Since the least squares estimator is a linear combination of normally distributed random variables $\mathbf{y}$, several useful properties follow. Its distribution is multivariate normal

$$\hat{\boldsymbol{\beta}} \sim N\left(\boldsymbol{\beta}, \sigma^2(\mathbf{X}^T\mathbf{X})^{-1}\right). \tag{7.9}$$

The estimator is symmetrically distributed around the true coefficients with variance matrix that becomes smaller if the error variance is smaller and an increasing number of observations n. The predicted outcomes $\hat{Y}_i = \mathbf{x}_i^T\hat{\boldsymbol{\beta}}$, for $i = 1, \cdots, n$ are important for several purposes. Setting Equation (7.4) to zero, inserting the solution to (7.3) gives the biased estimator $\hat{\sigma}^2_{ML} = \frac{1}{n}\|\mathbf{y} - \mathbf{X}\hat{\boldsymbol{\beta}}\|^2$, which does not attain the Cramer-Rao lower bound. It is custom to work with the unbiased estimator

$$\hat{\sigma}^2 = \frac{1}{n-p}\sum_{i=1}^{N}(Y_i - \hat{Y}_i)^2 = \frac{1}{n-p}\|\mathbf{y} - \mathbf{X}\hat{\boldsymbol{\beta}}\|^2, \tag{7.10}$$

which related to the Chi-squared distribution with $n-p$ degrees of freedom. To see this, we define the hat matrix

$$\mathbf{H} = \mathbf{X}(\mathbf{X}^T\mathbf{X})^{-1}\mathbf{X}^T \tag{7.11}$$

and note its projection property $\mathbf{HX} = \mathbf{X}$, so that

$$\begin{aligned}(n-p)\hat{\sigma}^2 &= \|\mathbf{y} - \mathbf{X}\hat{\boldsymbol{\beta}}\|^2 = \|\mathbf{y} - \mathbf{X}(\mathbf{X}^T\mathbf{X})^{-1}\mathbf{X}^T\mathbf{y}\|^2 = \|(\mathbf{I} - \mathbf{H})\mathbf{y}\|^2 \\ &= \|(\mathbf{I} - \mathbf{H})(\mathbf{X}\boldsymbol{\beta} + \varepsilon)\|^2 = \|(\mathbf{I} - \mathbf{H})\varepsilon\|^2 = \sigma^2\frac{\varepsilon^T}{\sigma}(\mathbf{I} - \mathbf{H})\frac{\varepsilon}{\sigma} \\ &\sim \sigma^2\chi^2_{n-p}\end{aligned}$$

TABLE 7.1
Eight temperature (^{o}C) and viscosity (mPa·s) measurements.

Observation	Temp.	Viscosity
1	24.9	1.1330
2	35.0	0.9772
3	44.9	0.8532
4	55.1	0.7550
5	65.2	0.6723
6	75.2	0.6021
7	85.2	0.5420
8	95.2	0.5074

using that the symmetric projection matrix $\mathbf{I} - \mathbf{H}$ equals its square and has rank $n - p$. Since the expected value of a χ^2_{n-p} distributed random variable equals its degrees of freedom $n - p$, it follows that $E[\hat{\sigma}^2] = \sigma^2$ the estimator is unbiased.

The proportion of variance of the outcome $\mathbf{y}$ explained by the model, is given by

$$R^2 = 1 - \frac{\sum(\hat{Y}_i - Y_i)^2}{\sum(\bar{Y}_i - Y_i)^2} = 1 - \frac{\text{RSS}}{\text{TSS}},$$

where $\sum(\bar{Y}_i - Y_i)^2$ is the total sum of square (TSS). R^2 is also called the coefficient of determination as it serves as a fit measure of the model to the data. It equals the squared correlation between the prediction $\hat{\boldsymbol{y}} = \mathbf{X}\hat{\boldsymbol{\beta}}$ and the outcome $\mathbf{y}$, that is $R^2 = \text{COR}(\hat{\boldsymbol{y}}, \boldsymbol{y})^2$. Since the value of R^2 increases with the number of variables p, it should be interpreted with some care. When n is small compared to p, it is recommended to use the R^2-adjusted defined as

$$R^2_a = 1 - \frac{\text{RSS}/(n-p)}{\text{TSS}/(n-1)}.$$

Example 7.1 *The impact of temperature on viscosity of toluene-tetralin blends was investigated in a study by Byers and Williams (1987), whose data is reported in Table 7.1. In the current setting the explanatory variable x is temperature and the outcome variable y is viscosity, so that the linear model simplifies to $\mathbf{y} \sim N(\mathbf{1}\beta_0 + \mathbf{x}\beta_1, \sigma^2\mathbf{I})$, with intercept β_0 and slope β_1; see also Equation 7.1. The estimated intercept and slope are given in Table 7.2. The*

TABLE 7.2
Results for regression analysis of temperature on viscosity of toluene-tetralin blends with estimated intercept $\hat{\beta}_0$ and slope $\hat{\beta}_1$, their standard error $SE(\hat{\beta}_j)$, corresponding t- and p-values.

	Estimate	Std. Error	t value	Pr(>|t|)
(Intercept)	1.2815	0.0469	27.34	< 0.0001
Temperature	−0.0088	0.0007	−12.02	< 0.0001

slope indicates a decrease of -0.0088 *viscosity with an increase in temperature of one degrees of Celsius. The residual standard error* $\hat{\sigma} = 0.04743$, *the coefficient of determination (multiple R-squared) = 0.9602, and the adjusted R-squared = 0.9535.*

7.2.2 Outcome prediction

Suppose $(1, x_{01}, x_{02}, \cdots, x_{0p}) = \boldsymbol{x}_0^T$ is a new observation for which we wish to predict Y_0. A prediction of such an unknown future observation Y_0 given $\boldsymbol{x}_0^T$ is provided by $\hat{Y}_0 = \boldsymbol{x}_0^T \hat{\boldsymbol{\beta}}$. Its **prediction interval** is

$$\hat{Y}_0 - t_{\alpha/2,n-p}\sqrt{\widehat{\sigma}^2(1 + \boldsymbol{x}_0^T(\boldsymbol{X}^T\boldsymbol{X})^{-1}\boldsymbol{x}_0)} \leq Y_0 \leq \hat{Y}_0 + t_{\alpha/2,n-p}\sqrt{\widehat{\sigma}^2(1 + \boldsymbol{x}_0^T(\boldsymbol{X}^T\boldsymbol{X})^{-1}\boldsymbol{x}_0)}. \tag{7.12}$$

It can be shown that a prediction interval is broader than a confidence interval due to the one under the square root.

Example 7.2 *For the temperature on viscosity data (Byers and Williams, 1987) we compute the prediction interval for Temperature 50 (^oC), so that* $\boldsymbol{x}_0^T = (1, 50)$. *For* $n = 8$, $p = 2$, $\alpha = 0.05$, *we obtain the quantile* $t_{\alpha/2,n-p} = -2.446912$. *From the linear regression we find* $\widehat{\sigma} = 0.04743$ *and* $\hat{Y}_0 = \boldsymbol{x}_0^T\hat{\boldsymbol{\beta}} = 0.8436195$. *Next, from* $\sqrt{(1 + \boldsymbol{x}_0^T(\boldsymbol{X}^T\boldsymbol{X})^{-1}\boldsymbol{x}_0)} = 1.071911$ *we find that the 95% prediction interval for viscosity of toluene-tetralin blends at Temperature 50°C is equal to* $[0.719, 0.968]$.

7.3 Model Diagnostics

For making valid inferences it is of importance that the model assumptions hold and that the estimators are sufficiently close to their true values. Since all estimators are continuous functions of the data, a single or a few extremely outlying data points, e.g., from contaminated chemical samples, may be highly influential on the values for the estimators, making these unreliable or scientifically useless. To detect these, we will briefly discuss the most important **regression diagnostics**.

7.3.1 Diagnostics for high leverage points

An important role is played by the n by n hat matrix $\mathbf{H} = \mathbf{X}(\mathbf{X}^T\mathbf{X})^{-1}\mathbf{X}^T$. Its ith diagonal value

$$h_{ii} = \boldsymbol{x}_i^T(\boldsymbol{X}^T\boldsymbol{X})^{-1}\boldsymbol{x}_i$$

is called **leverage**, for which we have $0 \leq h_{ii} \leq 1$ for all i. The leverage values generally decrease with the number of observations n. Since the sum of the leverages equals p, the average hat value $\overline{h} = p/n$. Values h_{ii} larger than $2 \cdot \overline{h} = 2p/n$ indicate that observation i is influential (Huber, 2009).

7.3.2 Diagnostics for outlying observations

For the diagnosis of outlying observations the predicted values $\hat{\mathbf{y}} = \mathbf{X}\hat{\boldsymbol{\beta}}$ and the residuals $\mathbf{y} - \hat{\mathbf{y}}$ are of importance. Note that the hat matrix projects the outcomes $\mathbf{y}$ onto the space spanned by the columns of the predictor matrix $\mathbf{X}$, that is

$$\hat{\mathbf{y}} = \mathbf{X}\hat{\boldsymbol{\beta}} = \mathbf{X}(\mathbf{X}^T\mathbf{X})^{-1}\mathbf{X}^T\mathbf{y} = \mathbf{H}\mathbf{y}.$$

Hence, the elements of $\hat{\mathbf{y}}$ are a linear combination of the outcomes in $\mathbf{y}$. From the law of large numbers, the consistency $\hat{\boldsymbol{\beta}} \xrightarrow{P} \boldsymbol{\beta}$ follows, so that the predicted values tend in probability to the structural part of the linear model $\mathbf{X}\boldsymbol{\beta}$, with increasing n. Moreover, the residuals $\mathbf{y} - \hat{\mathbf{y}} \xrightarrow{P} \varepsilon$, tend to the error terms. For the residuals, we have

$$\mathbf{y} - \hat{\mathbf{y}} = (\mathbf{I} - \mathbf{H})\mathbf{y},$$

so that we see that these are also a linear combination of the outcomes $\mathbf{y}$. Hence, the model assumption $\mathbf{y} \sim N(\mathbf{X}\boldsymbol{\beta}, \sigma^2\mathbf{I})$ implies that

$$\begin{bmatrix} \hat{\mathbf{y}} \\ \mathbf{y} - \hat{\mathbf{y}} \end{bmatrix} \sim N\left(\begin{bmatrix} \mathbf{X}\boldsymbol{\beta} \\ \mathbf{0} \end{bmatrix}, \begin{bmatrix} \sigma^2\mathbf{H} & \mathbf{0} \\ \mathbf{0} & \sigma^2(\mathbf{I} - \mathbf{H}) \end{bmatrix} \right).$$

That is, the predicted values and the residuals are stochastically independent. Hence, for practical purposes it makes sense to inspect a plot with horizontally the fitted values and vertically the residuals. Then we would expect to find a band of points, randomly distributed around the center zero, without any observable pattern.

To evaluate the size of the residuals $\hat{\varepsilon}_i = Y_i - \hat{Y}_i$, several normalizations are considered in the literature. The **standardized residuals**

$$d_i = \frac{\hat{\varepsilon}_i}{\sqrt{\hat{\sigma}^2}},$$

for $i = 1, \ldots, n$ have mean zero and approximately unit variance. Since, the residuals tend to the error terms, and $\hat{\sigma}^2$ to the error variance, it follows that the d_i random variables tend to the standard normal with increasing n. For practical purposes a rule is to see the i-th observation as outlying, if $|d_i| > 3$. Further more, note that, at least asymptotically, we know that the standardized residuals tend to standard normal deviates. Hence, if we for practical purposes inspect a plot with vertically the standardized residuals and horizontally the corresponding quantiles from the standard normal distribution, then we expect the points to appear close to the line. Such a quantile-quantile (QQ) plot is often informative.

An other way to normalize residuals comes from the so-called **studentized residuals**

$$r_i = \frac{\hat{\varepsilon}_i}{\sqrt{\hat{\sigma}^2 \cdot (1 - h_{ii})}}, \tag{7.13}$$

as these are adjusted for the influence of high leverage. However, note that the denominator and the denominator are dependent as both involve the i-th residual. Therefore, an other useful normalization to consider is the one based on

$$\hat{\sigma}^2_{(i)} = \sum_{j \neq i}^{n} (Y_j - \mathbf{x}_j^T \hat{\boldsymbol{\beta}}_{(i)})^2/(n - p - 1),$$

where the estimator $\hat{\boldsymbol{\beta}}_{(i)}$ is computed from all observations except the ith. This leaving-one-out idea plays an important role in diagnostics of linear model estimation. The **externally studentized residuals** have a t-distribution, more precisely for each i we have

$$\frac{\hat{\varepsilon}_i}{\sqrt{\hat{\sigma}^2_{(i)} \cdot (1 - h_{ii})}} \sim t_{n-(p+1)}.$$

On this fact a statistical test for the null-hypothesis of absence of outliers is constructed.

7.3.3 Diagnostics for influential observations

The influence of observations typically follow from influence functions defined as a derivative in the direction of an observation. Such a fundamental approach turn out to imply differences between estimates and those obtained by leaving out one observation. Two of the most important are given here. First, we define $\hat{\boldsymbol{\beta}}_{(i)}$ as the estimator obtained by leaving out the i-the observation, $\hat{\beta}_{j(i)}$ as its j-th element, C_{jj} as the jth diagonal element of $(\boldsymbol{X}^T\boldsymbol{X})^{-1}$, and $\boldsymbol{X}\hat{\boldsymbol{\beta}}_{(i)} = \widehat{\mathbf{y}}_{(i)}$. The **beta difference**, i.e., the normalized difference between the beta coefficient $\hat{\beta}_j$ and $\hat{\beta}_{j(i)}$ relative to its standard deviation, is given by

$$\text{DFBETA}_{ij} = \frac{\hat{\beta}_j - \hat{\beta}_{j(i)}}{\sqrt{\hat{\sigma}^2_{(i)} \cdot C_{jj}}}.$$

Observations for which an absolute DFBETA exceeds the cut-off $2/\sqrt{n}$ are seen as potentially influential.

An overall diagnostic for the influence of an observation on all estimates of the linear model is given by **Cook's distance**, for which there are several expressions

$$\begin{aligned} C_i &= \frac{(\hat{\boldsymbol{\beta}} - \hat{\boldsymbol{\beta}}_{(i)})^T \boldsymbol{X}^T \boldsymbol{X} (\hat{\boldsymbol{\beta}} - \hat{\boldsymbol{\beta}}_{(i)})}{p\hat{\sigma}^2} \\ &= \frac{(\widehat{\mathbf{y}} - \widehat{\mathbf{y}}_{(i)})^T (\widehat{\mathbf{y}} - \widehat{\mathbf{y}}_{(i)})}{p\hat{\sigma}^2} \\ &= \frac{1}{p}\frac{h_{ii}}{1 - h_{ii}} r_i^2. \end{aligned}$$

Cook's distance is proportional to the squared Euclidian distance between the predicted value with and without the ith observation. For practical purposes observations with $|C_i| > 1$ are seen as influential which require inspection. Furthermore, we may inspect a plot horizontally with the leverages and vertically with the standardized residuals within which contour line are depicted with with Cook's distances equal 0.5 and 1.0. Points beyond the latter are seen as influential.

7.3.4 Diagnostics for linear dependency among predictors

It seems clear that when the columns of $\boldsymbol{X}$ are nearly linear dependent, then the estimates may become numerically unstable or the statistical inference may become more difficult. Near linear dependency is known as "multi-collinearity" for which a **variance inflation factor** (VIF) is defined as

$$C_{jj} = \frac{1}{1 - R_j^2}.$$

Here R_j^2 is the coefficient of determination, when x_j is regressed on the remaining $p - 1$ regressors, and C_{jj} is element jj from the $(\boldsymbol{X}^T \boldsymbol{X})^{-1}$. If the columns of $\boldsymbol{X}$ are orthogonal, then VIF=1. If the VIF value is larger than 10, then this is a clear indication that the coefficients are poorly estimated due to multi-collinearity.

Although, at first sight the above may seem a somewhat lengthy list of influence measures, it is clear that models that pass an extensive diagnostic inspection are scientifically more sound in forming a basis for valid inferences. For applications, computer languages may conveniently be used to check for influential observations in conjunction with inspection of diagnostic plots.

Example 7.3 *In an Belle Ayr subbituminous coal liquefaction experiment (Cronauer et al., 1978) the outcome CO2 output was observed during 27 runs together with seven explanatory variables given in Table 7.5 below. Here we focus on the diagnostics from the model found by the Minimum AIC criterion which contains the explanatory variables Solvent total and Hydrogen consumption. The studentized residuals, the DFBETA values, the Cooks distances C_i, and the leverage values h_{ii} are given in Table 7.3. For the majority of the observations there are no indications for any of these to be largely influential*

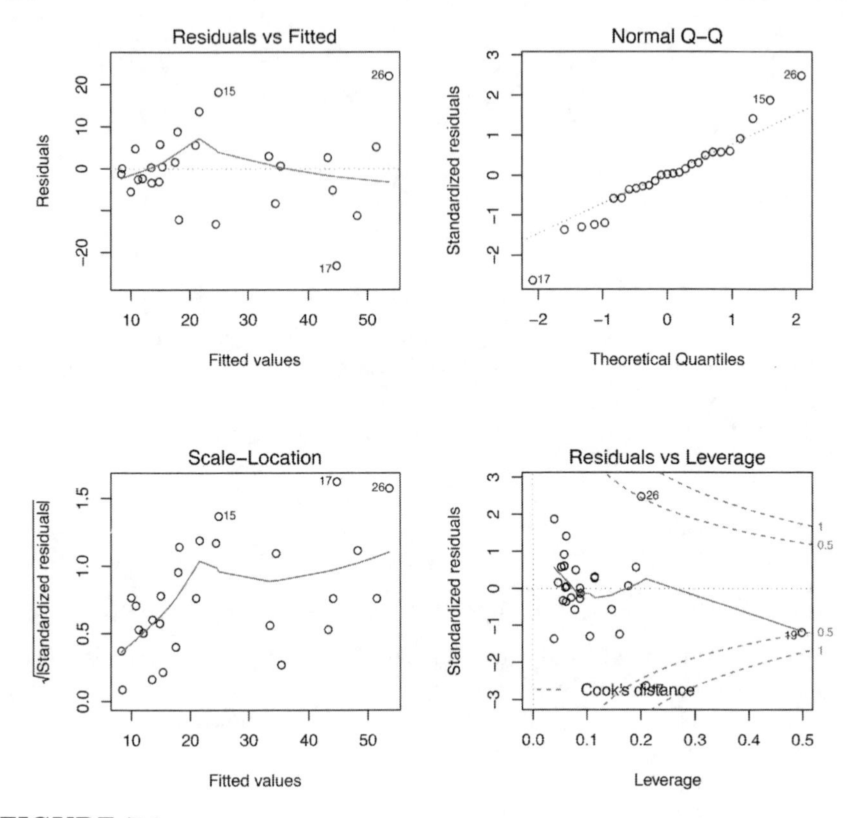

FIGURE 7.1
Diagnostic plot for regressions model on outcome CO_2 predicted from solvent total and hydrogen consumption.

as the absolute studentized residual $|r_i|$ smaller than 2, the DFBETA values are smaller than $2/\sqrt{27} = 0.38$, the Cook's distances C_i are smaller than 1, and the leverage values h_{ii} are smaller than $2 \cdot 2/27 = 0.148$. There are, however, three observations being potentially influential. That is, Observations 17, 19, and 26 have large DFBETA and large leverage values, and Observations 17 and 26 have large studentized residuals.

The two by two penal of diagnostic plots allows for a quick inspection. In Figure 7.1 the left upper panel gives the fitted values against the residuals, the right upper the quantiles from the residuals against the theoretical quantiles from the normal (Quantile-Quantile), the left lower fitted against the square root of the absolute residuals, and the right lower leverage points h_{ii} against the standardized residuals with a contour of Cook's distance. The large studentized residuals of Observation 17 and 26 are apparent from several diagnostic plots in Figure 7.1. The large leverage of Observation 19 appears at the right-hand side of the Residuals versus Leverage plot in Figure 7.1.

TABLE 7.3
Minimum AIC model Belle Ayr Liquefaction data with intercept and predictive variable Solvent total and Hydrogen consumption y_i, the predicted values $\hat{y}_i$, the studentized residuals r_i, influence measures on the coefficients (DFB), Cooks distance C_i, and the leverage values h_{ii}.

Observation	y_i	$\hat{y}_i$	r_i	**DFB**$_{i1}$	**DFB**$_{i2}$	**DFB**$_{i3}$	C_i	h_{ii}
1	36.98	48.28	−1.26	0.18	−0.48	0.07	0.10	0.16
2	13.74	13.49	0.03	0.01	−0.00	−0.00	0.00	0.06
3	10.08	13.56	−0.36	−0.09	0.04	0.03	0.00	0.06
4	8.53	8.46	0.01	0.00	−0.00	−0.00	0.00	0.09
5	36.42	33.49	0.31	0.02	0.07	−0.07	0.00	0.11
6	26.59	21.00	0.57	0.10	0.00	−0.07	0.01	0.05
7	19.07	17.53	0.16	0.03	−0.01	−0.00	0.00	0.05
8	5.96	18.17	−1.32	−0.17	0.26	−0.29	0.07	0.11
9	15.52	10.78	0.49	0.12	−0.10	0.01	0.01	0.08
10	56.61	51.48	0.57	−0.14	0.16	0.16	0.03	0.19
11	26.72	17.94	0.91	0.19	−0.03	−0.12	0.02	0.06
12	20.80	14.98	0.60	0.14	−0.06	−0.06	0.01	0.06
13	6.99	8.29	−0.13	−0.04	0.03	0.02	0.00	0.09
14	45.93	43.32	0.27	−0.02	0.08	−0.01	0.00	0.11
15	43.09	24.92	1.98	0.22	0.05	−0.10	0.05	0.04
16	15.79	15.35	0.05	0.01	−0.01	0.00	0.00	0.06
17	21.60	44.80	−3.05	0.19	−1.30	0.75	0.61	0.21
18	35.19	21.61	1.44	0.25	0.04	−0.23	0.04	0.06
19	26.14	34.54	−1.21	0.25	0.29	−1.15	0.47	0.50
20	8.60	11.25	−0.27	−0.08	0.03	0.05	0.00	0.09
21	11.63	14.82	−0.32	−0.07	0.04	0.00	0.00	0.06
22	9.59	12.01	−0.25	−0.06	0.05	−0.00	0.00	0.07
23	4.42	9.99	−0.58	−0.16	0.12	0.02	0.01	0.08
24	38.89	44.15	−0.57	0.04	−0.20	0.07	0.02	0.15
25	11.19	24.47	−1.39	−0.16	−0.02	0.06	0.03	0.04
26	75.62	53.65	2.81	−0.76	0.96	0.69	0.51	0.20
27	36.03	35.38	0.07	−0.01	−0.00	0.03	0.00	0.18

The outlier test on Observation 17 with externally studentized residual −3.05 *assigns an unadjusted p-value 0.0057 and Bonferroni corrected p-value 0.1533, which seems not very alarming. The diagnostic plots indicate potentially influential, but probably not extremely influential observations on the regression coefficients. A robust estimation approach to shed more light on this is undertaken in Chapter 12.*

7.4 Model Selection

The number of possible models having p explanatory variables equals 2^p, as each variable is part of the model or not. This exponential increase with p

may soon become too large for manual evaluation. Hence, automatic model selection by an information criterion that asymptotically minimizes the prediction error may be helpful, if the latter is deemed important with respect to the purpose of the research. If the estimation of the model parameters is of importance, then penalized inference is an option.

7.4.1 Marginal testing of parameters

Often, we wish to test the null hypothesis that a certain coefficient is of no importance, that is

$$H_0 : \beta_j = 0 \quad \text{against } H_A : \beta_j \neq 0.$$

Let C_{jj} be jth diagonal element of $(\mathbf{X}^T\mathbf{X})^{-1}$. A t-test for this null-hypothesis follows from Equation (7.9) as it suggests to use **standard error** $SE(\hat{\beta}_j) = \sqrt{\hat{\sigma}^2 C_{jj}}$ to normalize the difference between the estimator and its true value, under the null hypothesis. More specifically, we have

$$\hat{\beta}_j \sim N\left(\beta_j, \sigma^2 C_{jj}\right) \Rightarrow t = \frac{\hat{\beta}_j - \beta_j}{SE(\hat{\beta}_j)} = \frac{\hat{\beta}_j}{\sqrt{\hat{\sigma}^2 C_{jj}}} \sim t_{n-p-1}.$$

That is, the term $t = (\hat{\beta}_j - \beta_j)/SE(\hat{\beta}_j)$ has a t-distribution with $n - (p+1)$ degrees of freedom. If $|t| > t_{n-p-1,1-\alpha/2}$, then the decision is to reject the null hypothesis of zero slope. This is equivalent to rejecting the null-hypothesis $H_0 : \beta_j = 0$, if the **p-value**

$$P(t_{n-(p+1)} \geq |t|) < \alpha,$$

where α is the significance level of the test. The p-value in fact measures the probability of finding larger values than the observed value $|t|$, given that the null-hypothesis is true. Large t-values have low probability under H_0 and typically correspond to small p-values.

A $1 - \alpha$ **confidence interval** for β_j is given by

$$\left[\hat{\beta}_j - t_{1-\alpha/2,n-p} \cdot \sqrt{\hat{\sigma}^2 C_{jj}}, \hat{\beta}_j + t_{1-\alpha/2,n-p} \cdot \sqrt{\hat{\sigma}^2 C_{jj}}\right],$$

where $t_{1-\alpha/2,n-p}$ is the $1 - \alpha/2$ quantile of t_{n-p}. The confidence interval is small if the $SE(\hat{\beta}_j) = \sqrt{\hat{\sigma}^2 C_{jj}}$ is small. This obviously occurs when the residual variance $\hat{\sigma}^2$ is small or when the sample size n is large causing the term C_{jj} from the model matrix $\mathbf{X}$ to be small. The 95% percent confidence interval nicely summarizes our knowledge from the data of a chemical experiment about the unknown β_j.

Example 7.4 *In the example on the impact of temperature on viscosity of toluene/tetralin blends the t- and corresponding p-values from regression analysis in Table 7.2 suggest the decision to reject the null hypothesis of zero coefficients for the intercept as well as for the slope. The corresponding confidence* *intervals [1.1668, 1.3962] and* $[-0.0105, -0.0070]$, *respectively, for the intercept and slope seem fairly small in size.*

7.4.2 Testing a subset of parameters

Often, we wish to test

$$H_0 : \beta_1 = \beta_2 = \cdots = \beta_p = 0 \,\text{against}\; H_A : \beta_j \neq 0 \text{ for at least one}\, j.$$

To present a test for this null hypothesis by a single overall procedure we need to define the sum of squares due to the regression SS_R and the **residual sum of squares** SS_{Res}, as

$$\mathrm{SS}_R = \sum_{i=1}^{n} (\hat{Y}_i - \overline{Y})^2, \text{ and } \mathrm{SS}_{Res} = \sum_{i=1}^{n} (Y_i - \hat{Y}_i)^2,$$

respectively. Note that the SS_R value tends to be large, if the null hypothesis is false. The F-test for the null-hypothesis is based upon these

$$F = \frac{\mathrm{SS}_R/p}{\mathrm{SS}_{Res}/(n-p-1)} \sim F_{p,n-p-1},$$

where we

$$\text{reject } H_0 \text{ if p-value} = P(F_{p,n-p-1} > F) < \alpha = 0.05,$$

where α is the **significance level** of the test. The intuition is that when the SS_R/p is large relative to $\mathrm{SS}_{Res}/(n-p-1)$, then the p-value is small and H_0 gets rejected.

7.4.3 AIC

A way to proceed is by penalizing the maximum likelihood criterion in such a way that it corrects for some bias inherent in the estimation procedure when the data are used twice, that is, for estimation and for evaluation. **Akaike's Information Criterion** (AIC) is the estimated Kullback-Leibler divergence to the true model, defined as

$$AIC = -2(\text{maximum log-likelihood}) + 2(\text{number of free parameters}).$$

We may simplify this into

$$\begin{aligned} AIC &= -2\sum_{i=1}^{n} \log f(x_i, \widehat{\boldsymbol{\beta}}) + 2p \\ &= -2 \cdot \left(-\frac{n}{2} \log(2\pi\hat{\sigma}^2) - \frac{1}{2\hat{\sigma}^2} \sum_{i=1}^{n} (y_i - \mathbf{x}_i^T \widehat{\boldsymbol{\beta}})^2 \right) + 2p \\ &= n \log(2\pi + 1) + n \log(\hat{\sigma}^2) + 2(p+2) \end{aligned}$$

TABLE 7.4
Full model from linear regression and model found by minimum AIC.

Full model	**Estimate**	**Std. Error**	**t value**	**Pr(>|t|)**
(Intercept)	1.00210	0.00320	312.73414	0.00000
Fat content	−0.00004	0.00008	−0.53719	0.59267
Protein content	0.00009	0.00018	0.49705	0.62055
Casein content	0.00010	0.00022	0.46053	0.64642
Cheese dry substance factory	0.00008	0.00028	0.27628	0.78307
Cheese dry substance laboratory	−0.00010	0.00028	−0.34784	0.72890
Milk dry substance	0.00023	0.00005	5.06792	0.00000
Cheese product	−0.00027	0.00027	−1.01294	0.31422
Min AIC				
(Intercept)	1.00171	0.00301	333.26218	0.00000
Casein content	0.00017	0.00011	1.52385	0.13135
Milk dry substance	0.00019	0.00003	5.68039	0.00000

The model that gives the minimal value of this badness-of-fit criterion is best in the AIC sense. An increasing number of parameters p, would increase the maximum of the likelihood. The idea is to penalize this danger of over-fitting by the $2p$ term. A very fast stepwise type of algorithm to minimize AIC often leads to the best model in practical situations. It is known that minimizing the AIC is asymptotically equivalent to minimizing the prediction error for a new observation.

Example 7.5 *Daudin et al. (1988) give 8 readings on the composition of 86 containers of milk. We take for the outcome y the "density" of milk, and as explanatory variables fat content, protein content, casein content, cheese dry substance measured in the factory, cheese dry substance measured in the laboratory milk dry substance and cheese product. The resulting full linear model in Table 7.4 has several beta coefficients for which the marginal testing indicates these not to differ from zero, except for Milk Dry Substance. Its Multiple R-squared 0.5356 differs substantially from its Adjusted R-squared 0.4939, indicating too many estimated coefficients. The model found by minimum AIC consists of two explanatory variables of which Milk Dry Substance is significant. Its Multiple R-squared 0.5217 is almost equal to its Adjusted R-squared 0.5102, as the number of of explanatory variables is relatively small to the number of observations. The model found by minimum AIC has a better fit to the data according to the Adjusted R-squared.*

Example 7.6 *In an experiment the CO2 output was observed during 27 runs together with seven explanatory variables given in Table 7.5 below (Cronauer et al., 1978). From the full regression model with all explanatory variables*

TABLE 7.5
Full model from linear regression and model found by minimum AIC on the Belle Ayr Liquefaction data.

	Estimate	**Std. Error**	**t value**	**Pr(>|t|)**
(Intercept)	53.9370	57.4290	0.9392	0.3594
Space time (in min)	−0.1277	0.2815	−0.4535	0.6553
Temperature (in degrees Celsius)	−0.2292	0.2326	−0.9851	0.3370
Percent solvation	0.8249	0.7653	1.0779	0.2946
Oil yield (g/100g MAF)	−0.4382	0.3586	−1.2222	0.2366
Coal total	−0.0019	0.0097	−0.2007	0.8431
Solvent total	0.0199	0.0081	2.4586	0.0237
Hydrogen consumption	1.9935	1.0897	1.8294	0.0831
min AIC	**Estimate**	**Std. Error**	**t value**	**Pr(>|t|)**
(Intercept)	2.5265	3.6101	0.6998	0.4908
Solvent total	0.0185	0.0027	6.7420	0.0000
Hydrogen consumption	2.1858	0.9727	2.2471	0.0341

many coefficients are non-significant and the R-squared 0.73 differs considerably from the Adjusted R-squared 0.63. The AIC of the full model 212.6967 is considerably larger than 205.3722 that found for the model with minimum AIC. A minimum AIC approach yields a much smaller model with Solvent total and Hydrogen consumption as the two explanatory variables and Multiple R-squared 0.70 and Adjusted R-squared 0.67. The coefficient for Solvent total is highly significant and that of Hydrogen consumption is borderline significant.

7.4.4 SCAD penalized regression

In the search for a penalty function for the regression parameters one might require an estimator to have three properties.

1. Unbiasedness: The resulting estimator should avoid unnecessary bias when its value is sufficiently large.
2. Sparsity: The resulting estimator should conform a threshold rule in automatically setting small values to zero in order to reduce model complexity.
3. Continuity: The resulting estimator is a continuous function of the data to avoid instability in e.g., prediction.

We wish to minimize the **penalized least squares** function

$$\frac{1}{2}\|\mathbf{y} - \mathbf{X}\boldsymbol{\beta}\|^2 + \lambda \sum_{j=1}^{p} p_j(|\beta_j|),$$

where λ is a thresholding parameter and $p_j(|\beta_j|)$ a penalty function.

To explain the main ideas involved we assume that the model matrix $\mathbf{X}$ is column-wise orthonormal $\mathbf{X}^T\mathbf{X} = \mathbf{I}_p$. Then the least squares estimator $\hat{\boldsymbol{\beta}} = \mathbf{X}^T\mathbf{y} = \mathbf{z}$, and the predicted values can be written as $\hat{\mathbf{y}} = \mathbf{X}\mathbf{X}^T\mathbf{y} = \mathbf{X}\mathbf{z}$. It follows that

$$\begin{aligned}
\frac{1}{2}\|\mathbf{y} - \mathbf{X}\boldsymbol{\beta}\|^2 + \lambda\sum_{j=1}^{p} p_j(|\beta_j|) &= \frac{1}{2}\|\mathbf{y} - \hat{\mathbf{y}}\|^2 + \frac{1}{2}\|\hat{\mathbf{y}} - \mathbf{X}\boldsymbol{\beta}\|^2 + \lambda\sum_{j=1}^{p} p_j(|\beta_j|) \\
&= \frac{1}{2}\|\mathbf{y} - \hat{\mathbf{y}}\|^2 + \|\mathbf{z} - \boldsymbol{\beta}\|^2 + \lambda\sum_{j=1}^{p} p_j(|\beta_j|) \\
&= \frac{1}{2}\|\mathbf{y} - \hat{\mathbf{y}}\|^2 + \sum_{j=1}^{p}\Big(\frac{1}{2}(z_j - \beta_j)^2 + \lambda p_j(|\beta_j|)\Big),
\end{aligned}$$

so that we may continue component wise for each element of $\boldsymbol{\beta}$. We simplify the notation by writing θ for β_j, p for p_j, and compute the first order derivative

$$\frac{\partial}{\partial\theta}\frac{1}{2}(z_j - \theta)^2 + p_\lambda(|\theta|) = \theta - z + \text{sign}(\theta)\cdot\lambda p'(\theta).$$

Several types of penalty functions have been investigated in the past 50 years: $p_\lambda(|\theta|) = \lambda|\theta|^2$ gives ridge regression which is biased and not sparse, $p_\lambda(|\theta|) = \lambda|\theta|^q$ gives bridge regression which is not sparse, $p_\lambda(|\theta|) = \lambda|\theta|$ gives LASSO regression which is sparse and biased. For the **smoothly clipped absolute deviation** (SCAD) we take

$$\begin{aligned}
p_\lambda(|\theta|) = \lambda|\theta|\cdot I_{(-\infty,\lambda)}(|\theta|) &+ \left(\frac{a\lambda(|\theta| - \lambda) - (|\theta|^2 - \lambda^2)/2}{a-1} + \lambda^2\right)\cdot I_{[\lambda,a\lambda)}(|\theta|) \\
&+ \left(\frac{(a-1)\lambda^2}{2} + \lambda^2\right)\cdot I_{[a\lambda,\infty)}(|\theta|),
\end{aligned}$$

for a tuning parameter a often taken to be equal to 3.7 (Fan and Li, 2001), and $I_{(a,b)}(x)$ the indicator function equal to 1 if $x \in (a,b)$ and zero otherwise. Setting the first order derivative to zero separately for each of the three cases yields the continuous estimator

$$\widehat{\theta} = \begin{cases} \text{sign}(z)(|z| - \lambda)_+ & \text{for} \quad |z| \leq \lambda \\ ((a-1)z - \text{sign}(z)a\lambda)/(a-2) & \text{for} \quad \lambda \leq |z| < a\lambda. \\ z & \text{for} \quad |z| \geq a\lambda \end{cases}$$

When $|z|$ is smaller than λ, then the positive part $(|z| - \lambda)_+$ is zero, yielding $\widehat{\theta} = \widehat{\beta}_j = 0$ a zero estimator or sparsity. When $|z|$ is larger than $a\lambda$, then $\widehat{\theta} = \widehat{\beta}_j = z_j$ the least squares estimator, which is unbiased. The SCAD estimator follows from a non-convex loss function to be minimized for which effective algorithms have been developed.

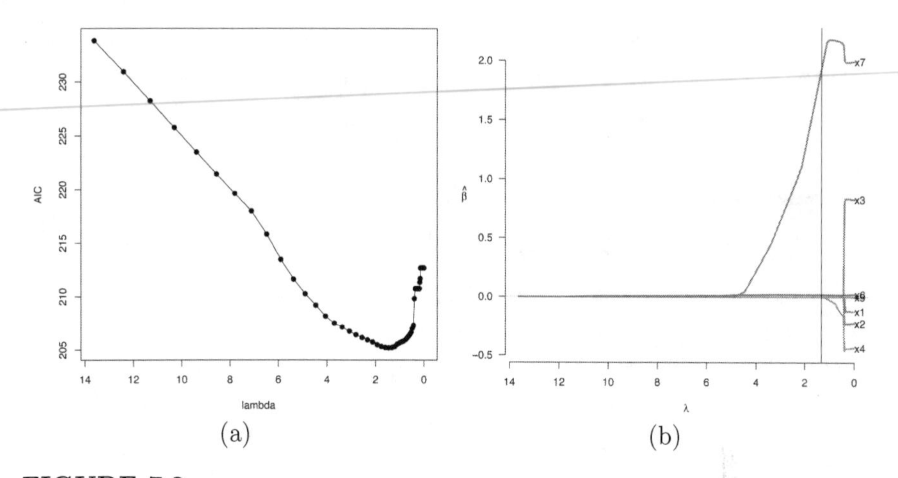

FIGURE 7.2
SCAD penalized regression with AIC values (a) and statistical learning of the regression parameters as a function of the penalty parameter λ.

The tuning parameter λ is chosen by a cross validation approach or by a function of the sample size n (Wang et al., 2014), in both cases it shrinks to zero as n increases without bound. In this way it is guaranteed that the SCAD estimator is consistent (asymptotically correct), also for small effects. The SCAD estimator has good asymptotic properties in terms of standard errors of the estimators and it outperformed several other estimators by simulation studies in terms of correctness of sparsity (Fan and Li, 2001; Wang et al., 2014).

In an application one may study the parameter estimates depending on λ values that tend from large to zero, where in the limiting case $|z|$ is larger than $a\lambda = 0$, so that SCAD equals least squares. A figure of the parameter estimates against values of λ is known as a statistical learning plot as it provides information on the predictive power of each explanatory variable under the "pressure" of the SCAD tuning penalty parameter.

Example 7.7 *For the Belle Ayr Liquefaction data with outcome CO_2 and explanatory variables numbered as x1 to x7 a SCAD analysis was undertaken. Figure 7.2 (a) gives the AIC values of the linear regression model depending on values of the tuning parameter λ tending from 14 down to zero. Its minimum occurs for $\lambda = 1.33073$ for which the regression parameters for Solvent total (x6) and Hydrogen consumption (x7) are 0.019 and 1.93, respectively. These results are very similar to those obtained before. The statistical learning plot in Figure 7.2 (b) reveals that the Hydrogen consumption (x7) and Solvent total (x6) are the most important and that all other predictive variables only become non-zero for very small values of the penalty parameter λ. A SCAD approach based on generalized cross validation reveals only Solvent total (x6) as a predictor for outcome CO_2. These findings are in line with the previous.*

To make the coefficients more comparable in a learning plot one might wish to scale all explanatory variables to have unit length.

7.5 Specific Linear Models

The linear model have many special cases and forms relevant for modeling and analysing chemical processes. Here two special cases are discussed, several others follow in Chapter 11.

7.5.1 Simple linear regression

The sub-model with a single predictive (explanatory) variable can be written as

$$Y = \beta_0 + \beta_1 x + \varepsilon, \ \varepsilon \sim N(0, \sigma^2), \tag{7.14}$$

where β_0 is the intercept and β_1 the slope. In chemistry the model is relevant for the construction of calibration curves, when e.g., the outcome Y can be linearized and depends on a single predictive variable such as concentration.

Example 7.8 *It is often useful to simulate outcomes of an experiment on a computer before actually conducting it. Suppose an aqueous solution of fluorescein is examined in a fluorescence spectrometer yielding fluorescence intensities (arbitrary units) measured at specific concentrations and that the fluorescence absorptions (y) are linearly related to the concentrations (x) with intercept $\beta_0 = 2.0$ and slope $\beta_1 = 2.5$. Let the error terms be normally distributed with mean zero and variance equal to 1. This yields the pseudo data in Table 7.6 giving the estimates, t-values and p-values and confidence intervals from linear regression in Table 7.7. The p-values are well below the 5% significance level suggesting the decision to reject the null-hypothesis of zero regression coefficients. The estimates do differ somewhat from the true, but the latter are contained in the 95% confidence intervals. From the confidence interval we are 95% certain that the true slope for the concentration is within* $[2.42; 2.81]$. *The residual standard error* $\hat{\sigma} = 0.9008$ *is fairly close to the true* $\sigma = 1$.

Various qualities to evaluate the presence of outlying data points are given in Table 7.6. All hat values h_{ii} are smaller than $2p/n = 2 \cdot 2/11 = 0.36$. Not all $|\text{DFBETA}_{ij}| < 2/\sqrt{11} = 0.60$, e.g., for observation 3 both are larger than the cut-off value. This seems due to the somewhat small observed intensity y_3, also indicated by its relatively large standardized residual. The largest Cook's distance occurs for observation 3, it is, however, well below the cut-off value 1.

From the two by two panel of diagnostic plots in Figure 7.3 the presence of outlying data points can be inspected. The residuals versus fitted values

TABLE 7.6
The concentrations and fluorescence intensities together with quantities from linear regression such as the predicted values $\hat{y}_i$, the residuals $y_i - \hat{y}_i$, the studentized residuals r_i, influence measures on the coefficients (DFB), Cooks distance C_i, and the leverage values h_{ii}.

	x_i	y_i	$\hat{y}_i$	$y_i - \hat{y}_i$	r_i	$\mathbf{DFB}_{i1}$	$\mathbf{DFB}_{i2}$	C_i	h_{ii}
1	0.00	2.18	2.25	−0.07	−0.09	−0.06	0.05	0.00	0.32
2	1.00	5.28	4.86	0.42	0.53	0.28	−0.22	0.04	0.24
3	2.00	5.65	7.48	−1.83	−2.24	−1.40	0.99	0.52	0.17
4	3.00	11.48	10.09	1.39	1.65	0.64	−0.38	0.20	0.13
5	4.00	13.24	12.71	0.53	0.62	0.15	−0.06	0.02	0.10
6	5.00	15.70	15.33	0.37	0.44	0.07	0.00	0.01	0.09
7	6.00	17.90	17.94	−0.04	−0.05	−0.00	−0.00	0.00	0.10
8	7.00	19.75	20.56	−0.80	−0.95	−0.00	−0.19	0.07	0.13
9	8.00	22.55	23.17	−0.62	−0.76	0.07	−0.23	0.06	0.17
10	9.00	26.40	25.79	0.62	0.78	−0.14	0.33	0.09	0.24
11	10.00	28.44	28.40	0.04	0.06	−0.02	0.03	0.00	0.32

TABLE 7.7
Results for regression analysis fluorescence intensities on Concentrations giving the estimates $\hat{\beta}_j$, their standard error $SE(\hat{\beta}_j)$, corresponding t- and p-values and 95% confidence intervals.

	Estimate	**Std. Error**	**t value**	**Pr(>|t|)**	**2.5 %**	**97.5 %**
(Intercept)	2.25	0.51	4.42	0.0017	1.10	3.40
Conc (slope)	2.62	0.09	30.45	< 0.0001	2.42	2.81

plot is fairly symmetric around zero with observation 3 clearly having the largest residual. The quantile-quantile (Q-Q) plot gives vertically the standardized residuals and horizontally their quantiles from the normal distribution. All residuals are close to the reference line except observation 3. In the scale by location plot the fitted values are given horizontally and the square root of the size of the standardized residuals vertically. There is no indication that the residuals increase with increasing sizes of the fitted values. The residuals versus leverage plot also has the contour lines of the Cook's distances at 0.5 and 1.0.

By decreasing the error variance σ^2 or increasing the number of observations n the precision of the estimates would increase.

7.5.2 Polynomial regression

In the special case where powers of x are taken as predictors in the linear model, we obtain the polynomial model, for example,

$$Y = \beta_0 + \beta_1 x + \beta_2 x^2 + \beta_3 x^3 + \varepsilon,\ \varepsilon \sim N(0, \sigma^2). \tag{7.15}$$

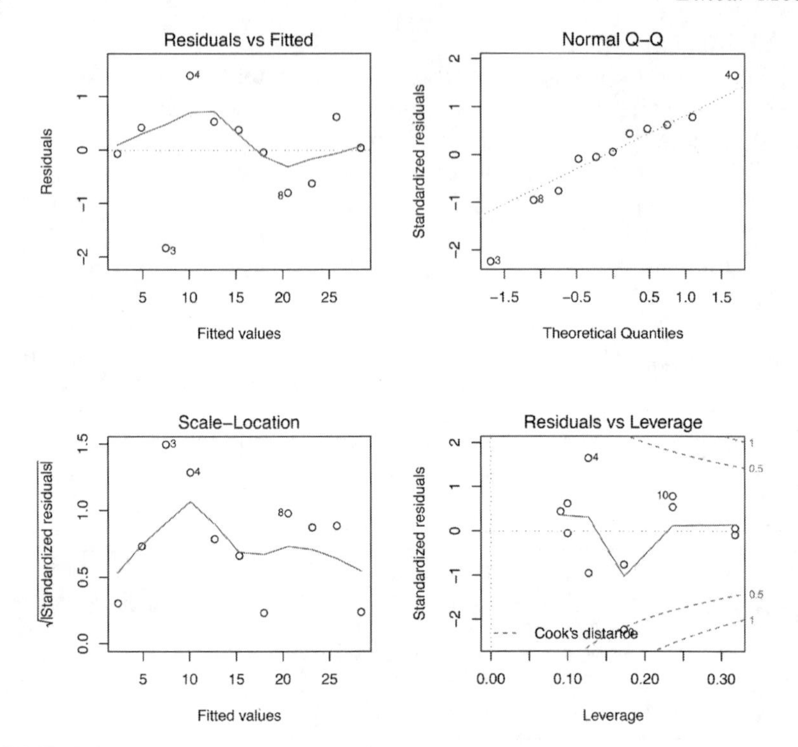

FIGURE 7.3
Diagnostic plots for linear regression of concentrations on fluorescence intensities, with some observation numbers.

Taking $\mathbf{x}_i^T = (1, x_i, x_i^2, x_i^3)$ to be row i of the n by p matrix $\mathbf{X}$, we can directly apply linear regression techniques described in above for the estimation of the parameters. The parameters to be estimated are the regression coefficients in order to find the best fitting cubic polynomial to the data.

The above F-test may also be used to test a reduced model (RM) against a full model (FM) according to

$$F = \frac{(\mathrm{SS}_{Res}(RM) - \mathrm{SS}_{Res}(FM))\,/r}{\mathrm{SS}_{Res}(FM)/(n-p)} \sim F_{r,n-p},$$

where r is the difference in the number of parameters in the models. Then we reject H_0 if p-value $= P(F_{p_1-p_0,N-p_1} > F) < \alpha$. An application of the test will be illustrated by the following example.

Example 7.9 *The data in Table 7.1 on the impact of temperature on viscosity of toluene-tetralin blends was re-analyzed by quadratic polynomial regression. The analysis of variance Table 7.8 clearly rejects the simple regression model.*

The estimated regression coefficients for the best fitting quadratic curve are given in Table 7.9. Each of the estimated coefficients is large relative to

TABLE 7.8
Analysis of variance table for testing the simple linear Model 1 against the quadratic polynomial Model 2 for the data on impact of temperature on viscosity of toluene-tetralin blends.

	Res.Df	RSS	Df	Sum of Sq	F	Pr(>F)
1	6	0.0135				
2	5	0.0003	1	0.0132	263.0795	0.0000

TABLE 7.9
Results for polynomial regression analysis of temperature (x) on viscosity of toluene-tetralin blends (y) giving the estimates $\hat{\beta}_j$, their standard error $SE(\hat{\beta}_j)$, corresponding t- and p-values.

	Estimate	Std. Error	t value	Pr(>|t|)
(Intercept)	1.5527	0.0181	85.64	0.0000
Temperature x	−0.0193	0.0007	−29.23	0.0000
Temperature^2 I(x^2)	0.0001	0.0000	16.22	0.0000

its standard error with p-values much smaller than the significance level. The resulting quadratic curve is given in Figure 7.4.

The residual standard error 0.007096 was considerable smaller than that obtained by simple regression, indicating that the latter overestimated the error in the outcome viscosity. The multiple R-squared 0.9993 as well as the Adjusted R-squared 0.999 are larger and indicate a better fit than those from the simple linear regression. The predicted value for temperature 50 degrees Celsius is 0.80 with 95 percent prediction interval (0.79, 0.83).

7.6 Notes and Comments

The least squares estimator of the linear model equals the maximum likelihood estimator is case the error terms are a random sample from the normal distribution. Estimation by maximum likelihood is attractive as it is asymptotically unbiased, consistent in tending in probability to the true parameter, and efficient in having the smallest possible standard errors (Ferguson, 1996). The Gauss-Markov theorem (Rao, 2009) tells us that the least squares estimator is best over all unbiased estimators of the linear model under the weak assumption that the error terms have a finite second moment (Jureckova and Picek, 2005), e.g., finite variance. The F-test for testing a sub-model is optimal if the data are normally distributed (Lehmann and Romano, 2005).

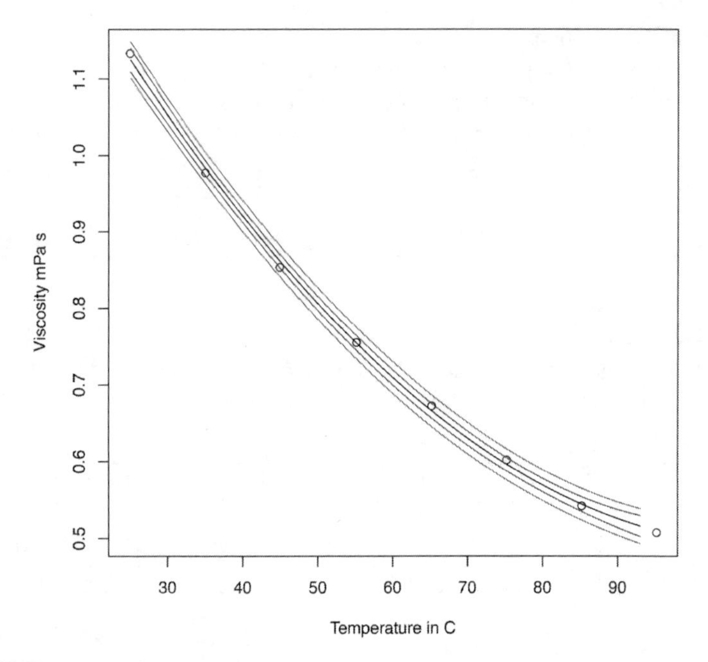

FIGURE 7.4
Temperature by viscosity of toluene-tetralin blends data with best fitting quadratic polynomial with inner 95% confidence lines and outer 95% prediction lines.

Some assumptions on the model matrix X are needed for nice asymptotic properties of the least squares estimator. If the smallest eigenvalue of $\mathbf{X}^T\mathbf{X}$ goes to infinity with n, then each of the C_{jj} will decrease to zero. This holds true for any reasonable model matrix $\mathbf{X}$. In that case, $\hat{\boldsymbol{\beta}} \xrightarrow{P} \boldsymbol{\beta}$ and consequently $\hat{Y}_i \xrightarrow{P} x_i^T\boldsymbol{\beta}$ for all i, and $\hat{\sigma}^2 \xrightarrow{P} \sigma^2$. If the error terms are only assumed to be independent with finite variance, then the condition that $\max_{1\leq i\leq n} \widehat{h}_{ii} \to 0$, as $n \to \infty$ is necessary and sufficient for the least squares estimator to be asymptotically normal (Huber, 1981). This emphasizes once more that small leverage values to be of importance, also for practical purposes.

The idea is to penalize the danger of over-fitting by the $2p$ term in the AIC approach follows from a deeper analysis of the estimation bias (Konishi and Kitagawa, 2008). The minimum AIC (Konishi and Kitagawa, 2008) procedure of model selection is popular, because it is approximately equivalent to minimizing of the final prediction error, as well as to cross-validation, and it is asymptotically efficient (Claeskens and Hjort, 2008). Minimum AIC seems the natural choice for chemical applications where prediction error is of main importance. If parameter selection is the main purpose of the analysis in a situation with a large number of predictive variables, then a SCAD type of analysis is recommended e.g., for its learning properties.

It is recommended to study additional diagnostics or influence measures if there is any doubt on drawing valid conclusions (Belsley et al., 2005; Fox, 1997; Cook and Weisberg, 1982; Williams, 1987; Chatterjee and Hadi, 1988).

The idea of polynomial regression up to order three (cubic) is one of the key ideas behind local regression splines, which are very flexible in fitting non-linear models in cases there is no specific theory available suggesting a class of functions behind the data generation process.

The above estimation methods yield an (asymptotic) variance matrix of the parameters. If this is combined with the delta method, then the standard error and a 95% confidence interval for a function of the true parameter vector can be computed. An example of such an approach is given in the next chapter.

7.7 Notes on Using R

Linear regression can easily be applied to chemical data by the `lm` function, which produces an extensive list on which further functions such as `summary`, `anova`, `confint`, and `influence.measures` are defined. Variance inflation factors are computed by `vif`. The `car` library has the `outlierTest` on the externally studentized residuals and a visualization technique `dfbetaPlots` for outliers. For simple regression `abline` conveniently adds the estimated line to a plot of the data. Model selection according to AIC is conducted by the `stepAIC` function from the `MASS` library and according to SCAD is available from the `grpreg` library.

7.8 Exercises

1. **Transform to linear**. Determine whether the following models can be transformed to a linear model.
   (a) $Y^* = \theta_1 e^{\theta_2 + \theta_3 \cdot x} \cdot \varepsilon^*$
   (b) $Y^* = \theta_1 + \theta_2 \cdot e^{\theta_3 \cdot x} + \varepsilon^*$
   (c) $Y^* = \theta_1 + \theta_2 / \theta_3 \cdot x + \varepsilon^*$
2. **Projection matrices**. The matrices $\mathbf{H} = \mathbf{X}(\mathbf{X}^T\mathbf{X})^{-1}\mathbf{X}^T$ in Equation (7.11) and $\mathbf{I} - \mathbf{H}$ are projection matrices with special properties.
   (a) Show that $\mathbf{H} = \mathbf{H}^2$, $\mathbf{H}\mathbf{X} = \mathbf{X}$, and $(\mathbf{I} - \mathbf{H})\mathbf{X} = \mathbf{0}$. Note that the latter property was used several times in the above.

(b) Use these properties to show that for all vectors $\mathbf{X}\boldsymbol{\beta}$ we have that

$$\|\mathbf{y} - \widehat{\mathbf{y}}\|^2 \leq \|\mathbf{y} - \mathbf{X}\boldsymbol{\beta}\|^2,$$

where $\widehat{\mathbf{y}} = \mathbf{X}\widehat{\boldsymbol{\beta}}$.

(c) Show for the diagonal elements h_{ii} of $\mathbf{H}$ that $h_{ii} \in [0, 1]$ for all $i = 1, \cdots, n$.

3. **Programming linear regression**. In the determination of lead in aqueous solution by electrochemical atomic absorption spectrometry with graphite-probe atomization (Giri et al., 1983), the following measurements were obtained. The data given are in Table 7.10 and can be read from the file `LeadConcentration.txt`. Use absorbance as outcome and concentration as predictive variable.

   (a) Construct the matrix $\mathbf{X} = [\mathbf{1}, \mathbf{x}]$ and compute the estimators $\hat{\boldsymbol{\beta}}$ according to Equation (7.7) and $\hat{\sigma}^2$ Equation (7.10). Compare with R its `lm` function.
   Hint: Use the R function `%*%` for matrix multiplication.

   (b) Compute the Standard Errors as the square root of the diagonal elements from to $\hat{\sigma}^2(\mathbf{X}^T\mathbf{X})^{-1}$ and compute the corresponding t and p-values. Compare your answer with that from summary of the `lm` function.
   Hint: Use the R function `x/y` for element-wise division.

   (c) Compute the hat values from the projection matrix in Equation (7.11), the studentized residuals in Equation (7.13), and Cook's distances from Equation (7.14). Use the R functions `hatvalues`, `rstandard`, and `cooks.distance` to check your answer.
   Hint: Use the R function `x*y` for element-wise multiplication.

4. **Lead concentration**. In the determination of lead in aqueous solution by electrochemical atomic absorption spectrometry with graphite-probe atomization (Giri et al., 1983), the following measurements were obtained. The data given are in Table 7.10 and can be read from the file `LeadConcentration.txt`.

   (a) Fit the linear model to the data and evaluate the results in terms of the size of the coefficients, the percentage of variance explained, and confidence intervals.

   (b) Are there influential observations? Would you advice to obtain more observations?

TABLE 7.10
Lead concentration and absorption.

Measurement	1	2	3	4	5	6
Lead concentration ng ml^{-1}	10	25	50	100	200	300
Absorbance	0.05	0.17	0.32	0.60	1.07	1.40

5. **Ibuprofen**. In a calibration study concentration by area measurements were collected for Ibuprofen. The ibuprofen data are available from the R library `quantchem` and from `ibuprofen.txt`.

   (a) Fit the simple linear regression model to the data and construct confidence intervals for the intercept and slope. Does the model fit well?

   (b) Investigate the diagnostics of the model and comment on your findings. It is useful to plot the data with the regression line.

6. **Wine quality**. From a testing experiment 38 observations became available on the quality of Pinot Noir wine. The data are available from the R library `MPV` under the name `table.b11`; see also `WineQuality.csv`.

   (a) Fit the outcome Quality linearly by the explanatory variables Clarity, Aroma, Body, Flavor, Oakiness and Region.
   Hint: Consider a factor for Region.

   (b) Find the best model according to the minimum AIC criterion and report its fit, diagnostics and give an interpretation of the estimated parameters.

7. **Tube-flow reactor**. The data `table.b6` from the R library `MPV` consists of 28 observations on a tube-flow reactor, where y is the $Nb0Cl_3$ concentration (g-mol/l), x1 the $COCl_2$ concentration (g-mol/l), x2 the Space time (s), x3 the Molar density (g-mol/l), and x4 the Mole fraction CO_2 (Graham and Stevenson, 1972).

   (a) Estimate the linear model from the data and give residual standard error $\hat{\sigma}^2$ and the fit of the model.

   (b) Check the diagnostics of the model by a plot and by testing for any outliers on the studentized residuals.

   (c) Compare the confidence interval from the linear model with that from the bootstrap.

   (d) Give an interpretation of the estimated parameters.

8. **Clausius-Clapeyron equation**. Eleven observations were collected on the vapor pressure (mm Hg) of water for various temperatures (K). From physical chemistry the Clausius-Clapeyron equation states that $\log(p_v) \propto 1/T$. The data are available from the R library `MPV` under the name `p5.2`, or as `ClausiusClapeyron.csv`.

   (a) Transform the data and plot these.

   (b) Fit a linear model and add the line to the plot.

   (c) Report the residual standard error $\hat{\sigma}$.

9. **Cement**. In an experiment on the heat evolved in the setting of each of 13 cements. The outcome y is heat (calories per gram of cement) with explanatory variables consisting of percentages of weight

of the constituents tricalcium aluminate x1, tricalcium silicate x2, tetracalcium alumumino ferrite x3 and dicalcium silicate x4 (Woods et al., 1932). The cement data are available from the MPV R library or from `cement.csv`. The data set is famous for its multicollinearity.

(a) Fit the regression model with the four explanatory variables and compute the variance inflation factors. Test the null hypothesis that the coefficients are zero.

(b) Use minimum AIC to fit a more parsimonious model and compute the variance inflation factors.
Remark: This is make the reader alert on the idea that for small data sets the attractive properties of AIC derived by asymptotic derivations (Konishi and Kitagawa, 2008) may not always to satisfactory modeling.

(c) Use x1 and x2 in a linear regression model, compare the variance inflation factors as well as the coefficient of determination with the previous models. Test the null hypothesis that the coefficients are zero.

10. **Corrosion**. In an experiment 13 specimen of 90/10 Cu-Ni alloys with varying percentages of iron content were submerged in seawater for 60 days. The weight loss due to corrosion was recorded in milligrams per square decimeter per day (Draper and Smith, 1998). The data corrosion available from the R library `faraway` and as `corrosion.csv`.

    (a) Fit the simple linear regression model to the data and test the null-hypothesis that the coefficients are zero.

    (b) Fit a degree 5 polynomial to the data and test this model against the linear model.

    (c) Plot the data with its best fitting degree 5 polynomial and reflect upon the scientific status of the model.

11. **Toluene**. In a calibration study Toluene amounts by GC/MS were measured (Rocke and Lorenzato, 1995). The `r195_toluene` data are available from the R library `chemCal` and from `r195_toluene.txt`.

    (a) Fit the simple linear regression model to the data and evaluate the estimated parameters.

    (b) Evaluate the diagnostics by plotting and testing. Comment on your findings.

    (c) Compare the bootstrap 95% confidence intervals with those from part (a). Construct a QQ plot for the empirical distribution of the slope. What do you conclude?

12. **Ash.** The dependent variable Softening Temperature (SOT) of ash is modeled by the elemental composition of the ash data by the

mass concentrations of the elements P2O5, SiO2, Fe2O3, Al2O3, CaO, MgO, Na2O. The data consist of 99 ash samples, originating from different biomass, with the comprise of an experimental SOT (630-1410 centigrades) (Varmuza and Filzmoser, 2016). The ash data are available from the R library `chemometrics` and from `ash.txt`. Note that it is not possible to add the element K2O to the model due to near singularity.

(a) Fit the outcome SOT in the linear model with the log transformed masses of the elements as explanatory variables.
Hint: R, and other software systems, are confused by the functional type of naming of the variables. Renaming the variables is recommended.

(b) Use the minimum AIC criterion for model selection.

(c) Construct a SCAD learning plot using the generalized cross validation for lambda.

8

Non-linear Models

A non-trivial fixed ODE with initial value condition has a unique non-linear function as its solution. An ODE with a number of free parameters may have a family of non-linear functions as its solution. In experimental settings the exact values of the parameters may be unknown to us, but often it is possible to obtain measurements of various concentrations across time. From these the best fitting maximum likelihood or least squares parameters can be estimated together with their uncertainty in terms of standard errors in order to make valid statistical inferences about the chemical process.

When some response y linearly depends on some concentration x, a simple linear regression model may fit the data well, as described in Section 7.5.1. In chemical practice we may want to use the estimated linear model as a calibration curve to determine the concentration for a newly measured response. This is known as an inverse regression problem since the dependent outcome y is given, and the independent predictive value x is to be estimated.

Thirdly, we have some idea about a class of fitting functions, but beforehand no explicit ODE is available to describe the process. In such a situation we may wish to use a smooth spline approach in order to fit the data and use it to predict future observations of the yield given the values on the predictive variables.

In case the outcome is a count or some binary response, we need a generalization of the linear model in order to find how the response depends on the predictor variables.

In this chapter you learn:

- In Section 8.1 how non-linear dependencies between x and y can be formulated in terms of non-linear models.
- In Section 8.2 how the parameters of non-linear model can be estimated the Gauss-Newton method on the basis of data from a chemical experiment.
- In Section 8.3 how to solve the inverse regression problem where an estimated line or curve is given together with a new value of y in order to estimate the responsible predictor value x.
- In Section 8.5 what cubic splines are and how these can be used to fit any unknown differentiable function underlying the data generation process.

DOI: 10.1201/9781003178194-8

- In Section 8.4 how to link a non-normal response to a set of predictor variables, using generalized linear models.

8.1 Some Non-linear Functions Modeling Chemical Processes

In terms of general parameters an ODE may have a family of non-linear functions as its solution. For example, in Chapter 4 we found that the ODE $y' = (\beta_2/x - \beta_3)y$ with initial condition $y(0) = 0$ has solution $y(x) = \beta_1 x^{\beta_2} e^{-\beta_3 x}$. The latter represents a class of concentration curves in time x

$$f(x, \beta_1, \beta_2, \beta_3) = \beta_1 x^{\beta_2} e^{-\beta_3 x}$$

over the positive parameters $\beta_1, \beta_2, \beta_3$. Each member of the class starts at zero, increases to a unique maximum, and than decreases to zero as time x tends to infinity.

The kinematic viscosity of a certain solvent system decreases exponentially with temperature x (Corradini et al., 1997), suggests the class of functions

$$f(x, \beta_1, \beta_2) = \beta_2 e^{-\beta_1 x}$$

defined over all positive parameters β_1 and β_2.

In a NIST study (Chwirut, 1979) an ultrasonic response decreases monotonically with metal distance x according to

$$f(x, \beta_1, \beta_2, \beta_3) = \frac{e^{-\beta_1 x}}{\beta_2 + \beta_3 x}.$$

The number of defects may increase with excited energy state x (Roszman, 1979) according to a function from the class

$$f(\beta_1, \beta_2, \beta_3, x) = \beta_1 - \frac{1}{\pi} \arctan\left(\frac{\beta_2}{x_i - \beta_3}\right),$$

with the parameters $\beta_1, \beta_2, \beta_3$.

Note that by collecting the parameters into a vector $(\beta_1, \cdots, \beta_p)^T = \boldsymbol{\beta}$ the function classes can be written as $f(\boldsymbol{\beta}, x)$. In many chemical research settings the class of functions is known, but the exact values of the parameters not. In order to estimate these an experiment is conducted yielding the outcome measurements y_i for each x_i, where $i = 1, \cdots, n$. It is often reasonable to assume that the measurements can be represented by a non-linear model with independent normally distributed measurement errors. That is, we have the model

$$Y_i = f(\boldsymbol{\beta}, x_i) + \varepsilon_i, \ \ \{\varepsilon_1, \cdots, \varepsilon_n\} \overset{IID}{\sim} N(0, \sigma^2)$$

where the measurement error terms $\varepsilon_1, \cdots, \varepsilon_n$ are independently normally distributed with mean zero and size given by the variance σ^2. The model decomposes the yield Y into a structural part $E[Y|x] = f(\boldsymbol{\beta}, x)$ and a random error part ε. In other words, the n measurements are $Y_i \sim N(f(\boldsymbol{\beta}, x_i), \sigma^2)$, where $i = 1, \cdots, n$. More succinctly we may write the nonlinear model as $\mathbf{y} \sim N(\mathbf{f}(\mathbf{x}, \boldsymbol{\beta}), \sigma^2\mathbf{I})$.

8.2 Non-linear Regression

Given the data (y_i, x_i), for $i = 1, \cdots, n$ from a chemical experiment, we need to estimate the parameters $\boldsymbol{\beta}$ and σ^2. Since the Y_i are independently distributed as $N(f(\boldsymbol{\beta}, x_i), \sigma^2)$, we have the density

$$p(\mathbf{y}, \boldsymbol{\beta}, \sigma^2) = (2\pi\sigma^2)^{-n/2} \cdot \exp\left(-\frac{1}{2}\sum_{i=1}^{n}\frac{[y_i - f(\boldsymbol{\beta}, x_i)]^2}{\sigma^2}\right).$$

Hence, the log likelihood becomes

$$\ell(\boldsymbol{\beta}, \sigma^2, \mathbf{y}) = -\frac{n}{2}\log(\sigma^2) - \frac{1}{2\sigma^2}\sum_{i=1}^{n}[y_i - f(\boldsymbol{\beta}, x_i)]^2.$$

Similar to linear models, the maximum likelihood estimator of the $\boldsymbol{\beta}$ parameters can be obtained by minimizing the squared difference between the outcomes Y_i and the function values $f(\boldsymbol{\beta}, x_i) = f_i(\boldsymbol{\beta})$. That is, the maximum likelihood estimator of the $\boldsymbol{\beta}$ parameters is equivalent to the least squares estimator. Given the estimator for $\boldsymbol{\beta}$, the error variance can be estimated from the residual sum of squares.

8.2.1 Non-linear least squares parameter estimation

Let's define the residual sum of squares function S with the defined as the sum of the squared differences between $f_i(\boldsymbol{\beta})$ and the data y_i

$$S(\boldsymbol{\beta}) = \sum_{i=1}^{n}(Y_i - f_i(\boldsymbol{\beta}))^2 = \|\mathbf{y} - \mathbf{f}(\boldsymbol{\beta})\|^2.$$

In order to develop a Gauss-Newton algorithm to minimize S over $\boldsymbol{\beta}$ we need a starting vector $\boldsymbol{\beta}^{(m)}$, for $m = 1$, which is to be updated to a vector $\boldsymbol{\beta}^{(m+1)}$. For this we use a first-order Taylor approximation

$$\mathbf{f}(\boldsymbol{\beta}) \approx \mathbf{f}(\boldsymbol{\beta}^{(m)}) + \mathbf{F}^{(m)}(\boldsymbol{\beta} - \boldsymbol{\beta}^{(m)}), \text{ where } \left.\frac{\partial\mathbf{f}}{\partial\boldsymbol{\beta}^T}\right|_{\boldsymbol{\beta}=\boldsymbol{\beta}^{(m)}} = \mathbf{F}^{(m)}.$$

The n by p Jacobian matrix $\mathbf{F}^{(m)}$, with the first order derivative of each row of $\left.\frac{\partial f_i}{\partial \boldsymbol{\beta}^T}\right|_{\boldsymbol{\beta}=\boldsymbol{\beta}^{(m)}}$ put one beneath the other, and is supposed to have full column rank. Such is reasonable since we would reduce the number of parameters without loss of generality is case of deficient rank. Then the residual vector equals

$$\begin{aligned}\mathbf{r}(\boldsymbol{\beta}) &= \mathbf{y} - \mathbf{f}(\boldsymbol{\beta}) \\ &\approx \mathbf{y} - \mathbf{f}(\boldsymbol{\beta}^{(m)}) - \mathbf{F}^{(m)}(\boldsymbol{\beta} - \boldsymbol{\beta}^{(m)}) \\ &= \mathbf{r}(\boldsymbol{\beta}^{(m)}) - \mathbf{F}^{(m)}(\boldsymbol{\beta} - \boldsymbol{\beta}^{(m)}).\end{aligned}$$

Using this into the residual sum of squares gives

$$S(\boldsymbol{\beta}) = \|\mathbf{r}(\boldsymbol{\beta})\|^2 \approx \|\mathbf{r}(\boldsymbol{\beta}^{(m)}) - \mathbf{F}^{(m)}(\boldsymbol{\beta} - \boldsymbol{\beta}^{(m)})\|^2.$$

From the linear regression Equation (7.7) we know that the right-hand side is minimal if we take

$$\boldsymbol{\beta} - \boldsymbol{\beta}^{(m)} = (\mathbf{F}^{(m)T}\mathbf{F}^{(m)})^{-1}\mathbf{F}^{(m)T}\mathbf{r}(\boldsymbol{\beta}^{(m)}) \equiv \boldsymbol{\delta}^{(m)}.$$

By simply solving for $\boldsymbol{\beta}$ we find the Gauss-Newton update

$$\boldsymbol{\beta}^{(m+1)} = \boldsymbol{\beta}^{(m)} + \boldsymbol{\delta}^{(m)}.$$

Repeating this, after starting with $m = 1$, generates a sequence of updating vectors $\{\boldsymbol{\beta}^{(1)}, \boldsymbol{\beta}^{(2)}, \boldsymbol{\beta}^{(3)}, \cdots\}$, which we would like to convergence in the sense of

$$\{\boldsymbol{\beta}^{(m)}\} \to \widehat{\boldsymbol{\beta}} = \underset{\boldsymbol{\beta}}{\operatorname{argmin}}\, S(\boldsymbol{\beta}), \quad \text{as } m \to \infty.$$

Although the latter is not completely certain, we note that convergence of $\boldsymbol{\beta}^{(m)}$ occurs if and only if $\boldsymbol{\delta}^{(m)} \to \mathbf{0}$. From this we have

$$\mathbf{F}^{(m)T}\mathbf{r}(\boldsymbol{\beta}^{(m)}) \to \widehat{\mathbf{F}}^T\mathbf{r}(\widehat{\boldsymbol{\beta}}) = \mathbf{0}, \quad \text{where } \widehat{\mathbf{F}} = \left.\frac{\partial \mathbf{f}}{\partial \boldsymbol{\beta}^T}\right|_{\boldsymbol{\beta}=\widehat{\boldsymbol{\beta}}}.$$

The latter is equivalent to the necessary condition for a stationary point, since setting the first order derivative of S to zero gives

$$\begin{aligned}\mathbf{0} &= \frac{\partial}{\partial \boldsymbol{\beta}^T}\|\mathbf{y} - \mathbf{f}(\boldsymbol{\beta})\|^2 \\ &= \frac{\partial}{\partial \boldsymbol{\beta}^T}\mathbf{y}^T\mathbf{y} - 2\mathbf{y}^T\mathbf{f}(\boldsymbol{\beta}) + \mathbf{f}(\boldsymbol{\beta})^T\mathbf{f}(\boldsymbol{\beta}) \\ &= -2\mathbf{y}^T\frac{\partial}{\partial \boldsymbol{\beta}^T}\mathbf{f}(\boldsymbol{\beta}) + 2\mathbf{f}(\boldsymbol{\beta})^T\frac{\partial}{\partial \boldsymbol{\beta}^T}\mathbf{f}(\boldsymbol{\beta}) \\ &= -2(\mathbf{y} - \mathbf{f}(\boldsymbol{\beta}))^T\mathbf{F} \\ &= -2\mathbf{r}(\boldsymbol{\beta})^T\mathbf{F}.\end{aligned}$$

This shows that Gauss-Newton iterators are very close to satisfying the necessary conditions of the optimization problem.

In order to define testing procedures for the estimated parameters we note that, under weak conditions such as the uniqueness of the global minimum of the residual sum of squares function S, it holds that the estimator tends in probability to the true parameter value $\boldsymbol{\beta}^*$, that is $\widehat{\boldsymbol{\beta}} \xrightarrow{P} \boldsymbol{\beta}^*$, as $n \to \infty$.

To develop statistical properties of the estimator we formulate the model in matrix terms as

$$\mathbf{y} = \mathbf{f}(\boldsymbol{\beta}^*) + \boldsymbol{\varepsilon}, \quad \text{where} \quad \boldsymbol{\varepsilon} \sim N(\mathbf{0}, \sigma^2\mathbf{I}).$$

Under the weak assumption that the matrix $\mathbf{F}$ is a continuous function of $\boldsymbol{\beta}$, it follows that

$$\widehat{\mathbf{F}} = \mathbf{F}(\widehat{\boldsymbol{\beta}}) = \left.\frac{\partial \mathbf{f}}{\partial \boldsymbol{\beta}^T}\right|_{\boldsymbol{\beta}=\widehat{\boldsymbol{\beta}}} \xrightarrow{P} \left.\frac{\partial \mathbf{f}}{\partial \boldsymbol{\beta}^T}\right|_{\boldsymbol{\beta}=\boldsymbol{\beta}^*} = \mathbf{F}_*$$

From reasoning, similar to the above, we now have from a first-order Taylor approximation that

$$\mathbf{f}(\widehat{\boldsymbol{\beta}}) \approx \mathbf{f}(\boldsymbol{\beta}^*) + \mathbf{F}_*(\widehat{\boldsymbol{\beta}} - \boldsymbol{\beta}^*),$$

or that the difference $\mathbf{f}(\widehat{\boldsymbol{\beta}}) - \mathbf{f}(\boldsymbol{\beta}^*)$ is approximately in the columnspace of $\mathbf{F}_*$. Using this into the residual sum of squares function

$$S(\widehat{\boldsymbol{\beta}}) = \|\mathbf{y} - \mathbf{f}(\boldsymbol{\beta}^*) + \mathbf{f}(\boldsymbol{\beta}^*) - \mathbf{f}(\widehat{\boldsymbol{\beta}})\|^2 \approx \|\boldsymbol{\varepsilon} - \mathbf{F}_*(\widehat{\boldsymbol{\beta}} - \boldsymbol{\beta}_*)\|^2.$$

From this we find that

$$\widehat{\boldsymbol{\beta}} - \boldsymbol{\beta}_* = (\mathbf{F}_*^T\mathbf{F}_*)^{-1}\mathbf{F}_*^T\boldsymbol{\varepsilon} \sim N(0, \sigma^2(\mathbf{F}_*^T\mathbf{F}_*)^{-1}),$$

since the left-hand side is a linear combination of the normally distributed error random variable $\boldsymbol{\varepsilon}$. The variance of the error term is estimated by

$$\hat{\sigma}^2 = \frac{1}{n-p}S(\widehat{\boldsymbol{\beta}}) = \frac{1}{n-p}\|\mathbf{y} - \mathbf{f}(\widehat{\boldsymbol{\beta}})\|^2.$$

In practical settings the matrix $\mathbf{F}_*$ is unknown. For this reason, it is approximated by $\widehat{\mathbf{F}}$ so that for practical purposes we use

$$\widehat{\boldsymbol{\beta}} \sim N(\boldsymbol{\beta}_*, \hat{\sigma}^2(\widehat{\mathbf{F}}^T\widehat{\mathbf{F}})^{-1}).$$

On this the statistical testing of chemical parameters is based as well as the definition of confidence intervals for the parameters.

Example 8.1 *In an experiment (Corradini et al., 1997) the kinematic viscosity of a certain solvent system was investigated dependent on temperature and four ratios of 2-methoxyethanol (ME) versus 1,2-dimethoxyethane (DME). The data are plotted in different colors depending on the ratios for kinematic*

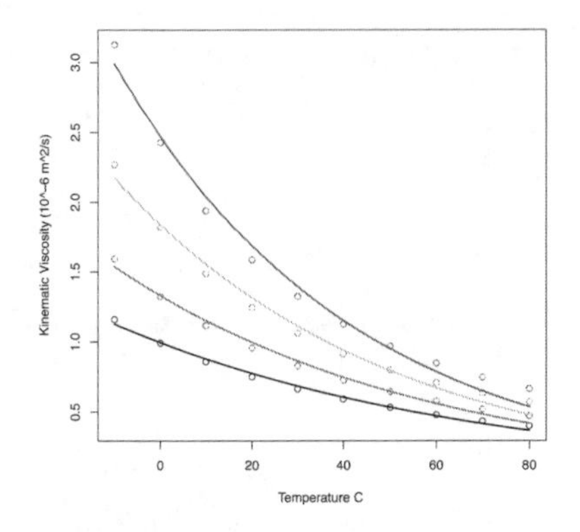

FIGURE 8.1
Kinematic viscosity $10^{-6}m^2/s$ by temperature ^{o}C horizontally for different ME:DME ratios.

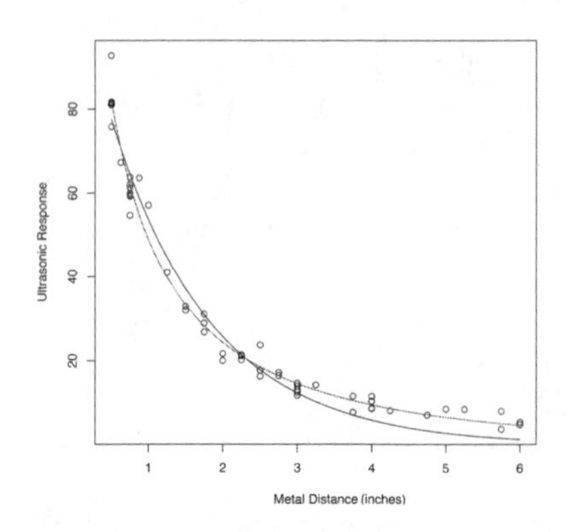

FIGURE 8.2
Two nonlinear models fitting the ultrasonic calibration data; f_1 for Model 1 and the better fitting f_2 for Model 2.

TABLE 8.1
Parameter estimates $\hat{\beta}_1$ and $\hat{\beta}_2$ with 95% confidence intervals from the exponential models of the kinematic viscosity $10^{-6}m^2/s$ values by temperature ^{o}C data for four different ratios of 2-methoxyethanol (ME) versus 1,2-dimethoxyethane (DME).

Ratio	Coefficient	Estimate	Std. Error	t value	Pr(>|t|)	2.5%	97.5%
0.36	$\hat{\beta}_1$	0.0124	0.0004	31.2694	0.0000	0.0115	0.0133
0.36	$\hat{\beta}_2$	0.9956	0.0118	84.5939	0.0000	0.9683	1.0228
0.56	$\hat{\beta}_1$	0.0144	0.0005	26.4990	0.0000	0.0131	0.0157
0.56	$\hat{\beta}_2$	1.3311	0.0200	66.5968	0.0000	1.2848	1.3775
0.75	$\hat{\beta}_1$	0.0167	0.0007	22.6004	0.0000	0.0150	0.0185
0.75	$\hat{\beta}_2$	1.8394	0.0344	53.4261	0.0000	1.7595	1.9192
0.91	$\hat{\beta}_1$	0.0191	0.0009	21.5470	0.0000	0.0170	0.0212
0.91	$\hat{\beta}_2$	2.4704	0.0506	48.8658	0.0000	2.3531	2.5875

viscosity against temperature in Figure 8.1. Since the relationship seems exponentially decreasing the model

$$y_i = \beta_2 e^{-\beta_1 x_i} + \varepsilon_i,\ \{\varepsilon_1, \cdots, \varepsilon_{10}\} \overset{IID}{\sim} N(0, \sigma^2)$$

was fitted to the data for each ratio, where y_i is the measured kinematic viscosity (0.000001 m^2/s) and x the corresponding temperature in $^o C$. From Table 8.1 it can be observed that the coefficients $\hat{\beta}_2$ at temperature 0 $^o C$ correspond with the different kinematic viscosity levels in Figure 8.1. The fit to the data is very good for the lowest, but less well for the highest in particular for larger values of temperature. From the summary statistics in Table 8.1, it

TABLE 8.2
Analysis of variance table for nonlinear models f_1 and f_2 for the ultrasonic calibration data.

	Res.Df	Res.Sum Sq	Df	Sum Sq	F value	Pr(>F)
Model 1 f_1	52	1011.33				
Model 2 f_2	51	513.05	1	498.28	49.53	< 0.0001

can be observed that the coefficients $\hat{\beta}_1$ that represent the speed of decrease of kinematic viscosity by one degrees of increase of the temperature ^oC do differ according to the four different graphs. From the confidence intervals it can be concluded that the data provides sufficient evidence for an increase of the ratio to cause an increase in size of the coefficients for the curves.

Example 8.2 *In a NIST study (Chwirut, 1979) involving ultrasonic calibration, measurements were obtained on ultrasonic response depending on metal distance. Two models that may be considered for modeling the process are*

$$\text{Model 1: } f_1(\beta_1, \beta_2, x) = \frac{e^{-\beta_1 x}}{\beta_2}, \quad \text{Model 2: } f_2(\beta_1, \beta_2, \beta_3, x) = \frac{e^{-\beta_1 x}}{\beta_2 + \beta_3 x}.$$

Note that the exponential decay function in Model is the solution to the ODE $y' = y$ with initial value condition $y(0) = 110$ in §4.2.2 as a special case. Both models tend to zero as x increases without bound, but for Model 2 this decrease is slower than for Model 1. The best fitting function to the data from Model 1 given in Figure 8.2 clearly improves the Runge-Kutta approximation obtained in Chapter 2. It can be observed that Model 1 is not fitting too well for larger values of metal distance on the right-hand side of the fitted curve. Model 2 was estimated by a Newton type of algorithm using the starting values $b_1 = 0.1, b_2 = 0.01, b_3 = 0.02$, the latter are easy to obtain by educated guessing based upon plotting the graph of a function together with the data points. Visual inspection of Figure 8.2 suggests that Model 2 fits the data better than Model 1 due to its higher level for larger metal distances. Since Model 1 is a sub-model of Model 2 (by setting β_3 to zero), we may use an F-test for testing Model 1 against Model 2. From Table 8.2 it can be observed that the residual sum of squares from Model 2 is about half in size of that from Model 1 and that the F-test rejects Model 1. The data provide strong evidence for rejecting the simpler Model 1. The parameter estimates in Table 8.3 are large relative to their standard error values. Also, the null hypothesis that the β_3 parameter from Model 2 is zero is rejected on the basis of its small p-value. The residual standard error $\hat{\sigma} = 3.172$.

8.2.2 Estimating a function of the parameters

With respect to non-linear models we may be interested in $g(\boldsymbol{\beta})$, the value of a function g of the parameters $\boldsymbol{\beta}$. After fitting a curve f to the data (y_i, x_i), for

TABLE 8.3
Parameter estimates for Model 2 their standard errors, t-values and corresponding p-values.

	Estimate	Std. Error	t value	Pr(>|t|)
$\hat{\beta}_1$	0.1666	0.0383	4.3489	0.0001
$\hat{\beta}_2$	0.0052	0.0007	7.7532	< 0.0001
$\hat{\beta}_3$	0.0122	0.0015	7.9390	< 0.0001

$i = 1, \ldots, n$ from an exepriment, we obtain the estimator $\widehat{\boldsymbol{\beta}}$ and its asymptotic variance matrix $\hat{\sigma}^2(\widehat{\mathbf{F}}^T\widehat{\mathbf{F}})^{-1}$. We may summarize the knowledge about $g(\boldsymbol{\beta})$ from the experiment by a confidence interval. To obtain this we apply the delta theorem, cf Section 5.3, to a differentiable function g of the parameters $\widehat{\boldsymbol{\beta}}$ in the sense that

$$g(\widehat{\boldsymbol{\beta}}) \xrightarrow{D} N\left(g(\boldsymbol{\beta}), \dot{g}(\boldsymbol{\beta})^T \sigma^2(\mathbf{F}^T\mathbf{F})^{-1}\dot{g}(\boldsymbol{\beta})\right). \quad (8.1)$$

From this we obtain its standard error and its 95 % confidence interval.

Example 8.3 *Suppose we are interested in the maximum of the concentration curve* $f(t) = \beta_1 t^{\beta_2}\exp(-\beta_3 \cdot t)$ *in time* t*. After setting the first order derivative to zero and solving we find that the maximum occurs at time point* $t = \beta_2/\beta_3$*, which provides us with a new differentiable function* $g(\boldsymbol{\beta}) = g(\beta_1, \beta_2, \beta_3) = \beta_2/\beta_3$ *of the parameters. In an experiment using naturally enriched* ^{13}C*-proteins the oxidation of protein was investigated by Isotope Ratio Mass Spectrometry producing measurements of change in* $^{13}CO_2 : {}^{12}CO_2$ *ratio over time (Reckman et al., 2019). The* $n = 34$ *measurements were conducted over 330 minutes each consecutive 10 minutes, taking and* $t_1 = 0, \ldots, t_n = 330$*, are given in Figure 8.3 for one of the subjects. The concentration in time model reads as*

$$y_i = \beta_1 t_i^{\beta_2} e^{-\beta_3 \cdot t_i} + \varepsilon_i, \quad \{\varepsilon_1, \ldots, \varepsilon_n\} \overset{IID}{\sim} N(0, \sigma^2).$$

After fitting the concentration curve f *to data* y_i, t_i*, for* $i = 1, \ldots, n$*, we obtain the estimates in Table 8.4 each of which is significantly different from zero. From the estimates we find the time point at which the maximum occurs* $g(\hat{\beta}_1, \hat{\beta}_2, \hat{\beta}_3) = \hat{\beta}_2/\hat{\beta}_3 = 127.51$*. The asymptotic variance matrix* $\hat{\sigma}^2(\widehat{\mathbf{F}}^T\widehat{\mathbf{F}})^{-1})$ *of the estimated parameters is given in Table 8.5. The first order derivative of the function* $\dot{g}(\widehat{\boldsymbol{\beta}})^T = (0, \hat{\beta}_3^{-1}, -\hat{\beta}_2 \cdot \hat{\beta}_3^{-2}) = (0.00000, 81.99426, -10455.13249)$*. Hence, from the Delta method we find that the standard error of the estimated time point of the maximum equals*

$$\sqrt{\dot{g}(\widehat{\boldsymbol{\beta}})^T\hat{\sigma}^2(\widehat{\mathbf{F}}^T\widehat{\mathbf{F}})^{-1})\dot{g}(\widehat{\boldsymbol{\beta}})} = \sqrt{7.715526} = 2.777684.$$

From the latter we find the 95% CI [122.06 ; 132.95].

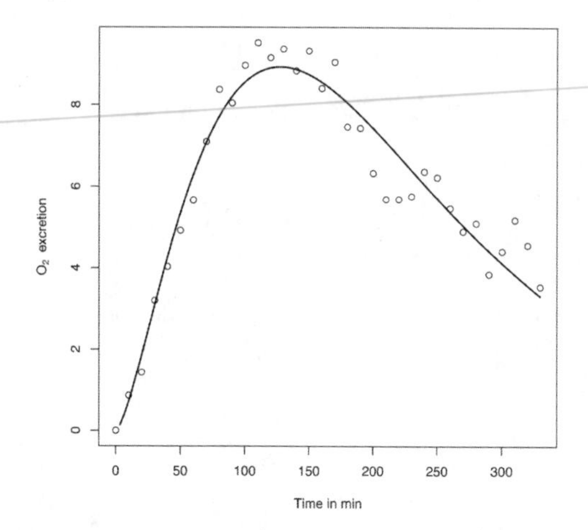

FIGURE 8.3
O2 excretion in time fitted by non-linear least squares concentration function.

TABLE 8.4
Coefficients of estimated concentration curve.

	Est	SE	t-val	p-val
$\hat{\beta}_1$	0.0225	0.0099	2.27	0.02
$\hat{\beta}_2$	1.5551	0.1128	13.78	0.00
$\hat{\beta}_3$	0.0122	0.0008	15.11	0.00

TABLE 8.5
Asymptotic variance matrix of estimates.

	$\hat{\boldsymbol{\beta}}_1$	$\hat{\boldsymbol{\beta}}_2$	$\hat{\boldsymbol{\beta}}_3$
$\hat{\beta}_1$	0.0000973	−0.0011078	−0.0000073
$\hat{\beta}_2$	−0.0011078	0.0127237	0.0000869
$\hat{\beta}_3$	−0.0000073	0.0000869	0.0000006

8.2.3 Using the bootstrap

Similar to linear regression one may want to consider the case where the error is independent and identically distributed, but that its distribution is unknown. Such is often denoted by

$$Y_i = f(\boldsymbol{\beta}, x_i) + \varepsilon_i, \ \ \{\varepsilon_1, \cdots, \varepsilon_n\} \sim IID(0, \sigma^2).$$

Under weak conditions, such as the uniqueness of the minimum of residual sum of squares function S, it holds that the estimator tends in probability to the true parameter value $\boldsymbol{\beta}^*$, that is $\widehat{\boldsymbol{\beta}} \xrightarrow{P} \boldsymbol{\beta}^*$. It follows that the least squares estimator is asymptotically normally distributed

$$\sqrt{n}(\widehat{\boldsymbol{\beta}} - \boldsymbol{\beta}^*) \stackrel{D}{\rightarrow} N(\mathbf{0}, \sigma^2 \boldsymbol{\Omega}^{-1}),$$

with

$$\frac{1}{n}\sum_{i=1}^{n} \frac{\partial f_i(\boldsymbol{\beta}^*)}{\partial \boldsymbol{\beta}} \frac{\partial f_i(\boldsymbol{\beta}^*)}{\partial \boldsymbol{\beta}^T} \stackrel{P}{\rightarrow} \boldsymbol{\Omega},$$

where the derivatives are evaluated at the true parameter point $\boldsymbol{\beta}^*$. In practical settings the unknown $\boldsymbol{\Omega}$ is replaced by

$$\frac{1}{n}\sum_{i=1}^{n} \frac{\partial f_i(\widehat{\boldsymbol{\beta}})}{\partial \boldsymbol{\beta}} \frac{\partial f_i(\widehat{\boldsymbol{\beta}})}{\partial \boldsymbol{\beta}^T} = \widehat{\boldsymbol{\Omega}}.$$

If the first order derivative of the f_i are a continuous function of $\boldsymbol{\beta}$, then $\widehat{\boldsymbol{\Omega}} \stackrel{P}{\rightarrow} \boldsymbol{\Omega}$ and the difference will be small if the sample size n is sufficiently large.

In many chemical estimation problems the estimators are asymptotically normally distributed and the 95% confidence intervals of the parameters are based on it. In case a researcher is somehow in doubt whether the sample is sufficiently large for the preciseness of the distributional approximation, it generally is a good idea to investigate the empirical distribution of the parameter estimates by the **bootstrap**. Such may very well reduce the uncertainty with respect to making valid inferences. In the example below we compare the asymptotic normality based 95% CI with those obtained from the bootstrap.

Example 8.4 *In a NIST study the outcome quantum defects was studied with the excited energy state as the explanatory variable (Roszman, 1979). Assuming that the dependency can be modeled by the arc tangent function, we obtain the model*

$$y_i = \beta_1 - \frac{1}{\pi}\arctan\left(\frac{\beta_2}{x_i - \beta_3}\right) + \varepsilon_i, \; \{\varepsilon_1, \cdots, \varepsilon_{25}\} \stackrel{IID}{\sim} N(0, \sigma^2).$$

The data with the best fitting curve are given in Figure 8.4. That all measurements are fairly close to the arc tangens curve is reflected in a small residual error standard deviation $\widehat{\sigma} = 0.0051$. *The non-parametric bootstrap procedure was used to investigate the degree of similarity between the normal 95% CI for the parameters and the 95% CI from the empirical distribution estimated from 2000 re-samples from the data with replacement. The bootstrap is slightly more general as it does not require the assumption of normality for the errors terms. From the resulting estimates and their 95% CI in Table 8.6 it can be observed that these are very similar. That is, the median from the 2000 bootstrap estimates and estimates from non-linear least squares are equal as well as the boundaries of the corresponding confidence intervals.*

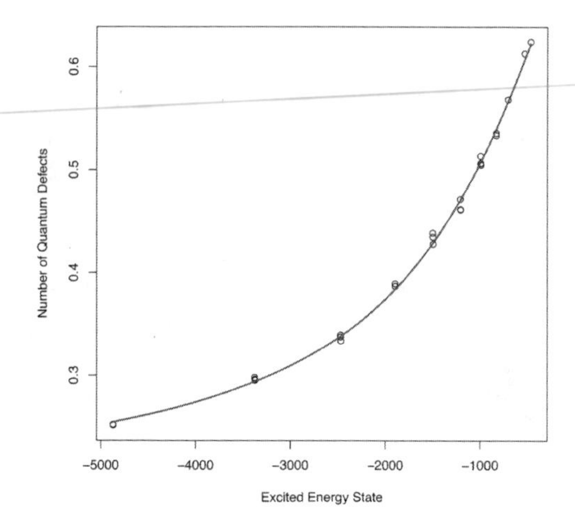

FIGURE 8.4
Number of quantum defects against excited energy state and its best fitting arctangens function.

TABLE 8.6
Parameter estimates from non-linear least squares at the left and from the bootstrap to the right with their 95% CI.

	Estimate	2.5%	97.5%	Median	2.5%	97.5%
$\hat{\beta}_1$	0.166	0.156	0.175	0.166	0.157	0.174
$\hat{\beta}_2$	1317.257	1231.310	1410.225	1316.833	1242.644	1398.098
$\hat{\beta}_3$	−282.992	−312.526	−253.477	−283.758	−310.663	−255.824

8.3 Inverse Regression

By inverse regression we wish to predict the value of x, given a new values of Y. Such frequently occurs when Y is absorbance of light and x the unknown concentration of a chemical substance.

8.3.1 Inverse linear regression

Recall the simple linear model

$$Y_i = \beta_0 + \beta_1 x_i + \varepsilon_i, \quad \{\varepsilon_1, \cdots, \varepsilon_n\} \overset{IID}{\sim} N(0, \sigma^2) \tag{8.2}$$

and suppose that x is the concentration, the amount of which is under the exact control of the chemical engineer, and that Y is absorbance of light, which is subject to some error ε. For a positive β_1 the absorbance increases with the level of the concentration. From the measurements $(x_1, Y_1), \cdots, (x_n, Y_n)$, the least square estimates for the linear model are special cases from the general estimation formula

$$b_1 = \frac{\sum_{i=1}^{n}(x_i - \overline{x})(Y_i - \overline{Y})}{\sum(x_i - \overline{x})^2}, \quad b_0 = \overline{Y} - b_1\overline{x}, \quad \hat{\sigma}^2 = s^2 = \frac{1}{n-2}\sum_{i=1}^{n}(Y_i - \widehat{Y}_i)^2.$$

From these we obtain the fitted straight line $\widehat{Y} = b_0 + b_1 x$, which will serve as a **calibration curve**.

After obtaining a new absorption observation Y_0, we wish to predict its corresponding concentration and to express the uncertainty associated with by an interval. The predicted concentration follows from solving $Y_0 = b_0 + b_1\widehat{X}_0$ with respect to the unknown concentration $\widehat{X}_0$. This yields

$$\widehat{X}_0 = (Y_0 - b_0)/b_1,$$

the so-called **inverse regression** as we calculated the unknown concentration from the calibration curve. The derivation of the limits of the confidence interval is somewhat lengthy, but not difficult, yielding the formula

$$\left.\begin{matrix} X_U \\ X_L \end{matrix}\right\} = \widehat{X}_0 + \frac{(\widehat{X}_0 - \overline{x})g \pm (ts/b_1)\cdot\left\{(\widehat{X}_0 - \overline{x})^2 S_{xx} + (1-g)\left(\frac{1}{n} + \frac{1}{m}\right)\right\}^{1/2}}{1-g},$$

where

$$S_{xx} = \sum_{i=1}^{n}(x_i - \overline{x})^2, \quad g = \frac{t^2 s^2}{b_1^2/S_{xx}}, \quad t = t(n-2, 1-\alpha/2).$$

The method to construct the confidence interval is called fiducial and gives non-symmetric intervals.

Example 8.5 *For various concentrations of Ibuprofen the peak areas in Table 8.7 were measured by chromatography. From the* $n = 14$ *data points we found* $b_0 = 7185.516$, $b_1 = 2436.725$, $S_{xx} = 61754.88$, *and* $s = 2421.491$. *The quantile* $t = t(0.975, df = n-2) = 2.1788$. *A new measurement yielded*

TABLE 8.7
Fourteen measurements of concentrations of ibuprofen and peak areas from chromatography.

Sample nr	Conc	Area	Sample nr	Conc	Area
1	103.9	265053.0	8	103.9	261357.0
2	139.3	345915.0	9	139.3	345669.0
3	180.1	445684.0	10	180.1	445753.0
4	200.3	494700.0	11	200.3	493846.0
5	219.9	540221.0	12	219.9	539610.0
6	278.1	683881.0	13	278.1	683991.0
7	305.7	755890.0	14	305.7	754901.0

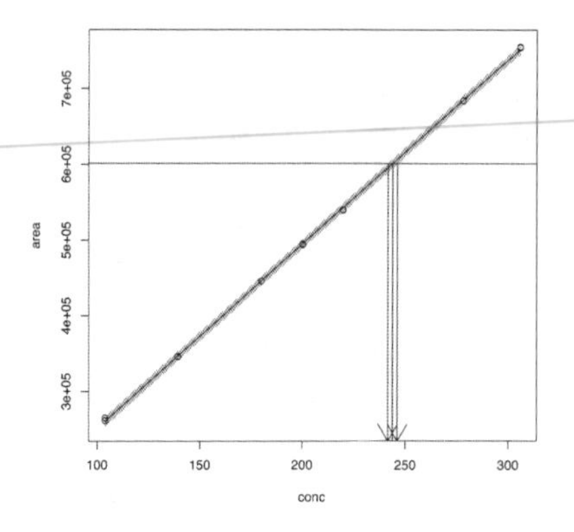

FIGURE 8.5
The graph of the prediction line, the data points and the prediction interval from concentrations of Ibuprofen and the corresponding peak areas from chromatography.

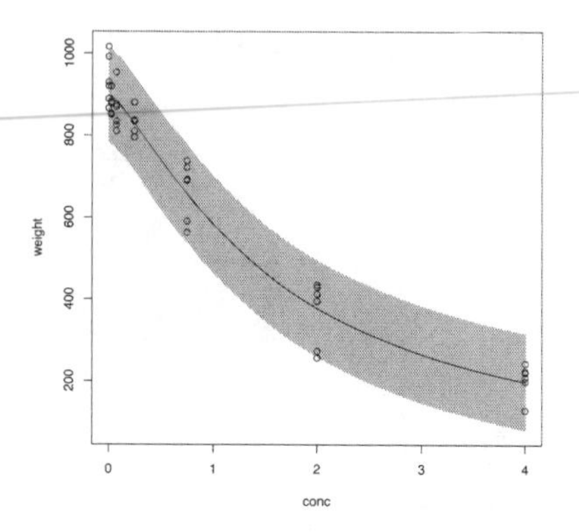

FIGURE 8.6
Weights of plants in (mg) versus concentrations (g/ha) of an agrochemical with fitted curve from non-linear least squares.

peak area $Y_0 = 601376$, *for which* $g = 7.591392 \cdot 10^{-5}$, *and the predicted concentration* $\widehat{X}_0 = 243.85$, *with 95% prediction interval* $(241.58, 246.12)$ *for the concentration. Figure 8.5 visualizes the idea of inversion by the graph and the vertical lines marking the prediction interval horizontally.*

8.3.2 Inverse non-linear regression

Note that inverse linear regression can be generalized to **inverse non-linear regression** if $f(x, \boldsymbol{\beta})$ is a monotonic function of x. In such a case the inverse function exists as a differentiable function of the parameters. If the data $(\mathbf{x}, \mathbf{y})$ are available, then non-linear least squares can be used to obtain the estimate $\widehat{\boldsymbol{\beta}}$ and the corresponding asymptotic variance matrix. If new observations on the outcome $y_{10}, \cdots, y_{n0}$ are obtained, then its mean $\bar{y}_0$ can be computed. The inverse function then solves $\bar{y}_0 = f(x, \widehat{\boldsymbol{\beta}})$ in terms of x which will be denoted by $\hat{x}_0$. Since the latter generally is a differentiable function of $\widehat{\boldsymbol{\beta}}$ and $\bar{y}_0$, the delta theorem can be used to compute the standard error of $\hat{x}_0$, from which a prediction interval can be constructed.

Example 8.6 *A Bioassay on Nasturtium gives weights y (mg) of plants after three weeks of growth depending on several concentrations (g/ha) of an agrochemical in the soil (Racine-Poon, 1988). Figure 8.6 illustrates that the weight decreases monotonically with an increase of concentration. A class of functions to consider is*

$$f(x, \boldsymbol{\beta}) = \frac{\beta_1}{1 + \exp(\beta_2 + \beta_3 \cdot \log(x))} = \frac{\beta_1}{1 + \exp(\beta_2) \cdot x^{\beta_3}},$$

TABLE 8.8
Estimated coefficients from non-linear least squares on Weights of plants in (mg) versus concentrations (g/ha) of an agrochemical.

	Estimate	Std. Error	t value	Pr(>|t|)
$\hat{\beta}_1$	897.86	13.71	65.47	< 0.0001
$\hat{\beta}_2$	−0.61	0.11	−5.75	< 0.0001
$\hat{\beta}_3$	1.35	0.11	12.41	< 0.0001

which indeed monotonically decreases in the concentration x, if β_3 is positive. Note that, $f(0, \boldsymbol{\beta}) = \beta_1$ is well-defined. From the summary statistics in Table 8.8 it can be concluded that the estimates are different from zero with $\hat{\beta}_3$ being positive. If new observations for y become available, then equation $\bar{y} = f(x, \boldsymbol{\beta})$ has the unique solution

$$\hat{x}_0 = \exp\left\{\frac{\log(\hat{\beta}_1/\bar{y} - 1) - \hat{\beta}_2}{\hat{\beta}_3}\right\},$$

a differentiable function of $\widehat{\boldsymbol{\beta}}$ and $\bar{y}$. The prediction and corresponding 95% prediction interval for the newly observed plant weights $(y_{10}, y_{20}, y_{30}) = (753, 804, 821)$ is $\hat{x}_0 = 0.35$ with 95% PI $(0.15; 0.56)$ and for the newly observed plant weights $(309, 296, 419)$ we have $\hat{x}_0 = 2.26$ with 95% PI $(1.77; 2.97)$. This illustrates that when the estimated curve has a larger slope than the prediction interval can be considerably smaller compared to those obtained for a more horizontal slope. The bootstrap gives the 95% prediction interval $(1.78; 2.86)$, which is fairly equal to the previous.

8.4 Generalized Linear Models

Besides the linearity assumption, another drawback of a linear is its reliance on the normal distribution for its response. In many experimental setting this assumption is inappropriate. A number of frequently encountered datatypes in engineering do not satisfy normality: consider for example strictly positive reaction times, discrete valued failure counts or binary outcomes such as success/failure, dead/alive or yes/no.

The two key ideas of generalized linear regression models are that

1. the outcome variable Y is distributed according to an exponential family distribution, and
2. the expected value of the outcome variable $\mu = EY$ is an invertible function of a linear combination of the explanatory variables X.

We suppose that the data $\{Y_1, \cdots, Y_n\}$ are independent and distributed according to a member of the exponential family, such as the Bernoulli, binomial, Poisson, exponential, gamma, normal, etc. The mean value for members of the exponential distribution is some set D. For example $D_{\text{Bernoulli}} = [0, 1]$, $D_{\text{Poisson}} = [0, \infty)$ or $D_{\text{normal}} = \mathbb{R}$. For all of the observations Y_i, we consider an associated set of explanatory variables $x_i \in \mathbb{R}^p$ $(i = 1, \ldots, n)$. As any linear combination of the explanatory variables might fall outside the domain of the mean value parameter D, we consider a monotone function $g : D \longrightarrow \mathbb{R}$, called the *link function*, with the key property that it links the mean value of the response to the linear combination of the explanatory variables,

$$g(\mu_i) = g[E(Y_i)] = \mathbf{x}_i^T \boldsymbol{\beta}. \tag{8.3}$$

Therefore, the expected value of Y_i is the inverse of g evaluated at the linear combination

$$\mu_i = g^{-1}(\mathbf{x}_i^T \boldsymbol{\beta}).$$

This gives, for example, the probability that insects die when given a certain dose of an insecticide, or the average failure count under various settings of some production process. The general aim in a generalized linear regression model is to estimate β given data (y, x) in order to determine what features x drive or are useful to predict the outcome y. In the remainder of this section we describe the computational algorithm that can be used to find the maximum likelihood estimate of β.

8.4.1 Estimation of a generalized linear model

Recall from Chapter 3 that the density of a member of the exponential family can be written as

$$0 \leq f(y, \theta) = \exp\left\{a(y)b(\theta) + c(\theta) + d(y)\right\}.$$

The function $a(y)$ is a sufficient statistic for θ and therefore is a vector of the same dimension of θ. In what follows, we consider only the mean parameter and we can take $a(y) = y$ without loss of generality.

Considering a random sample (Y, x) of size n, the log-likelihood function is given as,

$$\begin{aligned} \ell(\boldsymbol{\beta}) &= \sum_{i=1}^{n} \log f(Y_i, \theta_i) \\ &= \sum_{i=1}^{n} Y_i b(\theta_i) + c(\theta_i) + d(Y_i). \end{aligned} \tag{8.4}$$

The **score** U_j, which is its first order derivative of the log-likelihood function, equals

$$U_j = \frac{\partial}{\partial \beta_j} \ell(\boldsymbol{\beta}) = \sum_{i=1}^{n} \frac{\partial \ell_i}{\partial \theta_i} \frac{\partial \theta_i}{\partial \mu_i} \frac{\partial \mu_i}{\partial \beta_j} \tag{8.5}$$

This expression follows from the chain rule and that ℓ is a function of the canonical parameter θ_i, which is itself a function of the mean μ_i. We also have that

$$E[Y] = \mu_i = -\frac{c'(\theta)}{b'(\theta)} \tag{8.6}$$

and

$$\text{Var}[Y] = \frac{b''(\theta)c'(\theta) - b'(\theta)c''(\theta)}{(b'(\theta))^3}, \tag{8.7}$$

Without loss of generality we assume that $b'(\theta) \neq 0$ for all θ in the parameter set Θ. Since μ_i is an invertible function of θ_i, we have $\mu_i = q(\theta_i)$, and $\theta_i = q^{-1}(\mu_i)$, so that

$$\frac{\partial \theta_i}{\partial \mu_i} = 1/\frac{\partial \mu_i}{\partial \theta_i}.$$

Furthermore, from the quotient rule and equation 8.7, we find

$$\begin{aligned} \frac{\partial \mu_i}{\partial \theta_i} &= \frac{\partial}{\partial \theta_i} - \frac{c'(\theta_i)}{b'(\theta_i)} = -\left(\frac{c''(\theta_i)b'(\theta_i)c'(\theta_i)b''(\theta_i)}{b'(\theta_i)^2}\right) \\ &= b'(\theta_i)\text{Var}[Y]. \end{aligned}$$

If we write $g(\mu_i) = \eta_i = \mathbf{x}_i^T \boldsymbol{\beta}$, then

$$\frac{\partial \mu_i}{\partial \beta_j} = \frac{\partial \mu_i}{\partial \eta_i}\frac{\partial \eta_i}{\partial \beta_j} = \frac{\partial \mu_i}{\partial \eta_i} x_{ij}.$$

From equation (8.6), we find

$$\frac{\partial \ell_i}{\partial \theta_i} = y_i b'(\theta_i) + c'(\theta_i) = b'(\theta_i)(Y_i - \mu_i). \tag{8.8}$$

Now making the suggested substitutions we find that the first order derivative of the log-likelihood function equals

$$U_j = \sum_{i=1}^{n} \frac{(Y_i - \mu_i)}{\text{Var}[Y_i]} x_{ij} \frac{\partial \mu_i}{\partial \eta_i}.$$

It is immediate that its expected value $E(U_j) = 0$. Using independence between different random variables Y_i and Y_ℓ, the jk-th element of the Fisher information matrix equals

$$\begin{aligned} J_{jk} &= E[U_j U_k] = E\left[\sum_{i=1}^{n} \frac{(Y_i - \mu_i)}{\text{Var}[Y_i]} x_{ij} \frac{\partial \mu_i}{\partial \eta_j} \sum_{\ell=1}^{n} \frac{(Y_\ell - \mu_\ell)}{\text{Var}[Y_\ell]} x_{\ell k} \frac{\partial \mu_\ell}{\partial \eta_\ell}\right] \\ &= E\left[\sum_{i=1}^{n} \frac{(Y_i - \mu_i)^2}{\text{Var}[Y_i]^2} x_{ij} x_{ik} \left(\frac{\partial \mu_i}{\partial \eta_i}\right)^2\right] = \sum_{i=1}^{n} x_{ij} \frac{1}{\text{Var}[Y_i]} \left(\frac{\partial \mu_i}{\partial \eta_i}\right)^2 x_{ik} \\ &= \sum_{i=1}^{n} x_{ij} w_{ii} x_{ik} = \mathbf{x_j}^T \mathbf{W} \mathbf{x_k}, \end{aligned}$$

for all $j, k = 1, \cdots, p$, where $\mathbf{x_j}$ is column j of the matrix $\mathbf{X}$, and $\mathbf{W}$ is a diagonal matrix with elements indicated by the previous display. Hence, we find an expression for the information matrix

$$\mathbf{J} = \mathbf{X}^T\mathbf{W}\mathbf{X}.$$

As our aim is to find the MLE $\hat{\boldsymbol{\beta}}$ by solving the score equation $\mathbf{U} = \mathbf{0}$, we derive a first-order Taylor expansion of $\mathbf{U}$ using the Fisher information,

$$\mathbf{U}(\beta) \approx \mathbf{U}(\boldsymbol{\beta_0}) + \mathbf{J}(\boldsymbol{\beta} - \boldsymbol{\beta_0}).$$

Using this approximation, a Newton-Raphson algorithm can be used to iteratively find the MLE.

$$\boldsymbol{\beta}_{m+1} = \boldsymbol{\beta}_m + \mathbf{J}_m^{-1}\mathbf{U}_m.$$

A modified Newton-Raphson algorithm can be developed by noting that the previous equation implies

$$\mathbf{J}_m\boldsymbol{\beta}_{m+1} = \mathbf{J}_m\boldsymbol{\beta}_m + \mathbf{U}_m.$$

Now, by substitution, the j-th element $(j = 1, \cdots, m)$ of the right-hand side equals

$$\begin{aligned}
\sum_{k=1}^{p} J_{jkm}\beta_{km} + U_{jm} &= \sum_{k=1}^{p}\sum_{i=1}^{n} \frac{Y_i - \mu_i}{\text{Var}[Y_i]} x_{ij} \left(\frac{\partial\mu_i}{\partial\eta_i}\right)^2 \beta_{km} + \sum_{i=1}^{n} \frac{Y_i - \mu_i}{\text{Var}[Y_i]} x_{ij} \frac{\partial\mu_i}{\partial\eta_i} \\
&= \sum_{i=1}^{n} x_{ij} \frac{1}{\text{Var}[Y_i]} \left(\frac{\partial\mu_i}{\partial\eta_i}\right)^2 \cdot \left(\sum_{k=1}^{p} x_{ik}\beta_{km} + (Y_i - \mu_i)\frac{\partial\eta_i}{\partial\mu_i}\right) \\
&= \sum_{i=1}^{n} x_{ij} w_{ii} z_i = \mathbf{x}_j^T \mathbf{W}_m \mathbf{z}_m
\end{aligned}$$

where $\mathbf{z}_m = \sum_{k=1}^{p} x_{ik}\beta_{km} + (Y_i - \mu_i)\frac{\partial\eta_i}{\partial\mu_i}$ is sometimes referred to as the *working response.* This implies

$$\mathbf{X}^T\mathbf{W}_m\mathbf{X}\boldsymbol{\beta}_{m+1} = \mathbf{X}^T\mathbf{W}_m\mathbf{z}_m,$$

so that we find

$$\boldsymbol{\beta}_{m+1} = (\mathbf{X}^T\mathbf{W}_m\mathbf{X})^{-1}\mathbf{X}^T\mathbf{W}_m\mathbf{z}_m. \tag{8.9}$$

It is useful to see that the latter minimizes the Euclidian norm

$$\|\mathbf{W}_m^{1/2}\mathbf{z}_m - \mathbf{W}_m^{1/2}\mathbf{X}\boldsymbol{\beta}\|^2.$$

The iteratively re-weighted least squares algorithm reads:

- Step 1: Initialize a vector $\boldsymbol{\beta}_1$ and set the step number $m = 1$.
- Step 2:

1. Compute $\mathbf{W}_m$ and $\mathbf{z}_m$, and increase the step number to $m+1$.
2. Compute the updating vector $\boldsymbol{\beta}_{m+1}$ by equation (8.9).
3. While: $m \leq 10$, say, and $\|\boldsymbol{\beta}_m - \boldsymbol{\beta}_{m+1}\| > 0.0001$, repeat the previous 2 steps.

Almost always this generates a sequence of converging vectors

$$\{\boldsymbol{\beta}_m\} \to \widehat{\boldsymbol{\beta}}_{MLE} = (\mathbf{X}^T\widehat{\mathbf{W}}\mathbf{X})^{-1}\mathbf{X}^T\widehat{\mathbf{W}}\widehat{\mathbf{z}}$$

yielding the maximum likelihood estimator by iterative re-weighting least squares. From this we immediately see that the projection or hat matrix equals

$$\widehat{\mathbf{W}}^{1/2}\mathbf{X}(\mathbf{X}^T\widehat{\mathbf{W}}\mathbf{X})^{-1}\mathbf{X}^T\widehat{\mathbf{W}}^{1/2}.$$

The term $\mathbf{X}^T\widehat{\mathbf{W}}\mathbf{X}$ is known as the Fisher information matrix, the inverse of which gives the asymptotic variance matrix of the estimator. The maximum likelihood estimator is asymptotically normally distributed according to

$$(\mathbf{X}^T\mathbf{W}\mathbf{X})^{1/2}(\widehat{\boldsymbol{\beta}} - \boldsymbol{\beta}) \overset{D}{\to} N(\mathbf{0}, \mathbf{I}).$$

where $\mathbf{W}$ follows from Step 1 inserting the true parameter point $\boldsymbol{\beta}$. For practical purposes we use

$$\widehat{\boldsymbol{\beta}} \overset{D}{\approx} N(\boldsymbol{\beta}, (\mathbf{X}^T\mathbf{W}\mathbf{X})^{-1}),$$

as the basis to construct statistical tests and 95% confidence intervals for e.g., the effect of the dose.

8.4.2 Binary dose-response models

In dose-response models the aim is to describe the relationship between the amount of exposure to some stimulant or stressor and some response. In the case the response is binary, such as affected/non-affected or dead/alive, then these models are referred to as *binary dose-response models*. In binary dose-response experiments the outcome is the number of "successes" Y_i out of $n_i \geq 1$ trials, which needs to be linked to the chemical dose and possibly other covariates $\mathbf{x}_i$. A basic generalized linear model for Y_i can be constructed using the binomial distribution, i.e., for $i = 1, \ldots, n$,

$$\begin{cases} Y_i \overset{\text{IID}}{\sim} \text{Bin}(\pi_i, n_i) \\ \pi_i = g^{-1}(\mathbf{x}_i^t\beta) \end{cases}$$

The inverse of the link function $g : [0,1] \longrightarrow \mathbb{R}$ should map the linear predictor into the parameter space, which is a probability. There are a number of possibilities. A common choice for dose-response models is the **probit link**,

$$g(\pi) = \Phi(\pi), \quad g^{-1}(x) = \Phi^{-1}(x),$$

where Φ is the cumulative distribution function of the standard normal distribution. This choice results into a generalized linear model that is referred to as a **probit regression model** .

Alternatively, one can use the **logit link** function,

$$g(\pi) = \log\left(\frac{\pi}{1-\pi}\right), \quad g^{-1}(x) = \frac{e^x}{1+e^x}.$$

This is referred to as **logistic regression**. The logistic model is the most popular binary regression model, as it has a number of advantages. The main one is computational, as we will see below, which meant that it has been incorporated in software early on. Moreover, the logit itself has some interpretational advantages. The quantity

$$\text{Odds} = \frac{\pi}{1-\pi}$$

has a long history in gambling and connects quite closely how humans think about probability. It allows us to construct an easy interpretation for the parameters β. Consider a model with a single covariate x and we consider two individuals i and j that differ exactly by a value of 1 for that covariate, i.e., $x_i = x_j + 1$. Then using the logit link function we note that

$$\frac{\pi_i}{1-\pi_i} = e^{x_i\beta} = e^{(x_j+1)\beta} = e^{\beta}\frac{\pi_j}{1+\pi_j}.$$

So the odds of success for individual i are e^{β} times higher than the odds of individual j, where it should be noted that obviously e^{β} can be less than one, if β is negative.

Example 8.7 *In an experiment the number of beetle mortalities y_i out of n_i were counted after five hours of exposure to a gaseous carbon disulphide dosis x_i ($\log_{10} CS_2 mgl^{-1}$) in various concentrations (Bliss, 1935). Figure 8.7 contains the 8 different concentrations that were tested. The dose-response models aims to describe the number Y_i of insects killed out of N according to different doses x_i. Besides the dose, we also consider a general intercept in the model, which captures the expected mortality at a zero dose level. The model matrix $\mathbf{X}$ has its i-th row equal to $\mathbf{x}_i^T = (1, x_i)$, where x_i is the dose of the insecticide. The IRLS algorithm converged in 4 steps yielding the estimate $\widehat{\beta} = (-60.72, 34.27)^T$ with the asymptotic variance matrix*

$$(\mathbf{X}^T\widehat{\mathbf{W}}\mathbf{X})^{-1} = \begin{bmatrix} 26.84 & -15.08 \\ -15.08 & 8.48 \end{bmatrix}$$

from which we take the square root of the diagonal values to obtain the standard errors as 5.1807, 2.9121. The estimated coefficients and their standard errors from binomial regression given in Table 8.9 reveal that these are non-zero with large certainty. From the estimate $\widehat{\boldsymbol{\beta}}$ the

$$\hat{\pi}_i = \frac{e^{\mathbf{x}_i^T\widehat{\boldsymbol{\beta}}}}{1+e^{\mathbf{x}_i^T\widehat{\boldsymbol{\beta}}}}$$

x_i	n_i	y_i	$\hat{y}_i$	y_i/n_i	$\hat{\pi}_i$
1.69	59	6	3.46	0.10	0.06
1.72	60	13	9.84	0.22	0.16
1.76	62	18	22.45	0.29	0.36
1.78	56	28	33.90	0.50	0.61
1.81	63	52	50.10	0.83	0.80
1.84	59	53	53.29	0.90	0.90
1.86	62	61	59.22	0.98	0.96
1.88	60	60	58.74	1.00	0.98

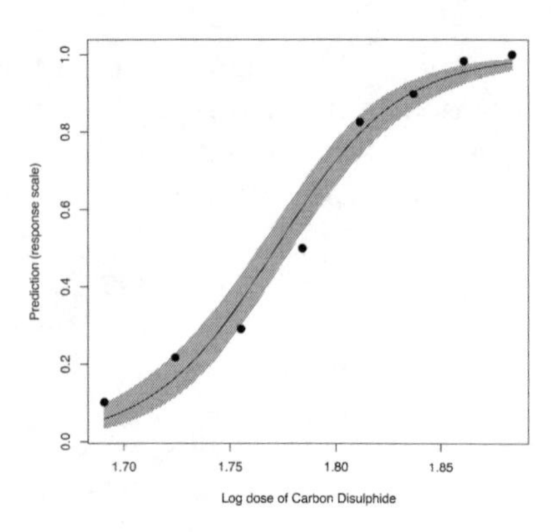

FIGURE 8.7
Number of beetle mortalities y_i out of n_i depending on dose x_i in $\log_{10} CS_2 mgl^{-1}$, with predicted mortalities $\hat{y}_i$, proportions y_i/n_i, and estimated probabilities $\hat{\pi}_i$ from binomial logistic regression.

TABLE 8.9
The estimated coefficients from generalized least squares for the beetle mortality data.

	Estimate	**Std. Error**	**z value**	**$\Pr(>\lvert z\rvert)$**
(Intercept)	−60.72	5.18	−11.72	< 0.0001
Dose x	34.27	2.91	11.77	< 0.0001

values are computed from which the predicted number of beetles killed $\hat{y}_i = n_i \cdot \hat{\pi}_i$ *follow. From Figure 8.7 it can be seen that the* $\hat{\pi}_i$ *values from the model are close to the proportions* y_i/n_i *from the experiment, and that the expected number of beetles killed* $\hat{y}_i$ *from the model are close to the number of observed beetles killed* y_i*. The estimated inverse link function*

$$\hat{\pi}_i = g^{-1}(\hat{\eta}_i) = \frac{e^{\mathbf{x}_i^T \widehat{\boldsymbol{\beta}}}}{1 + e^{\mathbf{x}_i^T \widehat{\boldsymbol{\beta}}}}$$

in Figure 8.7 fits the observed proportions y_i/n_i *well. Given the relative small sample size* $N = 8$*, the confidence curves are small.*

The monotonicity of the inverse link function makes it possible to obtain inverse regression predictions about the 50% lethal dose is 1.77 ($\log_{10} CS_2 mgl^{-1}$) *with 95% prediction interval* $(1.76, 1.78)$ *and the 90% lethal dose is 1.84 with 95% prediction interval* $(1.82, 1.85)$.

Computational estimation framework for logistic regression.

The binomial distribution is a member of the exponential family of distributions (see Chapter 3). We assume the $Y_1, \cdots, Y_N$ to be independently

$\text{Bin}(\pi_i, n_i)$ distributed, so that

$$\begin{aligned} f(y_i, n_i) &= \binom{n_i}{y_i} \pi_i^{y_i}(1 - \pi_i^{y_i}) \\ &= \exp\left\{ y_i \cdot \log\left(\frac{\pi_i}{1 - \pi_i}\right) + n_i \log(1 - \pi_i) + \log\binom{n_i}{y_i} \right\} \end{aligned}$$

The monotonically increasing logit function g has the closed interval $[0, 1]$ as its domain and the set of real numbers as its range. It links the success probability to the dose. Its inverse

$$\pi_i = g^{-1}(\eta_i) = \frac{e^{\mathbf{x}_i^T \boldsymbol{\beta}}}{1 + e^{\mathbf{x}_i^T \boldsymbol{\beta}}}$$

provides the success probability as a function of the dose. The maximum likelihood estimator can be found by the modified Newton-Raphson algorithm using the iterative re-weighted least squares (IRLS) algorithm. That is, given the data $\mathbf{X}, \mathbf{y}$ we start with an initial vector β_0, typically only including the intercept. Next, we proceed with the following iterative steps ($m = 1, 2, \ldots$):

Step 1: Calculate the individual probabilities,

$$\pi_{im} = \frac{e^{\mathbf{x}_i^T \beta_m}}{1 + e^{\mathbf{x}_i^T \beta_m}}, \quad \text{for } i = 1, \cdots, N,$$

Step 2: Calculate the diagonal "weight" matrix $\mathbf{W}_m$,

$$w_{m,ii} = n_i \pi_{im}(1 - \pi_{im}) \ \text{for,} \ i = 1, \cdots, N.$$

Step 3: Calculate the "working response" N-vector $\mathbf{z}_m$,

$$z_{im} = \mathbf{x}_i^T \boldsymbol{\beta}_m + \frac{y_i - n_i \pi_{im}}{n_i \pi_{im}(1 - \pi_{im})} \ \text{for,} \ i = 1, \cdots, N.$$

Step 4: Update the vector $\boldsymbol{\beta}_m$ as the argument that minimizes the weighted least squares problem

$$\begin{aligned} \text{Step 4: } \boldsymbol{\beta}_{m+1} &= \underset{\boldsymbol{\beta}}{\text{argmin}} \, ||\mathbf{W}_m^{1/2} \mathbf{z}_m - \mathbf{W}_m^{1/2} \mathbf{X} \boldsymbol{\beta}||^2 \\ &= (\mathbf{X}^T \mathbf{W}_m \mathbf{X})^{-1} \mathbf{X}^T \mathbf{W}_m \mathbf{z}_m. \end{aligned}$$

The algorithm consists of repeating Step 1 to 4 until convergence. This generates a sequence of vectors $\{\boldsymbol{\beta}_m\} \to \boldsymbol{\beta} = \widehat{\boldsymbol{\beta}}$ the maximum likelihood estimator for the parameters. Convergence usually takes only a few steps. After convergence we use $\widehat{\boldsymbol{\beta}}$ in Steps 1 to 3 to obtain $\hat{\pi}_i$, $\widehat{\mathbf{z}}$, and $\widehat{\mathbf{W}}$. Then according to Step 4, we obtain an expression for the estimator

$$\widehat{\boldsymbol{\beta}} = (\mathbf{X}^T \widehat{\mathbf{W}} \mathbf{X})^{-1} \mathbf{X}^T \widehat{\mathbf{W}} \hat{\mathbf{z}}$$

from which we immediately see that the projection or hat matrix equals

$$\widehat{\mathbf{W}}^{1/2}\mathbf{X}(\mathbf{X}^T\widehat{\mathbf{W}}\mathbf{X})^{-1}\mathbf{X}^T\widehat{\mathbf{W}}^{1/2}.$$

The term $\mathbf{X}^T\widehat{\mathbf{W}}\mathbf{X}$ is known as the Fisher information matrix, the inverse of which gives the asymptotic variance matrix of the estimator. The maximum likelihood estimator is asymptotically normally distributed according to

$$\widehat{\boldsymbol{\beta}} \stackrel{D}{\rightarrow} N(\boldsymbol{\beta}, (\mathbf{X}^T\mathbf{W}\mathbf{X})^{-1}),$$

where $\mathbf{W}$ follows from Step 1 inserting the true parameter point $\boldsymbol{\beta}$. On the basis of this statistical tests and confidence intervals can be constructed with respect to, e.g., the effect of the dose.

8.4.3 Count models

In many engineering applications, the response of interest is a count. For example, counting the daily failures of some kind of chemical production process or mortality counts as a result of the chemical composition of airborne particulate matter. In these cases traditional linear regression models are inappropriate for two reasons: first the response variable cannot be normally distributed and secondly the relationship between the mean counts and the covariates cannot be strictly linear, as it cannot become negative. In these situations we can consider another type of generalized linear regression model, the simplest and often the most appropriate of which is Poisson regression.

$$\begin{cases} Y_i \stackrel{\text{IID}}{\sim} \text{Poisson}(\mu_i) \\ \mu_i = g^{-1}(\mathbf{x}_i^T\boldsymbol{\beta}) \end{cases}$$

The link function $g : \mathbb{R}^+ \longrightarrow \mathbb{R}$ needs to map the positive real numbers into the real line. A natural choice is the natural logarithm,

$$g(\mu) = \log(\mu), \quad g^{-1}(x) = e^x$$

Example 8.8 Pollution induced mortality. *The National Morbidity, Mortality, and Air Pollution Study (NMMAPS) was a large US multi-city longitudinal study of the effects of ambient air pollution on daily mortality and morbidity (Peng and Welty, 2004). The study contains information on 108 U.S. cities for 14 years (1987–2000). The* **R** *package* **gamair** *(Wood, 2017) contains the data from 5114 days for the city of Chicago, consisting of the total mortality, various pollutants as well as the temperature in Fahrenheit. The measured pollutants are* PM_{10}, $PM_{2.5}$, SO_2, O_3. *These are the four "criteria pollutants" defined by the U.S. Environmental Protection Agency. Figure 8.8 shows the daily mortality and the daily ozone levels (low, medium, high) in Chicago in 1987.*

A main effects model $\mathcal{M}_1$ *in Table 8.10 shows that both fine inhalable particles* $PM_{2.5}$ *and* SO_2 *are positively related with mortality, whereas ozone* O_3

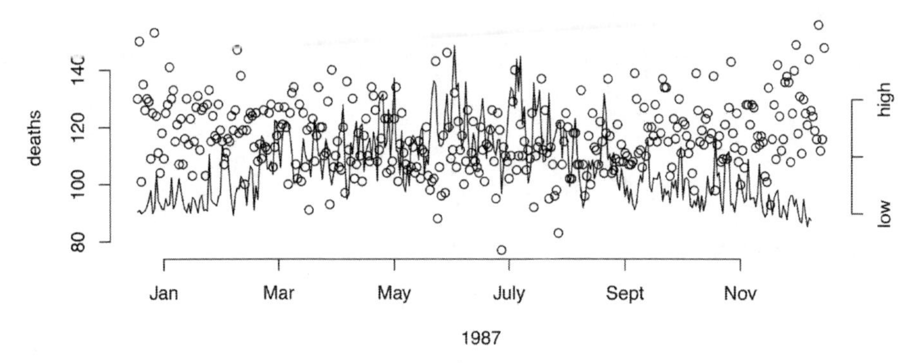

FIGURE 8.8
Subset of the NMMAPS pollution data for the city of Chicago over the year 1987, showing the daily deaths (left vertical axis) and the daily ozone O_3 levels (right vertical axis), showing a possible inverse relationship.

TABLE 8.10
Poisson regression coefficient estimates for the main effects and interaction models, showing that there is significant interactions between the pollutants on mortality.

Coefficient estimates						PM_{10}	PM_{10}	$PM_{2.5}$	O_3	
	Int	PM_{10}	$PM_{2.5}$	O_3	SO_2	$\times O_3$	$\times SO_2$	$\times O_3$	$\times SO_2$	**AIC**
$\mathcal{M}_1$	4.69		1.03	−2.98	5.51					5880.4
$\mathcal{M}_2$	4.68	0.00	0.87	−3.45	2.09	0.15	0.23	−0.19	−0.47	5834.4

is negatively correlated. However, as the interaction effect model $\mathcal{M}_2$ shows, the situation is probably somewhat more complicated, as there are significant interactions between the various pollutants, which means that also larger particles PM_{10} affect mortality.

Iterative Re-weighted Least Squares for Poisson regression.

In order to specify the iterative re-weighted least squares algorithm we need to derive the z_i and the w_{ii} for a $\mu_i = \exp(\mathbf{x}_i^T \boldsymbol{b})$. Note that

$$\frac{\partial \mu_i}{\partial \eta_i} = \frac{\partial}{\partial \eta_i} e^{\eta_i} = e^{\eta_i} = \mu_i \Rightarrow \frac{\partial \eta_i}{\partial \mu_i} = \frac{1}{\mu_i},$$

so that

$$w_{ii} = \frac{1}{\text{Var}[Y_i]} \cdot \left(\frac{\partial \eta_i}{\partial \mu_i} \right)^2 = \frac{1}{\mu_i} \mu_i^2 = \mu_i = \exp(\mathbf{x}_i^T \boldsymbol{b})$$

for $i = 1, \cdots, n$. Furthermore, we have

$$z_i = \mathbf{x}_i^T \boldsymbol{b} + (y_i - \mu_i)\left(\frac{\partial \eta_i}{\partial \mu_i}\right) = \mathbf{x}_i^T \boldsymbol{b} + (y_i - \mu_i)\frac{1}{\mu_i},$$

where $\exp(\mathbf{x}_i^T \boldsymbol{b})$.

8.5 Semi-parametric Models

There are research situations in which we do not have an idea about the underlying class of functions that determines the structural part of the model, but we have reasons to believe the error to be normally distributed. Such models are called semi-parametric since the parametric distribution of the error is known, but not the parameters that determine the underlying function. For such models an approximation by **cubic splines** may be used, as these have many favorable properties. Cubic splines are, for instance, very flexible in contrast to polynomials. The latter may even become arbitrarily large on a segment of the domain of the function if we would interpolate $f(x) = 1/(1+x^2)$ on $[-5, 5]$ by a polynomial of degree m (Schumaker, 2007, p. 102).

Suppose we have observed the data (y_i, x_i), for $i = 1, \cdots, n$ and that all x values are members of the interval $[x_1, x_k]$, where the left/right boundary are the smallest/largest observation among the x-values. Next, we define k equidistant knot points

$$x_1 < x_2 < \cdots < x_k,$$

where k is by default taken to be 10, or larger when many data points are available. Throughout we use the strictly positive $h_j = x_{j+1} - x_j$. With a slight abuse of language, the knots x_j will be indexed by $j = 1, \cdots, k$ and these points are to be distinguished from the observations x_i, where $i = 1, \cdots, n$. Cubic splines are a series of local third-order polynomials which are connected at the knot points over an interval.

Our goal is to minimize the squared loss with the penalty function

$$\sum_{i=1}^{n}(y_i - f(x_i))^2 + \lambda \int_{x_1}^{x_k} f''(x)^2 dx, \tag{8.10}$$

subject to f having a continuous first order derivative at the knot points. It is well-known that this is accomplished by a cubic spline function with continuous first-order derivatives. We will define the cubic spline function as

$$f(x) = \sum_{j=1}^{k} \left(a_j^-(x)\beta_j + a_j^+(x)\beta_{j+1} + c_j^-(x)\delta_j + c_j^+(x)\delta_{j+1}\right) \cdot I_{[x_j, x_{j+1}]}(x),$$

where the basis functions are defined as

$$\begin{aligned} a_j^-(x) &= (x_{j+1}-x)/h_j, \quad c_j^-(x) = [(x_{j+1}-x)^3/h_j - h_j(x_{j+1}-x)]/6, \\ a_j^+(x) &= (x-x_j)/h_j, \quad c_j^+(x) = [(x-x_j)^3/h_j - h_j(x-x_j)]/6. \end{aligned}$$

Note that the a functions are linear in x with zero second-order derivative and first order derivatives

$$a_j^{-\prime}(x) = -1/h_j; \; a_j^{+\prime}(x) = 1/h_j.$$

The c functions are cubic as these are of degree 3. Their first order derivatives are

$$c_j^{-\prime}(x) = [-3(x_{j+1}-x)^2/h_j + h_j]/6, \; c_j^{+\prime}(x) = [3(x-x_j)^2/h_j - h_j]/6.$$

Given the values h_j, $j = 1, \cdots, k$ from the knots, we define the matrices $\mathbf{D}$ and $\mathbf{B}$ by

$$\begin{array}{ccc} D_{j,j} = 1/h_j & D_{j,j+1} = -1/h_j - 1/h_{j+1} & D_{j,j+2} = 1/h_{j+1} \\ B_{j,j} = (h_j + h_{j+1})/3 & & j = 1, \cdots, k-2 \\ B_{j,j+1} = h_{j+1}/6 & B_{j+1,j} = h_{j+1}/6 & j = 1, \cdots, k-3 \end{array} \tag{8.11}$$

From the definition of f it is clear that it has continuous first order derivatives for all internal points of the intervals given by the knot points. To find continuous first order derivatives at the knots we note that

$$f'(x) = \sum_{j=1}^{k} \left(a_j^{-\prime}(x)\beta_j + a_j^{+\prime}(x)\beta_{j+1} + c_j^{-\prime}(x)\delta_j + c_j^{+\prime}(x)\delta_{j+1}\right) \cdot I_{[x_j, x_{j+1}]}(x).$$

For interval $[x_j, x_{j+1}]$, we obtain at the left-hand boundary point

$$f'(x_j) \;=\; -h_j^{-1}\beta_j + h_j^{-1}\beta_{j+1} - h_j\delta_j/3 - h_j\delta_{j+1}/6$$

and for interval $[x_{j-1}, x_j]$ we find for the right-hand boundary point

$$f'(x_j) \;=\; -h_{j-1}^{-1}\beta_{j-1} + h_{j-1}^{-1}\beta_j + h_{j-1}\delta_{j-1}/6 + h_{j-1}\delta_j/3.$$

To obtain continuity of f' at the knots x_j we set the last two equations equal and collect terms to find

$$\begin{aligned} &[h_j/6, (h_j + h_{j+1})/6, h_{j+1}/6] \begin{bmatrix} \delta_j \\ \delta_{j+1} \\ \delta_{j+2} \end{bmatrix} \\ &= [1/h_j, -(1/h_j + 1/h_{j+1}), 1/h_{j+1}] \begin{bmatrix} \beta_j \\ \beta_{j+1} \\ \beta_{j+2} \end{bmatrix}. \end{aligned}$$

From this and the matrices defined in Equation (8.11) we find the necessary condition

$$\mathbf{B}\delta_- = \mathbf{D}\beta,$$

which allows for the substitution $\boldsymbol{\delta}_- = \mathbf{B}^{-1}\mathbf{D}\boldsymbol{\beta}$ below.

To find an expression for f'' and the integral of its square we proceed with expressing f'' as a linear combination with the delta coefficients. The a functions vanish, because these are of degree 1. If x is a member of the interval $[x_j, x_{j+1}]$, it follows that

$$\begin{aligned} f''(x) &= \left(c_j^{-\prime\prime}(x)\delta_j + c_j^{+\prime\prime}(x)\delta_{j+1}\right) \cdot I_{[x_j,x_{j+1}]}(x) \\ &= \left(\delta_j h_j^{-1}(x_{j+1} - x) + \delta_{j+1} h_j^{-1}(x - x_j)\right) \cdot I_{[x_j,x_{j+1}]}(x), \end{aligned}$$

so that after collecting the δ_j terms we can write

$$\begin{aligned} f''(x) &= \sum_{j=2}^{k-2} \delta_j \left(h_j^{-1}(x - x_{j-1}) \cdot I_{[x_{j-1},x_j]}(x) + h_j^{-1}(x_{j+1} - x) \cdot I_{[x_j,x_{j+1}]}(x)\right) \\ &= \sum_{j=1}^{k} \delta_j d_j(x) = \boldsymbol{\delta}_-^T \mathbf{d}(x), \end{aligned} \tag{8.12}$$

where, as suggested by the notation, the elements of the functions $d_j(x)$, $j = 2, \cdots, k-2$ are collected in the vector $\mathbf{d}(x)$. The latter provides us with the relatively straightforward expression

$$\begin{aligned} \int_{x_1}^{x_k} f''(x)^2 dx &= \int_{x_1}^{x_k} (\boldsymbol{\delta}_-^T \mathbf{d}(x))^2 dx \\ &= \boldsymbol{\delta}_-^T \int_{x_1}^{x_k} \mathbf{d}(x)\mathbf{d}(x)^T dx \boldsymbol{\delta}_- \\ &= \boldsymbol{\delta}_-^T \mathbf{B} \boldsymbol{\delta}_- \end{aligned}$$

The latter two expressions show that the matrix $\mathbf{B}$ is symmetric and positive definite. To gain more insight into the equality, we use a property from indicator functions

$$I_{[x_{j-1},x_j]}(x) \cdot I_{[x_j,x_{j+1}]}(x) = I_{\{x_j\}}(x),$$

so that we can conclude that integrals in which multiplications of such indicator functions are involved vanish. We find from the definition of $d_j(x)$ in equation (8.12), that the integration of its square comes down to

$$\begin{aligned} \int_{x_1}^{x_k} d_j(x)^2 dx &= \int_{x_{j-1}}^{x_j} h_{j-1}^{-2}(x - x_{j-1})^2 dx + \int_{x_j}^{x_{j+1}} h_j^{-2}(x_{j+1} - x)^2 dx \\ &= (h_{j-1} + h_{j-1})/3 = B_{j,j} \end{aligned}$$

Next, using integration by parts, we obtain

$$\begin{aligned}
\int_{x_1}^{x_k} d_{j-1}(x)d_j(x)dx &= \int_{x_1}^{x_k} h_{j-1}^{-1}(x_j - x)h_{j-1}^{-1}(x - x_{j-1}) \cdot I_{[x_{j-1},x_j]}(x)dx \\
&= \int_{x_{j-1}}^{x_j} -h_{j-1}^{-2}(x - x_j)(x - x_{j-1})dx \\
&= h_{j-1}/6 = B_{j-1,j}
\end{aligned}$$

Furthermore,

$$\int_{x_1}^{x_k} d_{j-1}(x)d_{j+1}(x)dx = 0,$$

which also holds for all other products of d functions with indices that differ larger than 2. We now reformulate the problem in terms of $\boldsymbol{\beta}$ coefficients so that we become able to actually minimize the **penalized regression** function. That is, for the penalty term, we find, from Equation (8.13), using $\delta_- = \mathbf{B}^{-1}\mathbf{D}\boldsymbol{\beta}$, that

$$\begin{aligned}
\int_{x_1}^{x_k} f''(x)^2 dx &= \boldsymbol{\beta}^T\mathbf{D}\mathbf{B}^{-1}\mathbf{B}\mathbf{B}^{-1}\mathbf{D}\boldsymbol{\beta} \\
&= \boldsymbol{\beta}^T\mathbf{D}^T\mathbf{B}^{-1}\mathbf{D}\boldsymbol{\beta} = \boldsymbol{\beta}^T\mathbf{S}\boldsymbol{\beta}
\end{aligned}$$

To redefine the function f we use

$$\begin{bmatrix} 0 \\ \boldsymbol{\delta}_- \\ 0 \end{bmatrix} = \begin{bmatrix} 0 \\ \mathbf{B}^{-1}\mathbf{D}\boldsymbol{\beta} \\ 0 \end{bmatrix} = \begin{bmatrix} \mathbf{0}^T \\ \mathbf{B}^{-1}\mathbf{D} \\ \mathbf{0}^T \end{bmatrix} \boldsymbol{\beta} = \mathbf{F}\boldsymbol{\beta}$$

so that we may write $\delta_j = \mathbf{f}_j^T\boldsymbol{\beta}$, where $\mathbf{f}_j^T$ is row j of the matrix $\mathbf{F}$. Let $\mathbf{e}_j^T$ be row j of the identity matrix. Then we have

$$\begin{aligned}
f(x) &= \sum_{j=1}^{k} \left(a_j^-(x)\beta_j + a_j^+(x)\beta_{j+1} + c_j^-(x)\delta_j + c_j^+(x)\delta_{j+1}\right) \cdot I_{[x_j,x_{j+1}]}(x). \\
&= \sum_{j=1}^{k} \left(a_j^-(x)\mathbf{e}_j^T + a_j^+(x)\mathbf{e}_{j+1}^T + c_j^-(x)\mathbf{f}_j^T + c_j^+(x)\mathbf{f}_{j+1}^T\right) \cdot I_{[x_j,x_{j+1}]}(x)\boldsymbol{\beta} \\
&= \sum_{j=1}^{k} b_j^T(x)\boldsymbol{\beta}.
\end{aligned}$$

If we now define row i of matrix $\mathbf{X}$ as $\sum_{j=1}^{k} b_j^T(x_i)$, for $i = 1, \cdots, n$, then the loss function (8.10) can be written as

$$\|\mathbf{y} - \mathbf{X}\boldsymbol{\beta}\|^2 + \lambda\boldsymbol{\beta}^T\mathbf{S}^{1/2}\mathbf{S}^{1/2}\boldsymbol{\beta} = \left\| \begin{bmatrix} \mathbf{y} \\ 0 \end{bmatrix} - \begin{bmatrix} \mathbf{X} \\ \lambda^{1/2}\mathbf{S}^{1/2} \end{bmatrix} \boldsymbol{\beta} \right\|^2 \tag{8.13}$$

which is globally minimized by

$$\widehat{\boldsymbol{\beta}} = (\mathbf{X}^T\mathbf{X} + \lambda\mathbf{S})^{-1}\mathbf{X}^T\mathbf{y}. \tag{8.14}$$

We immediately see that the predicted $\widehat{\mathbf{y}}$ are a linear combination of the outcomes $\mathbf{y}$, since

$$\widehat{\mathbf{y}} = \mathbf{X}\widehat{\boldsymbol{\beta}} = \mathbf{X}(\mathbf{X}^T\mathbf{X} + \lambda\mathbf{S})^{-1}\mathbf{X}^T\mathbf{y} = \mathbf{A}\mathbf{y}.$$

The matrix $\mathbf{A}$ tends to the hat matrix $\mathbf{X}(\mathbf{X}^T\mathbf{X})^{-1}\mathbf{X}^T$ as λ tends to zero. In analogy with this the **Effective Degrees of Freedom** (EDF), defined as trace $\mathbf{A}$, indicates the degree in which the estimated curve differs from the linear. The residual error variance is estimated by

$$\hat{\sigma}^2 = \frac{\|\mathbf{y} - \mathbf{A}\mathbf{y}\|^2}{n - tr\mathbf{A}}.$$

If the number of knots are sufficient to represent f by the basis functions, then choosing more knots does not have a large influence on the model. However, the flexibility of the model is determined by the value of the smoothing parameter λ. If it is chosen to be very small, then the model becomes overly flexible in over-fitting the data and making it dependent upon error in the data. If λ is chosen to be too large, then the model becomes overly inflexible and would under-fit the true underlying model. A very general way to solve for this over/under fitting problem is to minimize the **mean squared error**

$$MSE = \frac{1}{n}\sum_{i=1}^{n}(\hat{f}(x_i) - f(x_i))^2,$$

where the predicted function value at x_i is defined as

$$\hat{f}(x_i) = \sum_{j=1}^{k} b_j^T(x)\widehat{\boldsymbol{\beta}} = \mathbf{x}_i^T\widehat{\boldsymbol{\beta}} = \hat{y}_i.$$

Obviously, the true function f is unknown making direct estimation of MSE unfeasible. A general solution to this is provided by the idea of cross-validation by leaving one observation out. That is, if we estimate the model leaving out observation (y_i, x_i), then we obtain $\widehat{\boldsymbol{\beta}}^{[-i]}$ and for each i we can compute

$$\hat{f}_i^{[-i]} = \hat{f}^{[-i]}(x_i) = \hat{f}(x_i)^{[-i]} = \mathbf{x}_i^T\widehat{\boldsymbol{\beta}}^{[-i]},$$

where the left expression defined turns out top be convenient. Note that $\hat{f}_i^{[-i]}$ depends on the data (y_j, x_j) for $j \neq i$ and on the smoothing parameter λ.

Then we may consider the **Ordinary Cross Validation** criterion, defined as

$$
\begin{aligned}
OCV(\lambda) &= \frac{1}{n}\sum_{i=1}^{n}(\hat{f}_i^{[-i]} - y_i))^2 = \frac{1}{n}\sum_{i=1}^{n}(\hat{f}_i^{[-i]} - f(x_i) - \varepsilon_i)^2 \\
&= \frac{1}{n}\sum_{i=1}^{n}(\hat{f}_i^{[-i]} - f(x_i))^2 - 2\frac{1}{n}\sum_{i=1}^{n}\left(\hat{f}_i^{[-i]} - f(x_i))\varepsilon_i\right) + \frac{1}{n}\sum_{i=1}^{n}\varepsilon_i^2 \\
&\xrightarrow{P} \frac{1}{n}E\sum_{i=1}^{n}(\hat{f}_i^{[-i]} - f(x_i))^2 - 2\frac{1}{n}\sum_{i=1}^{n}E\left(\hat{f}_i^{[-i]} - f(x_i)) \cdot E\varepsilon_i\right) \\
&\quad + E\frac{1}{n}\sum_{i=1}^{n}\varepsilon_i^2 \\
&= \frac{1}{n}E\sum_{i=1}^{n}(\hat{f}_i^{[-i]} - f(x_i))^2 + \sigma^2
\end{aligned}
$$

using that $E\varepsilon_i = 0$ as well as that $\hat{f}_i^{[-i]}$ and ε_i are independent for each i. The last equality gives the sum of the squared bias and the variance. Recalling that the **smoothing parameter** λ determines $\hat{f}_i^{[-i]}$, it follows that the value of λ that minimizes OCV is best in approximating the true underlying function.

It seems clear that when n is large, repeated model estimation to obtain $\hat{f}_i^{[-i]}$ may be time consuming. By using A_{ii} as the i-th diagonal element of $\mathbf{A}$, it follows that

$$
OCV(\lambda) = \frac{1}{n}\sum_{i=1}^{n}\frac{(\hat{f}_i - y_i))^2}{1 - A_{ii}}.
$$

The latter is often replaced by the computationally more convenient Generalized Cross Validation criterion. The idea is to replace the $1 - A_{ii}$ terms by their mean $\text{tr}(\mathbf{I} - \mathbf{A})/n$. This yields the minimization of

$$
GCV(\lambda) = \frac{1}{n}\sum_{i=1}^{n}\frac{(\hat{f}_i - y_i)^2}{\text{tr}(\mathbf{I} - \mathbf{A})/n}.
$$

over the smoothing parameter as a feasible criterion to estimate the flexibility of the model. It is known that **cross-validation** is consistent in approximating the true model function.

Example 8.9 *Suppose we have the true model* $f(x) = ax^b\exp(-Kx)$ *with* $a = 0.07$, $b = 2.15$, $K = 0.04$ *and that the data come from the model* $y_i = f(x_i) + \varepsilon_i$, *where the latter are independently normally distributed* $N(0, 1/2^2)$. *25 data points were generated according to the model with* x *data values between 5 and 245 using step-size 10. The true curve, the data, and the cubic approximation are given in Figure 8.9. It can be observed that the cubic approximates the true concentration curve quite well.*

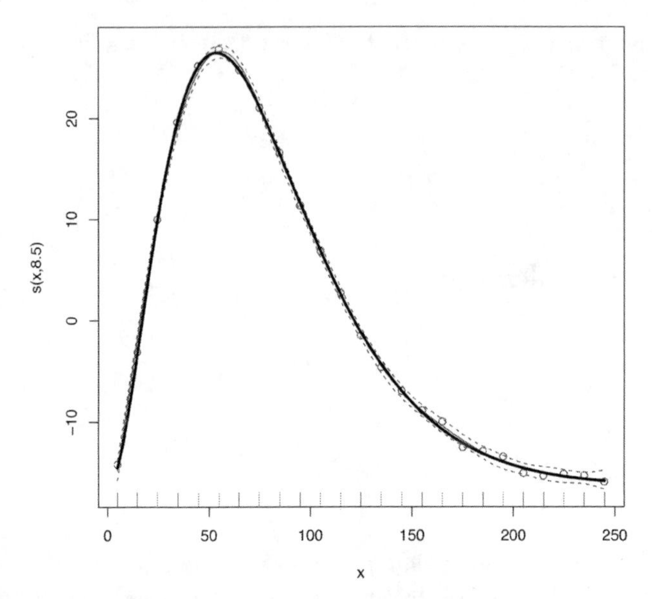

FIGURE 8.9
Cubic spline approximation (thin curve) to a known, thick concentration curve.

Example 8.10 *In a NIST study (Chwirut, 1979) involving ultrasonic calibration, several measurements were obtained on ultrasonic response depending on metal distance. The best fitting cubic spline through the data has intercept 31.5641 and an effective degrees of freedom of 8.417, largely different from the expected value 1 under a linear model. The estimated spline is presented as a smooth curve in Figure 8.10. The adjusted R-square 0.984 indicates a good fit to the data. Its GCV = 13.023, and its Scale estimator* $\widehat{\sigma}^2 = 10.752$ *are in line with the spread of the data around the fitted smoothed cubic spline. The latter are in line with the size of the error in the observations of ultrasonic response for identical metal distances.*

8.6 Notes and Comments

In the literature there exist examples with very small n for which there exist multiple minima of the function S or for which the Gauss-Newton algorithm yields a diverging sequence. Although such cases are exceptional in practice, but if such occurs then it may be worthwhile to reconsider the model parameters or increasing the number of runs in the experiment. In addition, if the observations are expensive to obtain a simulation study may be of great help

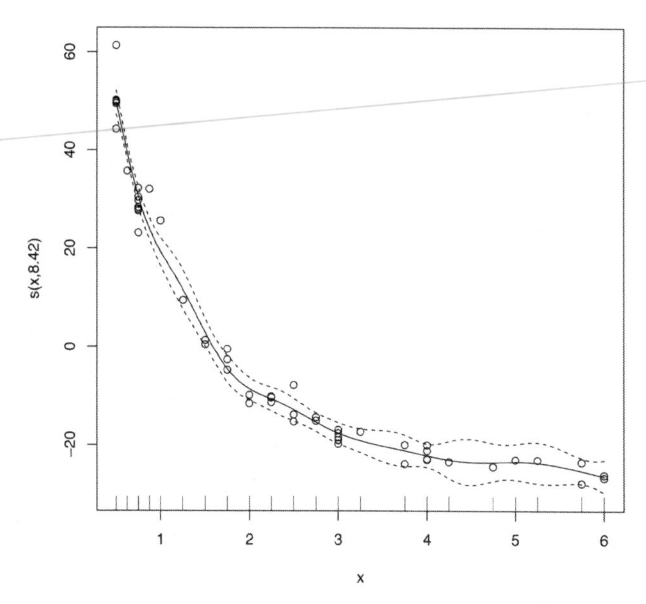

FIGURE 8.10
Cubic spline approximation of the Chwirut data from a NIST study on Ultrasonic response depending on Metal distance (inches).

in determining the sample size n and the range of x-values from which an analysis will become conclusive.

An account of inverse regression covering applied s well as theoretical issues is given by Draper and Smith (1998).

It may be noted that for in regression analysis we have

$$E|f(X) - Y| = \int_{R^d} |f(x) - m(x)|\mu(dx) + E|m(X) - Y|,$$

where $m(x) = E[Y|X = x]$. It follows that the optimal approximation of Y by a function of X is given by $m(x)$, cf Chapter 3. The cubic splines f do converge to this $m(x)$ as the sample size increases, although at a slower rate than non-linear least squares assuming its class to be correct (Györfi et al., 2002; van de Geer, 2000). Wahba (1990) gives many mathematical details on spline type of modeling.

On the other hand we have seen in the above examples that a relative small sample size may be sufficient for a close approximation. Obviously, a cubic spline approach may either confirm or disconfirm a fitted model from a class to the data e.g., depending on model fit expressed as R-squared adjusted. The regression theory of cubic splines can be generalized to several additive splines to fit a chemical outcome as well as into directions of a mixed model when repeated measurements are obtained.

A detailed account of non-linear estimation and its properties is given by Seber and Wild (1989) which goes back to e.g., work of Jennrich (1969). A statistical overview of the bootstrap is given by Davison and Hinkley (1997) and a thorough mathematical exposition by Hall (2013).

A breakthrough in generalized linear modeling was accomplished by Nelder and Wedderburn (1972). It is the heading of a large class of link functions specific for types of outcomes such as binomially, binary, or Poisson distributed (Myers et al., 2010; Dobson and Barnett, 2018; Venables and Ripley, 2002). Generalized mixed linear modeling can be applied if the data come from a chemical experimental setting with repeated measurements.

In the above we applied non-linear regression to a single predictive variable x, this may obviously be generalized to more predictive variables. This is illustrated by the **Viscosity** exercise below.

8.7 Notes on Using R

Non-Linear regression can be applied in to data by the R function `nls` (Venables and Ripley, 2002). For a book length in depth treatment of estimation of nonlinear models with many examples, see (Seber and Wild, 1989). Inverse regression is accomplished by the `calibrate` and the `invest` function of the library `investr` (Greenwell and Schubert Kabban, 2014). The generic `vcov` function applied to an object from the `nls` function gives the asymptotic variance matrix $\widehat{\mathbf{\Omega}}$ in non-linear least squares. The `deriv` function can be used to compute the matrix $\widehat{\mathbf{F}}$. From a given function g the derivative $\dot{g}(\widehat{\boldsymbol{\beta}})$ is often easily constructed so that the computation of the SE is straight forward from the delta theorem. Such type of computations are implemented in the R library `car`.

Cubic splines can be fitted to data by the `gam` function, the `gamm` function is useful for fitting smooth functions in a setting of repeated observations on different objects (Wood, 2017).

Generalized linear modeling can be applied with the `glm` function (Venables and Ripley, 2000) and generalized mixed linear modeling with the `lme` or `lmer` function (Pinheiro and Bates, 2006).

8.8 Exercises

1. **Gauss-Newton.** With the Kinematic viscosity data from Example 8.1 it is possible to compute a few updates Gauss-Newton for the function $f(x) = e^{-ax} \cdot b$, taking $a = \beta_1$ and $b = \beta_2$. Extract the data `table.b10` from the R library `MPV` for ME:DME ratio `x1==0.9189`.

The aim of this question is to model the viscosity y as the function $y = ae^{bx_2}$ of temperature x2. Use starting values a = beta[1] = 0.05; b = beta[2] = 1.0

(a) Derive an expression for the Jacobian matrix $\mathbf{F}^{(m)}$ matrix.

(b) Compute the update of the vector $\boldsymbol{\beta}^{(m+1)} = \boldsymbol{\beta}_{(m)} + \boldsymbol{\delta}_{(m)}$, where

$$\boldsymbol{\delta}^{(m)} = (\mathbf{F}^{(m)T}\mathbf{F}^{(m)})^{-1}\mathbf{F}^{(m)T}\mathbf{r}(\boldsymbol{\beta}^{(m)}),$$

where $\mathbf{r}(\boldsymbol{\beta}^{(m)}) = \mathbf{y} - \mathbf{f}(\boldsymbol{\beta}^{(m)})$ and repeat the updating in a while loop. Compare with the solution from a non-linear least squares function using your favorite software.
Hint: In R use the `nls` function.

2. **Puromycin.** Treloar (1974) measured the counts per minute of some radioactive product as a function of the substrate concentration in parts per million (ppm). From these counts the initial rate (or velocity) of the reaction was calculated (counts/min/min). The experiment was conducted once with the enzyme treated with Puromycin, and once with the enzyme untreated. The data are available from the file `Puromycin.txt`.

   (a) Fit the Michaelis-Menten model $Y = V_m \cdot x/(K + x) + \varepsilon$ to the data separately for the treated and untreated condition. Report on the fitted models to the data.
   Hint: Use e.g., starting values $V_m = 200, K = 0.05$.

   (b) Use confidence intervals to test the hypothesis that the parameters are equal.

   (c) Construct a plot of the data including the two functions corresponding to the two models.

3. **Quantum defects.** In a NIST (Roszman, 1979) study Roszman investigated the number of quantum defects within as time frame in iodine atoms is investigated depending on the excited energy state. The data are available from the file `Roszman.txt` or from `library(NISTnls)`.

   (a) Fit a simple linear model to the data. Does it seem realistic?

   (b) Test which of the models fits the data best

$$f_1(x) = b_1 - \arctan(b_3/(x - b_4))/\pi$$

$$f_2(x) = b_1 - b_2 x - \arctan(b_3/(x - b_4))/\pi$$

   use starting values $b1 = 0.1, b2 = -0.00001, b3 = 1000, b4 = -100$ to fit the model. Plot the data and the evaluate the model fit. Are the parameters significant?

   (c) Test the arctan model against the linear.

TABLE 8.11
Lead concentration and absorption.

Measurement	1	2	3	4	5	6
Lead concentration ng ml^{-1}	10	25	50	100	200	300
Absorbance	0.05	0.17	0.32	0.60	1.07	1.40

4. **Lead concentration**. In the determination of lead in an aqueous solution by electrochemical atomic absorption spectrometry with graphite-probe atomization (Giri et al., 1983), several measurements were obtained. The data given are in Table 7.10 as well in Table 8.11 and can be read from the file `LeadConcentration.txt`.
   (a) Estimate the simple linear model (see also §4.13) and Construct a prediction interval for Lead concentration 75ng ml^{-1}.
   (b) Use inverse regression for the construction of a prediction interval for absorbance 0.50.
   (c) Use inverse regression to predict the concentration from the following repeated absorbance values (0.509, 0.539, 0.432, 0.599, 0.562) intended to be associated with a unique concentration. Give the confidence interval and evaluate its length against that found in the previous question based on a single absorbance. How to obtain an even smaller interval?
5. **Ultrasonic response**. In §4.11 we found for the analyzed the Chwirut data from an experiment measuring ultrasonic response depending on metal distance and found that the

$$\text{Model 2: } f_2(\beta_1, \beta_2, \beta_3, x) = \frac{e^{-\beta_1 x}}{\beta_2 + \beta_3 x},$$

   fits the data best from a non-linear regression based on the Newton algorithm with starting values $b_1 = 0.1, b_2 = 0.01, b_3 = 0.02$. The data can be read from the file `Chwirut.txt`.
   (a) Use inverse non-linear regression for a confidence interval on Metal Distance given a measurement for Ultrasonic Response of 40.
   (b) Answer the same question for Ultrasonic Response 12 and explain the large difference in length of the confidence interval.
6. **Mitcherlich equation**. The Mitcherlich equation account for a non-linear model

$$y_i = \beta_1 - \beta_2 \cdot \exp(-\beta_3 * x_i)) + \varepsilon_i,$$

   where the error terms areas ample from the normal distribution.

The data `p12.11` in the library `MPV` contain fractions of active chlorine in a chemical product. Note that there is an error in row 6 which should read 6, 10, 0.47. It is conveniet to rename the columns to x and y, and to use the starting values $b1 = 0.38, b2 = -5.0, b3 = 0.5$.

(a) Fit the model to the data. Report the residual standard error $\hat{\sigma}$.
(b) Plot the estimated curve together with the data.
(c) Compare the confidence interval from NLS with that from the bootstrap. Study the pairs plot with the covariations of the bootstrap estimates. Are the estimated asymptotically normal?

7. **A simulation.** Construct data according to the concentration function $f(x) = ax^b \exp(-Kx)$ with $a = 0.07$, $b = 2.15$, $K = 0.04$ in time x. Suppose we have some error in the measurement which is represented by noise in the model $y_i = f(x_i) + \varepsilon_i$, where the latter are independently normally distributed $N(0, \sigma^2)$.

   (a) Take $\sigma = 5$, time points between zero to 250 with equidistances 5 and generate 1 measurement per time point. Generate the data according to the model and fit the model by nls. Repeat this 1000 times and collect the parameter estimates as rows in a matrix. Construct qq-plots for each of the parameter estimates and apply the Shapiro-Wilk normality test.
   (b) Repeat the previous question with the x value b/k at which the maximum occurs and the arie underthe curve for x over the domain 0 to 200.
   (c) Does the normality of the parameter estimates increase if you increase the number of measuremnets per time point? Or if you decrease the variance σ^2?

8. **Secalonic acids.** In an experiment assessing the inhibitory effect of secalonic acids on plant growth seven measurements were obtained for the dose (mM) and the root length (cm). The data can be found under the name `secalonic` from the `drc` library.

   (a) Give the formula for the 5-parameter Log-logistic model. Hint: Use the help from the library.
   (b) Test the 4-parameter Log-logistic model against the 3-parameter Log-logistic model. Hint: Conveniently use the `drm` function from the `drc` library together with the `anova` function.
   (c) Ptot the function and the data to the screen.
   (d) Does the function tend to zero as x increases without bound? What is the role of the parameter d?

9. **Dose response.** In a small experiment (Collett, 2003) different doses of pyrethroid trans-cypermethrin in μg were given to the tobacco budworm Heliothis virescens to protect in order to protect it against the moth. Batches of 20 moths for the female sex yielded the number dead 0, 2, 6, 10, 12, 16 for the dosis 1, 2, 4, 8, 16, 32. Use the number dead and the number alive as input for a binomial logistic regression analysis with the $\log_2$ dosis as predictive variable.
   (a) Test the null hypothesis that the parameters are zero.
   (b) Construct a table such as 8.7 to summarize the proportions and their predictions by the model. Extract the predicted proportions from the estimated model.
   (c) Construct a plot with the proportions and the 95% confidence band fromthe model.
   Hint: Use the `visreg` from the library that goes under the same name.
10. **Viscosity.** The Stormer viscometer measures the viscosity of a fluid by the time for an inner cylinder to perform a fixed number of revolutions in the response to an actuating weight (Williams, 1959). The viscometer is calibrated by measuring the time taken with various weights, while the mechanism is suspended in fluids of known viscosity. The data come from a calibration experiment and contains 23 measurements of time t, weight w and viscosity v. Theoretical considerations suggest a non-linear relation ship between time, weight and viscosity

$$t = f(\beta_1, \beta_2, v, w) + \varepsilon = \frac{\beta_1 v}{w - \beta_2} + \varepsilon,$$

where β_1 and β_1 are unknown parameters. The data `stormer` are available from the library MASS.
   (a) Fit the model to the data and test the hypothesis that the parameters are equal to zero. Compute the 95 percent confidence intervals for the parameters.
   Hint: Use starting values c(b1=29,b2=2).
   (b) Use the bootstrap to compute the confidence intervals and compare these with the previous. Are the parameters symmetrically distributed?
   (c) Plot the data with the fitted curve.

9

Chemodynamics and Stoichiometry

In this chapter we describe several modeling and inference techniques that can be used to study the dynamics of chemical reactions. A chemical reaction is a process in which one or more substances, i.e., the *reactants*, are converted to one or more different substances, called the *products*. These substances are either chemical elements or chemical compounds. The most common chemical reactions are combustion and oxidation, which is the reaction with oxygen. A reaction is described by means of a chemical equation. For example, the chemical equation for the burning of octane (C_8H_{18}) can be given as

$$2\ C_8H_{18} + 25\ O_2 \longrightarrow 16\ CO_2 + 18\ H_2O,$$

which results in generating carbon dioxide (CO_2) and water (H_2O). The constants in front of the substance names in the reaction are the stoichiometric constants, which describe the combination of substances so that matter is conserved: on both the left and right-hand side of the equation there are 16 carbon (C) atoms, 36 hydrogen (H) atoms and 50 oxygen (O) atoms. A reaction always need to be balanced.

Consider a closed system that contains 2 octane molecules and 25 oxygen molecules. How quickly will they react to form 16 carbon dioxide and 18 water molecules? First of all, this depends on the temperature and pressure of the system. The higher the temperature or pressure, the faster the reaction occurs; or rather, *tends* to occur, because reactions occur, for all practical purposes, *randomly*. A common choice for modeling the reaction time is by means of an exponential distribution,

$$T \sim \text{Exp}(\theta),$$

i.e., $P(T < t) = 1 - e^{-\theta t}$. The quantity θ describes the rate of the reaction, which more precisely we should have written as $\theta(T, P)$, depending on temperature T and pressure P. The average reaction time is given as

$$ET = 1/\theta.$$

Therefore, the higher the reaction rate, the faster the reaction tends to occur. Under thermal equilibrium, constant pressure and a fixed volume, a biochemical reaction shows which species and in what proportions react together and what they produce.

In this chapter you will learn about the following problems and solution techniques:

DOI: 10.1201/9781003178194-9

- In Section 9.1, you will see how to extend the single chemical reaction set-up to a reaction system containing multiple, interacting reactions.
- In Sections 9.2.1 and 9.2.2, you will learn how to use the Gillespie algorithm and the Euler-Maruyama approximation to simulate systems of chemical reactions.
- In Section 9.3, you will learn how to use a local linear approximation to estimate the reaction rates $\boldsymbol{\theta}$, when observing time course data from a system of chemical reactions. In this section we focus on systems in which the total number of molecules is large and therefore particularly affected by stochastic fluctuations.
- In Section 9.4, you will learn how to write the mean of the stochastic process as the solution of a linear differential equation.
- In Section 9.4.2, you will learn how to infer the reaction rates $\boldsymbol{\theta}$ for systems with large number of molecules, in which the observations are mainly affected by measurement error, but not by stochastic fluctuations.

9.1 Stoichiometry of Systems of Reactions

In this section we will look a system of r reactions involving p different substances, that can be described by $j = 1, \ldots, r$ reaction equations:

$$k_{1j}Q_1 + k_{2j}Q_2 + \ldots + k_{pj}Q_p \xrightarrow{\theta_j} s_{1j}P_1 + s_{2j}P_2 + \ldots + s_{pj}P_p. \tag{9.1}$$

The terms on the left-hand side, denoted as Q, are the *reactants*, whereas the terms on the right-hand side, denoted as P, are the *products*. The coefficients k_{ij} and s_{ij} $(i = 1, \ldots, p)$ represent the *stoichiometric coefficients* associated with the ith reactant Q_i and product P_i, respectively. The chemical interpretation of this equation is that k_{ij} molecules of type Q_i are converted into s_{ij} molecule of type P_i across all the $i = 1, \ldots, p$ substrates. How quickly does reaction j occur? First note, if we have $Y_{01}, \ldots, Y_{0p}$ molecules of the p substrates and we need $k_{j1}, \ldots, k_{jp}$ molecules to execute a reaction, then there are $\prod_{j=1}^{p} \binom{Y_{0i}}{k_{ij}}$ combinations of molecules for the reaction to occur. If each single combination has rate θ_j, then the overall rate for reaction j is:

$$h_j(\mathbf{Y}_0, \boldsymbol{\theta}) = \theta_j \prod_{i=1}^{p} \binom{Y_{0i}}{k_{ij}}, \tag{9.2}$$

where $\binom{x}{y} = 0$ for $x < y$. The waiting time for reaction j will therefore be

$$T_j \sim \text{Exp}(h_j(\mathbf{Y}_0, \boldsymbol{\theta})).$$

We will denote with Y the overall state of the system, in particular

$$Y_{ti} = \text{number of molecules of substrate } i \text{ at time } t.$$

So, if we start the system at time 0 with Y_0 molecules and if after time t, reaction j is the first reaction to be triggered, then the new state Y_t for the p substrates can be calculated as follows,

$$Y_{ti} = Y_{0i} + s_{ij} - k_{ij}.$$

The difference $v_{ij} = s_{ij} - k_{ij}$ is called the net-change. For a set of r reactions and p species, the molecular transfer from reactant to product species is according toa net change of $\mathbf{V} = \mathbf{S} - \mathbf{K}$, where $\mathbf{S}$ denotes the $p \times r$ dimensional matrix of stoichiometric coefficients of the products, $\mathbf{K}$ is the $p \times r$ dimensional matrix of stoichiometric coefficients of the reactants, and $\mathbf{V}$ is called the $p \times r$ dimensional *stoichiometric* or *net effect* matrix.

The stoichiometric matrix is an important matrix to study the dynamics of a reaction system. For example, the mean change of the state $\mathbf{Y}_t$ at time t is given by

$$E[\mathbf{Y}_{t+dt} - \mathbf{Y}_t | \mathbf{Y}_t] = \mathbf{V}\mathbf{h}(\mathbf{Y}_t, \boldsymbol{\theta})dt. \tag{9.3}$$

Therefore, the system is *stationary* in mean if

$$\mathbf{V}\mathbf{h}(\mathbf{y}, \boldsymbol{\theta}) = 0.$$

It is possible to find the mean stationary states $\mathbf{y}$ by solving the above equation. In particular, one needs to find the null space of $\mathbf{V}$ and then match it with $\mathbf{h}(\mathbf{y}, \boldsymbol{\theta})$.

Example 9.1 *We consider a fictive four reaction system (Wilhelm, 2009) consisting of four substrates* $Y_1, \ldots, Y_4$*:*

$$Y_2 + Y_4 \xrightarrow{\theta_1} 2Y_3 \tag{9.4}$$

$$2Y_3 \xrightarrow{\theta_2} Y_3 + Y_4 \tag{9.5}$$

$$Y_3 + Y_4 \xrightarrow{\theta_3} Y_1 + Y_4 \tag{9.6}$$

$$Y_3 \xrightarrow{\theta_4} Y_1 \tag{9.7}$$

Equation 4 is called a unary *reaction, as its occurrence depends on the presence of only one molecule of one type of substrate. The other reactions are* binary *reactions, since they depend on two substrates. In particular, reaction 2 is a* homeo binary *reactions, whereas reactions 1 and 3 are* hetero binary *reactions, since they involve one and two types of substrates, respectively. The hazards of the four reactions are given as*

$$\mathbf{h}(\mathbf{Y}, \boldsymbol{\theta}) = \begin{bmatrix} Y_2 Y_4 \theta_1 \\ Y_3(Y_3 - 1)\theta_2/2 \\ Y_3 Y_4 \theta_3 \\ Y_3 \theta_4 \end{bmatrix},$$

whereas the stoichiometric matrix is given as

$$\mathbf{V} = \begin{bmatrix} 0 & 0 & 1 & 1 \\ -1 & 0 & 0 & 0 \\ 2 & -1 & -1 & -1 \\ -1 & 1 & 0 & 0 \end{bmatrix}$$

You can try to find the null space of $\mathbf{V}$, *but as* Y_2 *is only consumed and* Y_1 *is only produced, it can be shown that only* $\mathbf{y} = (0,0,0,0)$ *is the only stationary point for the overall system. However, it is more interesting if we only concentrate on the "main" substrates* Y_3 *and* Y_4*. The null space of*

$$\mathbf{V}_{3:4} = \begin{bmatrix} 2 & -1 & -1 & -1 \\ -1 & 1 & 0 & 0 \end{bmatrix}$$

can be found by using the reduced row echelon form, i.e.,

$$\left[\begin{array}{cccc|c} 2 & -1 & -1 & -1 & 0 \\ -1 & 1 & 0 & 0 & 0 \end{array}\right] \Longrightarrow \left[\begin{array}{cccc|c} 0 & 1 & -1 & -1 & 0 \\ -1 & 1 & 0 & 0 & 0 \end{array}\right]$$

$$\Longrightarrow \left[\begin{array}{cccc|c} 0 & 1 & -1 & -1 & 0 \\ -1 & 0 & 1 & 1 & 0 \end{array}\right].$$

The last matrix means

$$\begin{aligned} h_2 - h_3 - h_4 &= 0 \\ -h_1 \quad + h_3 + h_4 &= 0 \end{aligned}$$

and therefore, the null space of $V_{3:4}$ *is*

$$N(\mathbf{V}_{3:4}) = \{\mathbf{v} \mid \mathbf{v} = h_3 \begin{bmatrix} 1 & 1 & 1 & 0 \end{bmatrix}^T + h_4 \begin{bmatrix} 1 & 1 & 0 & 1 \end{bmatrix}^T\}.$$

This means that besides the trivial $(0,0)$, *there are other stationary states* $\mathbf{y}$, *given by the solution of*

$$N(\mathbf{V}_{3:4}) = \mathbf{h}(\mathbf{y}, \boldsymbol{\theta}).$$

In particular, this is the solution of

$$\begin{cases} y_4 y_2 \theta_1 = y_3(y_3 - 1)\theta_2/2 \\ y_4 y_2 \theta_1 = y_3 y_4 \theta_3 + y_3 \theta_4 \end{cases}$$

Simplifying the above system, we get that the stationary solutions y *are given by*

$$\begin{cases} y_4 = -\frac{\theta_2/2+\theta_4}{\theta_3} + \frac{\theta_2}{2\theta_3} y_3 \\ \frac{\theta_2\theta_3}{2} y_3^2 - \left(\frac{\theta_2\theta_3}{2} + \frac{\theta_1\theta_2}{2} y_2\right) y_3 + y_2(\theta_2/2 + \theta_4)\theta_1 = 0 \end{cases}$$

Solving the quadratic equation, we obtain

$$\begin{cases} y_4 = -\frac{\theta_2/2+\theta_4}{\theta_3} + \frac{\theta_2}{2\theta_3} y_3 \\ y_3 = \frac{1+\frac{\theta_1}{\theta_3} y_2 \pm \sqrt{(1+\frac{\theta_1}{\theta_3} y_2)^2 - 8\theta_1(\theta_2/2+\theta_4) y_2}}{2} \end{cases}.$$

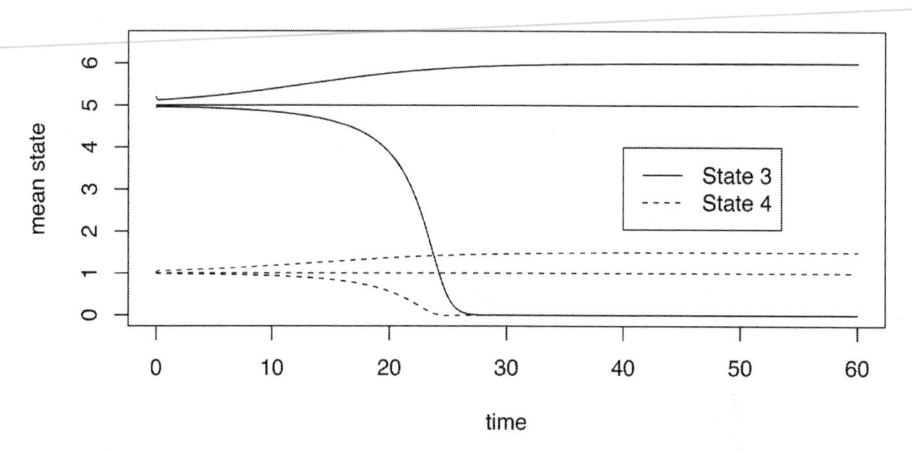

FIGURE 9.1
For $\boldsymbol{\theta} = (1, 1, 1, 1)$ and y_2 fixed at 10, there are three stationary states. The stationary state $(y_3 = 5, y_4 = 1)$ is unstable and slight deviations from that state will result in the convergence of the mean to either $(0, 0)$ or $(6, 1.5)$.

We note that eventually $y_2 \downarrow 0$ as it only appears on the left-hand side of the chemical reactions and therefore there are no additional positive *stationary points for (y_3, y_4). However, if we could keep the amount of y_2 constant due to outside intervention, then we can have 2 additional stationary states. For example, if $\boldsymbol{\theta} = (1, 1, 1, 1)$ and $y_2 \equiv 10$, then we get the stationary states $(y_3 = 6, y_4 = 1.5)$ and $(y_3 = 5, y_4 = 1)$. By simulating the mean process,*

$$\boldsymbol{\mu}_{t+dt} = \boldsymbol{\mu}_t + dt\mathbf{V}\mathbf{h}(\boldsymbol{\mu}_t, \boldsymbol{\theta}),$$

as shown in Figure 9.1 it is clear that the second equilibrium state is unstable. Given that the system is stochastic, the unstable system is in practice invisible.

9.2 Stochastic Models for Particle Dynamics

Whereas until now in this book we have encountered uncertainty typically as as measurement uncertainty, chemical reactions are themselves stochastic events that introduces an additional level of stochasticity in the system. In this section we present a simple simulation scheme that describes a realistic system of chemical reactions affecting the concentrations of multiple particles. Although it is simple to apply this so-called *Gillespie algorithm*, it turn out that its continuous approximation, called the *Euler-Maruyama algorithm*, is more useful from a statistical point of view.

9.2.1 Gillespie algorithm for simulating reactions

In this section, you will learn how to simulate a reaction system of r reactions with p substrates. This algorithm is known as the *Gillespie algorithm* (Gillespie, 1977) for simulating coupled chemical reactions. The simulation is stochastic and therefore will always result in different patterns, especially if the total number of molecules in the system is low.

Given a starting point $\mathbf{Y}_0$, the reaction equations and the reaction rates, the algorithm iterates the following steps:

1. Start at $t = 0$ with $\mathbf{Y}_0$ molecules of the p substrates.
2. Simulate an exponential time

$$\Delta T \sim \text{Exp}\left(\sum_{j=1}^{r} h_j(\mathbf{Y}_t, \boldsymbol{\theta})\right)$$

   for a reaction to occur.
3. Select the reaction of interest via multinomial draw,

$$J \sim \text{multinomial}\left((h_1(\mathbf{Y}_t, \boldsymbol{\theta}), \ldots, h_r(\mathbf{Y}_t, \boldsymbol{\theta})) \,/ \sum_{j=1}^{r} h_j(\mathbf{Y}_t, \boldsymbol{\theta})\right).$$

4. Update the state with the net-effect of the Jth reaction,

$$\mathbf{Y}_{t+\Delta T} = \mathbf{Y}_t + V_{.J}$$

5. Set $t \leftarrow t + \Delta t$.
6. Return to step 2.

The algorithm makes use of the fact that the minimum of a set of exponential distributions is again exponentially distributed (step 2), the location of the minimum is multinomially distributed (step 3) and that the exponential distribution is memoryless (step 6). In Exercise 3 you will explore this further.

Example 9.2 *Figure 9.2 shows a single simulation using the Gillespie algorithm of the system described above (Wilhelm, 2009). The system parameter are taken as* $\boldsymbol{\theta} = (1, 1, 1, 1)$, y_2 *fixed at* 10 *and we start the system at the instable mean equilibrium state* ($\mu_3 = 5, \mu_4 = 1$). *Although the instable state is invisible in the simulation, the process first hoovers around the "stable" stationary mean state* ($\mu_3 = 6, \mu_4 = 1.5$). *However, due to mere randomness, eventually the process converges to the only stochastically stable state* ($y_3 = 0, y_4 = 0$).

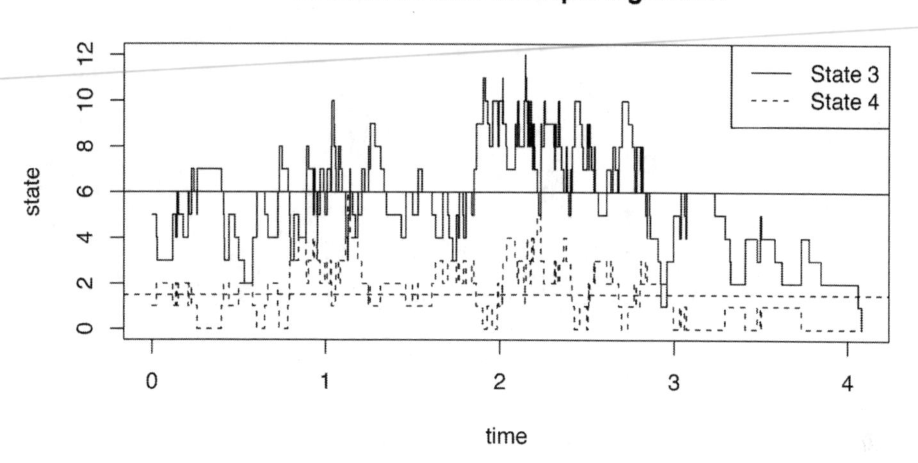

FIGURE 9.2
For $\boldsymbol{\theta} = (1, 1, 1, 1)$ and y_2 fixed at 10, the stochastic process, simulated here by the Gillespie algorithm, initially fluctuates around the mean stationary state $(\mu_3 = 6, \mu_4 = 1.5)$ before collapsing in the stochastically stationary state $(y_3 = 0, y_4 = 0)$.

9.2.2 Euler-Maruyama approximation

Although the true reaction process is a discrete system, where each state is represented as the number of molecules of a certain type, it is convenient sometimes to consider a continuous approximation. In this section we describe this so-called *diffusion approximation* as a tool for simulation. However, given the exact and simple Gillespie algorithm, which we described in the previous section, this is not really the importance of the diffusion approximation. Instead, its real use is in inference, as we will discover in section hereafter.

Let dt be an infinitesimal amount of time, then the change from $\mathbf{Y}_t$ to $\mathbf{Y}_{t+dt}$ can be described in terms of mean and variance, i.e.,

$$\begin{aligned} E(\mathbf{Y}_{t+dt} - \mathbf{Y}_t | \mathbf{Y}_t)/dt &= \boldsymbol{\mu}(\mathbf{Y}, \boldsymbol{\theta}) \\ &= \mathbf{V}\mathbf{h}(\mathbf{Y}_t, \boldsymbol{\theta}) \\ V(\mathbf{Y}_{t+dt} - \mathbf{Y}_t | \mathbf{Y}_t)/dt &= \boldsymbol{\beta}(\mathbf{Y}, \boldsymbol{\theta}) \\ &= \mathbf{V}\text{diag}\{\mathbf{h}(\mathbf{Y}, \boldsymbol{\theta})\}\mathbf{V}' \end{aligned}$$

The continuous *Euler-Maruyama approximation* of the true discrete reaction process considers possibly larger time increments Δt and approximates the increments of the process by means of a multivariate normal variable,

$$\mathbf{Y}_{t+\Delta t} = \mathbf{Y}_t + N(\boldsymbol{\mu}(\mathbf{Y}_t, \boldsymbol{\theta})\Delta t,\ \boldsymbol{\beta}(\mathbf{Y}_t, \boldsymbol{\theta})\Delta t).$$

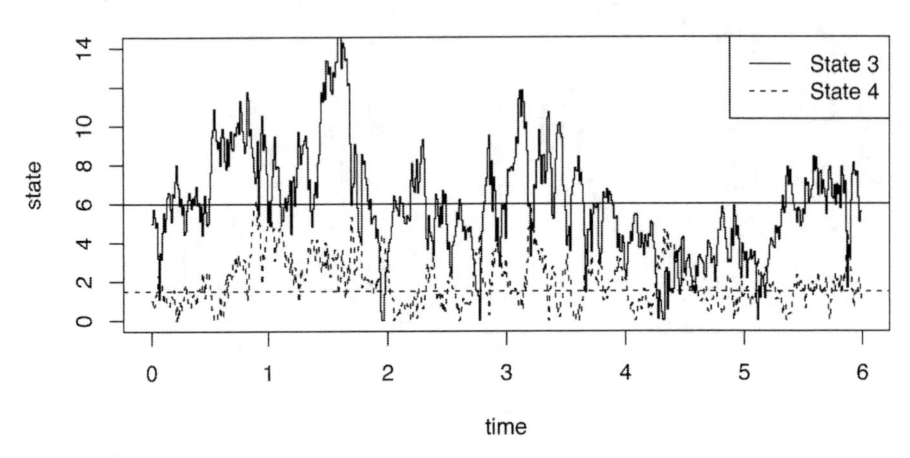

FIGURE 9.3
For $\boldsymbol{\theta} = (1, 1, 1, 1)$ and y_2 fixed at 10, the stochastic process, simulated here by the Euler-Maruyama approximation, fluctuates around the mean stationary state ($\mu_3 = 6, \mu_4 = 1.5$) and never collapses to the stationary state ($y_3 = 0, y_4 = 0$).

Special care has to be paid to the fact that the states cannot become negative. In practice, any update that results in a negative value can be truncated at zero.

Example 9.3 *Figure 9.3 shows a single simulation using the Euler-Maruyama approximation of the system described above (Wilhelm, 2009). We take again the same system parameters as before,* $\boldsymbol{\theta} = (1, 1, 1, 1)$, *fix state* y_2 *at* 10 *and we start the system at the instable mean equilibrium state* ($\mu_3 = 5, \mu_4 = 1$). *Again, the instable state is completely invisible in the simulation, as expected. This time the process fluctuates around the "stable" stationary mean state* ($\mu_3 = 6, \mu_4 = 1.5$). *Due to mere continuous approximation, the process never converges exactly to the only stable state* ($y_3 = 0, y_4 = 0$), *as it did do in the Gillespie algorithm.*

9.3 Estimating Reaction Rates

In this section you will learn how to estimate the reaction rates of the chemical system when observing a time-series of the substrates involved. By combining the Euler-Maruyama approximation with the (approximate) method of moments, the procedure can be formulated as a linear regression problem, which you encountered in Chapter 7. Using the standard linear regression tools, it is possible to extract the reaction rates from this highly non-linear system. This method is called the *local linear approximation (LLA)*.

Consider conducting an experiment of a reaction system where at time points $t_0, \dots, t_n$ the system is measured, resulting in measurements $\mathbf{Y} = \{\mathbf{Y}_{t_0}, \dots, \mathbf{Y}_{t_n}\}$ of the system. Each measurement $\mathbf{Y}_t$ consists of p measurements of the individual substrates. Consider the $k = 1, \dots, n$ increments in the system,

$$\Delta\mathbf{Y}_k = \mathbf{Y}_{t_k} - \mathbf{Y}_{t_{k-1}}.$$

From the Euler-Maruyama approximation and the fact that the hazard function is linear in $\boldsymbol{\theta}$ it is possible to write

$$\begin{aligned} E\Delta\mathbf{Y}_k &= (t_k - t_{k-1})\mathbf{V}\mathbf{h}(\mathbf{Y}_{t_{k-1}}, \boldsymbol{\theta}) \\ &= \underbrace{(t_k - t_{k-1})\mathbf{V}\mathbf{H}(\mathbf{Y}_{t_{k-1}})}_{\mathbf{X}_k}\boldsymbol{\theta}, \end{aligned}$$

since according to (9.2) the rate $h_j(\mathbf{Y}, \boldsymbol{\theta}) = H_j(\mathbf{Y})\theta_j$, where $H_j(\mathbf{Y}) = \prod_{i=1}^{p} \binom{Y_i}{k_{ij}}$. Therefore, we can write the following regression equation

$$\Delta\mathbf{Y} = \mathbf{X}\boldsymbol{\theta} + \epsilon$$

where $\Delta\mathbf{Y} = \begin{bmatrix} \Delta\mathbf{Y}_1 \\ \vdots \\ \Delta\mathbf{Y}_n \end{bmatrix}$ is a $np \times 1$ response vector, $\mathbf{X} = \begin{bmatrix} \mathbf{X}_1 \\ \vdots \\ \mathbf{X}_n \end{bmatrix}$ is a $np \times p$ model matrix and ϵ a $np \times 1$ vector of residuals. The least squares solution for $\boldsymbol{\theta}$ is given by

$$\hat{\boldsymbol{\theta}} = (\mathbf{X}^T\mathbf{X})^{-1}\mathbf{X}^T\Delta\mathbf{Y}.$$

Uncertainty estimates for the parameters can be obtained from the regression output.

A sanity check on the estimates are important, however. Given that the reaction rates cannot be negative, one needs to adjust the estimation procedure, if the above estimation procedure results in negatives estimates. In that case, the least squares criterion needs to be supplemented with a linear constraint,

$$\hat{\boldsymbol{\theta}} = \arg\min_{\boldsymbol{\theta}} (\Delta\mathbf{Y} - \mathbf{X}\boldsymbol{\theta})^T(\Delta\mathbf{Y} - \mathbf{X}\boldsymbol{\theta}) \text{ subject to } \boldsymbol{\theta} \geq 0.$$

This is a so-called quadratic programming problem and can be solved using, e.g., the `nnls` or `quadprog` packages in `R`.

Example 9.4 (Wilhelm, 2009) *Consider the data in Table 9.1, visualized in Figure 9.4. It is a subsample of data simulated by Gillespie algorithm using* $\boldsymbol{\theta} = (1, 1, 1, 1)$.

TABLE 9.1
3 longitudinal measurements of the Wilhelm (2009) system.

Time ($\times 10^{-3}$)	**0**	**12**	**30**
Y_1	100	138	176
Y_2	100	60	33
Y_3	50	69	49
Y_4	50	33	42

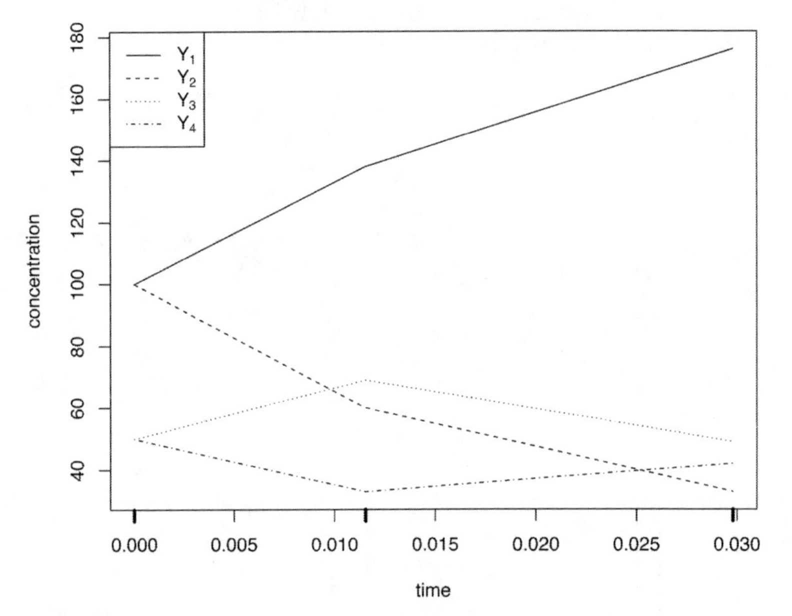

FIGURE 9.4
Subsample of 3 longitudinal observations of the Wilhelm (2009) system for $\boldsymbol{\theta} = (1, 1, 1, 1)$.

The local linear approximation $\Delta \mathbf{Y} = \mathbf{X}\boldsymbol{\theta} + \boldsymbol{\epsilon}$ *can be written as*

$$\begin{bmatrix} 38 \\ -40 \\ 19 \\ -17 \\ 38 \\ -27 \\ -20 \\ 9 \end{bmatrix} = \begin{bmatrix} 0.0 & 0.0 & 28.8 & 0.6 \\ -57.7 & 0.0 & 0.0 & 0.0 \\ 115.3 & -14.1 & -28.8 & -0.6 \\ -57.7 & 14.1 & 0.0 & 0.0 \\ 0.0 & 0.0 & 41.6 & 1.3 \\ -36.2 & 0.0 & 0.0 & 0.0 \\ 72.3 & -42.8 & -41.6 & -1.3 \\ -36.2 & 42.8 & 0.0 & 0.0 \end{bmatrix} \begin{bmatrix} \theta_1 \\ \theta_2 \\ \theta_3 \\ \theta_4 \end{bmatrix} + \begin{bmatrix} e_1 \\ e_2 \\ e_3 \\ e_4 \\ e_5 \\ e_6 \\ e_7 \\ e_8 \end{bmatrix}.$$

This results in a least squares estimate of $\hat{\boldsymbol{\theta}} = \begin{bmatrix} 0.61 & 0.74 & 2.40 & -51.04 \end{bmatrix}^T$, *which falls outside the feasible range. The quadratic programming solution with the additional constraint that* $\boldsymbol{\theta} \geq 0$, *results in* $\hat{\boldsymbol{\theta}} = \begin{bmatrix} 0.55 & 0.60 & 1.00 & 0.00 \end{bmatrix}^T$, *which is not unreasonable given the small amount of observations available.*

9.4 Mean-Field Approximation of Reaction System

There are two ways one can approximate the non-linear stochastic reaction system. In the local linear approximation the aim is to keep the underlying stochastic nature of the system, but to compromise on the non-linear dynamics by means of piece-wise linear forecasts. This approach is particularly useful in small systems with few molecules where the stochastic component important. However, larger systems with many particles it often turns out to be more important to keep the non-linear interactions, but to drop any consideration of the underlying stochastic interactions. This is what we explore in this section.

9.4.1 Chemical reaction system as ODE

We are interested in how a chemical reaction system, such as the one described in (9.1), develops *on average* when started at $\mathbf{Y}(0)$. Defining the mean function $\boldsymbol{\mu}_{\boldsymbol{Y}}(t) = E[\boldsymbol{Y}_t | \boldsymbol{Y}_0]$, it is clear from (9.3) that $\boldsymbol{\mu}_{\boldsymbol{Y}}(t)$ is the solution of the ordinary differential equation

$$\begin{aligned} \frac{d\boldsymbol{\mu}_{\boldsymbol{Y}}(t)}{dt} &= \boldsymbol{V}\boldsymbol{h}(\boldsymbol{\mu}_{\boldsymbol{Y}}(t), \boldsymbol{\theta}) \\ &= \underbrace{\boldsymbol{V}\text{diag}\{\boldsymbol{\theta}\}}_{\boldsymbol{P}_{\boldsymbol{\theta}}} \boldsymbol{H}(\boldsymbol{\mu}_{\boldsymbol{Y}}(t)). \end{aligned} \tag{9.8}$$

Given the polynomial nature of $\boldsymbol{H}$, where $H_j(\boldsymbol{\mu}_{\boldsymbol{Y}}) = \prod_{i=1}^{p} \binom{\mu_{Y i}}{k_{ij}}$, in general there is no analytical solution. However, in the special case of unitary reactions, an analytical solution is possible.

Example 9.5 (Circular reaction system) *Consider the following circular unitary reaction system,*

$$\begin{aligned} Y_1 &\xrightarrow{\theta_1} Y_2 \\ Y_2 &\xrightarrow{\theta_2} Y_3 \\ Y_3 &\xrightarrow{\theta_3} Y_1 \end{aligned}$$

Following (9.8) we get the following differential equation,

$$\frac{d\boldsymbol{\mu}_{\boldsymbol{Y}}(t)}{dt} = \boldsymbol{P}_{\boldsymbol{\theta}} \boldsymbol{\mu}_{\boldsymbol{Y}}(t),$$

since $\mathbf{H}(\boldsymbol{\mu}_{\mathbf{Y}}(t)) = \boldsymbol{\mu}_{\mathbf{Y}}(t)$ *with*

$$\boldsymbol{P}_{\boldsymbol{\theta}} = \begin{bmatrix} -\theta_1 & 0 & \theta_3 \\ \theta_1 & -\theta_2 & 0 \\ 0 & \theta_2 & -\theta_3 \end{bmatrix}.$$

Bringing the right-hand side over to the left and multiplying by the standard integrating factor $I(t) = e^{\int -\boldsymbol{P}_{\boldsymbol{\theta}} dt} = e^{-\boldsymbol{P}_{\boldsymbol{\theta}} t}$, *where the matrix exponential is defined as* $e^{\mathbf{A}} = \sum_{i=0}^{\infty} \frac{\mathbf{A}^i}{i!}$, *we obtain*

$$e^{-\boldsymbol{P}_{\boldsymbol{\theta}} t} \left(\frac{d\boldsymbol{\mu}_{\mathbf{Y}}(t)}{dt} - \boldsymbol{P}_{\boldsymbol{\theta}} \boldsymbol{\mu}_{\mathbf{Y}}(t) \right) = 0.$$

Integrating both sides, we get a nice simplification on the left-hand side,

$$e^{-\boldsymbol{P}_{\boldsymbol{\theta}} t} \boldsymbol{\mu}_{\mathbf{Y}}(t) = \boldsymbol{\mu}_0,$$

where $\boldsymbol{\mu}_0$ *is an arbitrary constant. By multiplying both sides with* $e^{\boldsymbol{P}_{\boldsymbol{\theta}} t}$ *we get*

$$\boldsymbol{\mu}_{\mathbf{Y}}(t) = e^{\boldsymbol{P}_{\boldsymbol{\theta}} t} \boldsymbol{\mu}_0.$$

Plugging in $t = 0$, *we get that* $\boldsymbol{\mu}_0 = \boldsymbol{\mu}_{\mathbf{Y}}(0)$. *For example, if* $\boldsymbol{\mu}_{\mathbf{Y}}(0) = (100, 100, 100)$ *and* $\boldsymbol{\theta} = (.7, 1, 1.1)$, *we obtain the mean dynamics as shown in Figure 9.5.*

9.4.2 Estimating reaction rates

The stochastic fluctuations that make every run different in small systems get evened out in larger systems. The main source of variation in these systems is measurement error. In particular, we consider n observations at locations $t_1 \leq t_2 \leq \ldots \leq t_n$ and that each observation is given as

$$\mathbf{Y}_{t_i} = \boldsymbol{\mu}_{\mathbf{Y}}(t_i) + \epsilon_i,$$

with $i = 1, \ldots, n$ and $V(\epsilon_i) = \sigma^2$. Just as before, our aim is to estimate the underlying dynamics of the system. Using smoothing of the data $\mathbf{y}$ you can obtain an estimate $\hat{\boldsymbol{\mu}}_{\mathbf{Y}}(t)$ of the p mean trajectories $\mu_Y(t)$ on $[t_1, t_n]$. Given that function $\hat{\boldsymbol{\mu}}_{\mathbf{Y}}(t)$, the idea is then to find the best possible $\boldsymbol{\theta}$, so that one obtains

$$\begin{aligned} \frac{d\hat{\boldsymbol{\mu}}_{\mathbf{Y}}(t)}{dt} &\approx \mathbf{V}\mathbf{h}(\hat{\boldsymbol{\mu}}_{\mathbf{Y}}(t), \boldsymbol{\theta}) \\ &= \underbrace{\mathbf{V}\text{diag}\{\mathbf{H}(\hat{\boldsymbol{\mu}}_{\mathbf{Y}}(t))\}}_{\mathbf{g}(t)} \boldsymbol{\theta}. \end{aligned} \tag{9.9}$$

which is linear in $\boldsymbol{\theta}$. By integrating both sides, we obtain

$$\hat{\boldsymbol{\mu}}_{\mathbf{Y}}(t) \approx \boldsymbol{\mu}_0 + \underbrace{\int_{t_1}^{t} \mathbf{g}(s) ds}_{\mathbf{G}(t)} \boldsymbol{\theta},$$

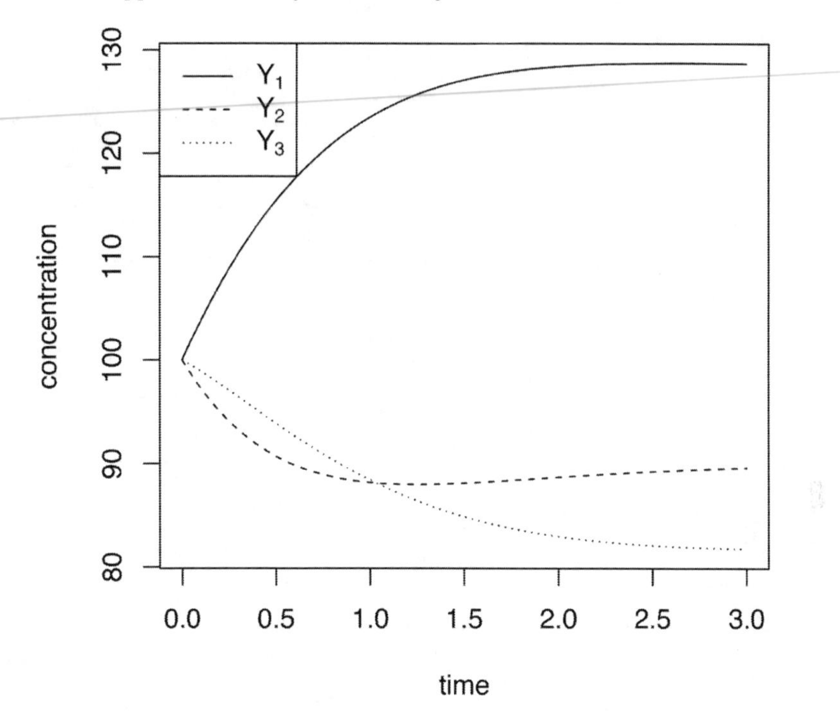

FIGURE 9.5
Analytic solution of the ODE for $\mu_Y(t)$ for example system 9.5 with $\boldsymbol{\theta} = (0.7, 1, 1.1)$ and $\boldsymbol{\mu_Y}(0) = (100, 100, 100)$.

where $\boldsymbol{\mu_0} = \boldsymbol{\mu_Y}(t_1)$ is the "initial value" of the ODE. The "best" is defined as the one that minimizes the integrated difference between the left and right-hand sides.

$$
\begin{aligned}
(\hat{\boldsymbol{\theta}}, \hat{\boldsymbol{\mu}}_{\mathbf{0}}) &= \arg\min_{\boldsymbol{\theta},\boldsymbol{\mu_0}} \int_{t_1}^{t_n} ||\hat{\boldsymbol{\mu}}_{\boldsymbol{Y}}(t) - \boldsymbol{\mu_0} - \boldsymbol{G}(t)\boldsymbol{\theta}||^2 dt \\
&= \arg\min_{\boldsymbol{\theta},\boldsymbol{\mu_0}} \begin{bmatrix} \boldsymbol{\theta} \\ \boldsymbol{\mu_0} \end{bmatrix}^T \underbrace{\begin{bmatrix} \int_{t_1}^{t_n} \mathbf{G}^T(t)\mathbf{G}(t)dt & \int_{t_1}^{t_n} \mathbf{G}^T(t)dt \\ \int_{t_1}^{t_n} \mathbf{G}(t)dt & (t_n - t_1)\mathbf{I}_p \end{bmatrix}}_{\mathbf{D}} \begin{bmatrix} \boldsymbol{\theta} \\ \boldsymbol{\mu_0} \end{bmatrix} \\
&\quad - 2 \underbrace{\begin{bmatrix} \int_{t_1}^{t_n} \hat{\boldsymbol{\mu}}_{\boldsymbol{Y}}(t)^T \mathbf{G}(t)dt \\ \int_{t_1}^{t_n} \hat{\boldsymbol{\mu}}_{\boldsymbol{Y}}(t)^T dt \end{bmatrix}}_{\mathbf{d}} \begin{bmatrix} \boldsymbol{\theta} \\ \boldsymbol{\mu_0} \end{bmatrix}
\end{aligned}
$$

In principle, one has, again, an explicit least squares estimate for $\boldsymbol{\theta}$,

$$
\begin{bmatrix} \hat{\boldsymbol{\theta}} \\ \hat{\boldsymbol{\mu}}_{\mathbf{0}} \end{bmatrix} = \boldsymbol{D}^{-1}\boldsymbol{d}.
$$

However, given that negative estimates for $\boldsymbol{\theta}$ and $\boldsymbol{\mu_0}$ are not meaningful, in practice one may have to solve a quadratic programming problem. For

TABLE 9.2
Noisy data from the unitary circular reaction system encountered in Example 9.5.

Time	0	0.03	0.55	0.92	1.77	2.21	2.59
y_1	98	101	114	124	125	135	129
y_2	103	95	87	83	94	88	88
y_3	100	88	94	89	83	88	82

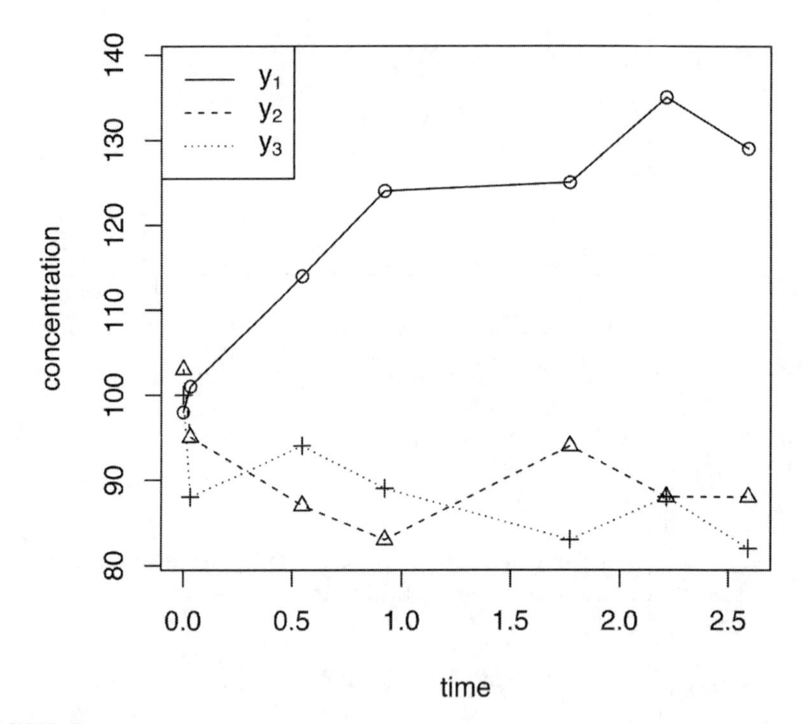

FIGURE 9.6
Seven observations of the unitary circular system, introduced in Example 9.5, together with a linear interpolation $\hat{\boldsymbol{\mu}}_{\boldsymbol{Y}}(t)$.

obtaining the matrix $\boldsymbol{D}$ and the vector $\boldsymbol{d}$, you may have to do some numerical integration. However, it is important to note that it is not necessary to obtain the solution of the system of differential equations. This is important, especially for non-unitary reaction systems that do not have an explicity solution.

Example 9.6 (Circular reaction system) *We consider the unitary circular reaction system described in Example 9.5. An experiment is performed gathering 7 observations, shown in Table 9.2. In this example, we consider a very simple estimation of the underlying mean μ_Y, namely by a linear interpolation of the data, as shown in Figure 9.6.*

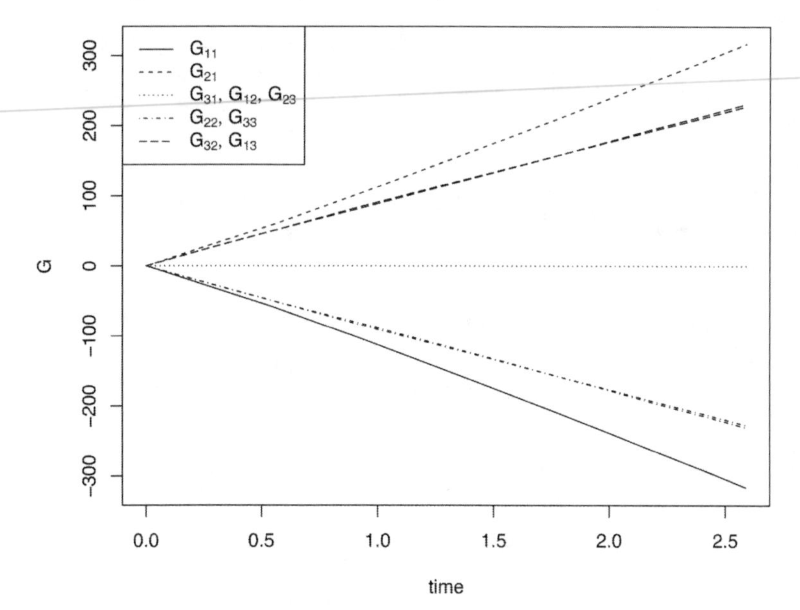

FIGURE 9.7
Calculation of intermediate matrix valued function $\mathbf{G}(t) = \int_{t_1}^{t} \mathbf{g}(s)\, ds$.

First, we have to define the $p = 3 \times 3 = r$ matrix-valued function $\mathbf{g}(t)$, which is different for every reaction system:

$$\begin{aligned}\mathbf{g}(t) &= \mathbf{V}\, diag\{\mathbf{H}(\hat{\boldsymbol{\mu}}_{\mathbf{Y}}(t))\} \\ &= \begin{bmatrix} -\hat{\mu}_{Y_1}(t) & 0 & \hat{\mu}_{Y_3}(t) \\ \hat{\mu}_{Y_1}(t) & -\hat{\mu}_{Y_2}(t) & 0 \\ 0 & \hat{\mu}_{Y_2}(t) & -\hat{\mu}_{Y_3}(t) \end{bmatrix}\end{aligned}$$

Plugging in the data (or, alternatively, a more sophisticated smoother of the data) and integrating it from t_1 subsequently to $t_1 \ldots t_7$ we obtain the matrix valued function $\mathbf{G}(t)$ as shown in Figure 9.7. Using $\mathbf{G}$ and $\hat{\boldsymbol{\mu}}_{\mathbf{Y}}(t)$ we construct the $(r+p) \times (r+p)$ matrix $\mathbf{D}$ and $(r+p)$-vector $\mathbf{d}$,

$$\mathbf{D} = \begin{bmatrix} 168488 & -63016 & -62722 & -394 & 394 & 0 \\ -63016 & 94354 & -46969 & 0 & -299 & 299 \\ -62722 & -46969 & 93538 & 299 & 0 & -299 \\ -394 & 0 & 299 & 3 & 0 & 0 \\ 394 & -299 & 0 & 0 & 3 & 0 \\ 0 & 299 & -299 & 0 & 0 & 3 \end{bmatrix}, \quad \mathbf{d} = \begin{bmatrix} -15066 \\ -927 \\ 12301 \\ 316 \\ 231 \\ 227 \end{bmatrix},$$

resulting in the estimates

$$\hat{\boldsymbol{\theta}} = \begin{bmatrix} 0.719 & 1.023 & 1.107 \end{bmatrix} \text{ and } \hat{\boldsymbol{\mu}}_{\mathbf{0}} = \begin{bmatrix} 103.7 & 97.8 & 97.2 \end{bmatrix}.$$

Given that these were simulated according to the system in Figure 9.5 with values $\boldsymbol{\theta} = (0.7, 1, 1.1)$ and $\boldsymbol{\mu}_{\mathbf{Y}}(0) = (100, 100, 100)$, the fit is remarkably good.

9.5 Exercises

1. **Stoichiometric coefficients.** In the following reactions a, b, c, d are the *stoichiometric coefficients* so that the equation is balanced. Find the smallest possible values to make the equation balanced.

   (a) Lead (IV) hydroxide and sulfuric acid react to produce lead (IV) sulfate and water. The reaction is given as

   $$a\ Pb(OH)_4 + b\ H_2SO_4 \longrightarrow c\ Pb(SO_4)_2 + d\ H_2O$$

   (b) Propane burns (i.e., reacts with oxygen) to produce water and carbon dioxide,

   $$a\ C_3H_8 + b\ O_2 \longrightarrow c\ CO_2 + d\ H_2O$$

2. **Ozone chemistry.** Ozone is probably best known as the constituent of layer in the upper atmosphere that protects life on earth from harmful UV radiation. It is also a powerful oxidant that has many industrial applications, despite being dangerous to animals and plants in concentrations over 0.1 ppm. In this exercise we consider the ozone chemistry in a pure oxygen environment, that can be described by the following 4 reactions.

   $$\begin{aligned} O_2 &\xrightarrow{\theta_1} 2O \\ O + O_2 &\xrightarrow{\theta_2} O_3 \\ O_3 &\xrightarrow{\theta_3} O + O_2 \\ O + O_3 &\xrightarrow{\theta_4} 2O_2 \end{aligned}$$

   (a) Derive the stoichiometric matrix $\mathbf{V}$. $V = \begin{bmatrix} 2 & -1 & 1 & -1 \\ -1 & -1 & 1 & 2 \\ 0 & 1 & -1 & -1 \end{bmatrix}$.

   (b) Find the null space $N(\mathbf{V})$ of $\mathbf{V}$. $N(V) = \{x \mid x = a\begin{bmatrix}1 & 1 & 0 & 1\end{bmatrix}^t + b\begin{bmatrix}0 & 1 & 1 & 0\end{bmatrix}^t\}$.

   (c) Find the hazard function $\mathbf{h}(\mathbf{O}, \boldsymbol{\theta})$ for the state (O_1, O_2, O_3). $h(o, \boldsymbol{\theta}) = \begin{bmatrix}\theta_1 o_2 & \theta_2 o_1 o_2 & \theta_3 o_3 & \theta_4 o_1 o_3\end{bmatrix}$.

   (d) Find an expression for a stable state besides the trivial. Hint: set the null space equal to the hazard function and derive two constraints for the stable state (O_1, O_2, O_3). ANSWER: $0_1 = \frac{\theta_1}{2\theta_2}(1+\sqrt{1+4\frac{\theta_2\theta_3}{theta_1\theta_4}})$, o_3 can be free and $o_2 = \frac{\theta_4}{\theta_1}o_1o_3$.

(e) Let $\boldsymbol{\theta} = (0.5, 0.25, 12, 1)$ and assume that the at the start the system consists only of 47 O_3 molecules. Calculate around which state the system will eventually start to fluctuate. You can either solve this analytically with the answer to the previous question or by means of a simulation. ANSWER: $o = (6, 60, 5)$.

3. **Properties of exponential distribution.** In the quasi-reaction process, r reactions compete for the molecules in order to be executed. Each reaction j is assumed to have an exponential waiting time T_j with rate h_j, i.e.,

$$P(T_j < t) = 1 - e^{-h_j t}.$$

(a) **Hazard function.** The formal definition of the hazard function of random variable T is

$$h(t) = \lim_{dt \downarrow 0} \frac{P(T < t + dt | T > t)}{dt}.$$

Show that the hazard function of the exponential random variable T_j with rate h_j is equal to h_j.

(b) **Minimum of exponentials.** Given that r reactions are competing to occur, show that the first reaction to occur has an exponential waiting time T with rate

$$h = \sum_{j=1}^{r} h_j$$

(c) **Multinomial reaction probability.** If we know that the first reaction occurred at $T = s$, show that the probability that it was reaction j, i.e., $T_j = s$, is given by:

$$P(T_j = s \mid T = s) = h_j / h.$$

(d) **Memoryless property.** If the first reaction to occur at time s is not j, show that the additional time for reaction j to occur is still distributed as an exponential distribution with rate h_j, i.e.,

$$P(T_j < s + t \mid T_j > s) = 1 - e^{-h_j t}.$$

This is known as the memoryless property of the exponential distribution.

4. **Simulating reaction system.** The quasi-reaction process $\{\mathbf{Y}_t\}$ with r reactions and p proteins is described by the stochastic differential equation

$$\frac{dP(\mathbf{Y};t)}{dt} = \sum_{j=1}^{r} \{h_j(\mathbf{Y}_t - \mathbf{V}_{\cdot j}, \boldsymbol{\theta}) P(\mathbf{Y} - \mathbf{V}_{\cdot j}; t) - h_j(\mathbf{Y}_t, \boldsymbol{\theta}) P(\mathbf{Y}; t)\},$$

where $h_j(\mathbf{Y}, \boldsymbol{\theta})$ is the hazard of reaction j in state Y and $\mathbf{V}$ the $p \times r$ stoichiometric or net effect matrix. The original process starts with an initial state $\mathbf{Y}_0 = \mathbf{y}_0$.

   (a) Write a function that implements the Gillespie algorithm. In particular, the function should take as input the two $r \times p$ matrices of stoichiometric coefficients for the reactants $\mathbf{K}$ and the products $\mathbf{S}$, rates $\boldsymbol{\theta}$ and initial value $\mathbf{y}_0$. Using the properties of the exponential distribution (see Exercise 3) simulate (i) each waiting time between reactions according to an exponential distribution with rate given by the sum of the hazards $\mathbf{h}$ of all reactions; (ii) the reaction according to a multinomial distribution with probability proportional to the hazard h.

   (b) Write a function that simulates an Euler-Maruyama approximation of the reaction network, taking as additional input an updating time increment Δt, where each increment of the process $\Delta \mathbf{Y}_t$ is simulated according to

$$\Delta \mathbf{Y}_t \sim N(\boldsymbol{\mu}(\mathbf{Y}_t, \boldsymbol{\theta})\Delta t, \boldsymbol{\beta}(\mathbf{Y}_t, \boldsymbol{\theta})\Delta t),$$

   where $\boldsymbol{\mu}(\mathbf{Y}_t, \boldsymbol{\theta}) = \mathbf{V}\mathbf{h}(\mathbf{Y}_t, \boldsymbol{\theta})$ and $\boldsymbol{\beta}(\mathbf{Y}, \boldsymbol{\theta}) = \mathbf{V}\text{diag}\{\mathbf{h}(\mathbf{Y}, \boldsymbol{\theta})\}\mathbf{V}'$.

5. **Simulating ozone process.** Consider the ozone system that we considered in Exercise 2 above.

   (a) Assuming that the reaction rates are given as $\boldsymbol{\theta} = (0.5, 0.25, 12, 1)$ and the initial value of the system as $\mathbf{o} = (0, 0, 47)$, use the two functions from the previous exercise to simulate the development of the process for 500 iterations.

      i. How does the Gillespie algorithm behave in the long run and how does it compare to the stationary state that you found in Exercise 2?

      ii. Compare the simulations. How good is the Euler-Maruyama approximation (make sure to adjust the dt to make the simulations comparable)?

6. **Local linear approximation (LLA).** Write a function in `R`, that takes as input

   - `tm`: n-dimensional vector of observation times
   - `dat`: $n \times p$ matrix of observations Y_{ti} of substrate i at time t

- •V: $p \times r$ dimensional stoichiometric matrix $\mathbf{V}$
- •K: $p \times r$ dimensional reactants matrix $\mathbf{K}$

and that gives as output the LLA estimate of the reaction rates $\boldsymbol{\theta}$ using either unconstrained least squares or constraint least squares.

7. **Ozone LLA.** Observations from the ozone system from Exercise 2 are given in the table below. Estimate the reaction rates of the four reactions, including their standard errors.

time (10^{-3})	62	67	73	78	83	87	89	91
O	7	9	10	8	6	8	8	6
O_2	429	434	432	436	437	439	439	437
O_3	45	41	42	40	40	38	38	40

8. **Mean reaction dynamics.** Consider the following unitary reaction system consisting of two substrates and three reactions:

$$\begin{aligned} \emptyset &\xrightarrow{\theta_1} Y_1 \\ Y_1 &\xrightarrow{\theta_2} Y_2 \\ Y_2 &\xrightarrow{\theta_3} \emptyset \end{aligned}$$

whereby the first reaction stands for the spontaneous appearance of the first type of substrate and the third reaction stands for the disintegration of the substrate of the second type.

(a) Write down the differential equation of the mean of the process in the following form,

$$\frac{d\boldsymbol{\mu}_{\boldsymbol{Y}}(t)}{dt} = \boldsymbol{P}_{\boldsymbol{\theta}}\boldsymbol{\mu}_{\boldsymbol{Y}}(t) + \mathbf{r}_{\boldsymbol{\theta}},$$

where $\boldsymbol{P}_{\boldsymbol{\theta}}$ is a 2×2 matrix and $\mathbf{r}_{\boldsymbol{\theta}} \in \mathbb{R}^2$ is a remainder vector.

(b) Use the *integration factor* to write

$$e^{-\boldsymbol{P}_{\boldsymbol{\theta}}t}\left(\frac{d\boldsymbol{\mu}_{\boldsymbol{Y}}(t)}{dt} - \boldsymbol{P}_{\boldsymbol{\theta}}\boldsymbol{\mu}_{\boldsymbol{Y}}(t)\right) = e^{-\boldsymbol{P}_{\boldsymbol{\theta}}t}\mathbf{r}_{\boldsymbol{\theta}}.$$

(c) Integrate on both sides to obtain:

$$e^{-\boldsymbol{P}_{\boldsymbol{\theta}}t}\boldsymbol{\mu}_{\boldsymbol{Y}}(t) = -e^{-\boldsymbol{P}_{\boldsymbol{\theta}}t}\boldsymbol{P}_{\boldsymbol{\theta}}{}^{-1}\mathbf{r}_{\boldsymbol{\theta}} + \mathbf{c},$$

where $\mathbf{c}$ is an arbitrary constant. By multiplying both sides with $e^{\boldsymbol{P}_{\boldsymbol{\theta}}t}$ derive an explicit solution $\boldsymbol{\mu}_{\boldsymbol{Y}}(t)$ of the ordinary differential equation.

(d) For reaction rates $\boldsymbol{\theta} = (15, 2, 1)$ and initial state $\boldsymbol{\mu}_{\boldsymbol{Y}}(0) = (30, 20)$, obtain an explicit expression of the solution of the differential equation, including the constant $\mathbf{c}$.

(e) Plot the solution of the differential equation on the interval $[0, 5]$.

9. **Function for inferring reaction rates.** In this exercise, you will programme an `R` function or a function in another computational environment that is able to infer the reaction rates ∂ from noisy observations $\mathbf{y}(t_1), \ldots, \mathbf{y}(t_n)$ from a reaction system,

$$\mathbf{y}(t_i) = \boldsymbol{\mu}_{\boldsymbol{Y}}(t_i) + \boldsymbol{\epsilon_i},$$

where $\boldsymbol{\mu}_{\boldsymbol{Y}}(t)$ is the solution of the differential equation $\frac{\boldsymbol{\mu}_{\boldsymbol{Y}}(t)}{dt} = \mathbf{V}\text{diag}\{H(\boldsymbol{\mu}_{\boldsymbol{Y}}(t))\}\boldsymbol{\theta}$.

(a) Write a function that estimates the kinetic parameters with the following two input and output consideration. The input of the function needs to be:

- `dat`: $n \times p$ observation matrix, where n is the number of observations and p the number of substrates.
- `tm`: a vector of length n indication the time points $t_1, \ldots, t_n$, corresponding to the rows of `dat`.
- `V`: The net-effect matrix of size $p \times r$, where r are the number of reactions.
- `K`: If the $p \times r$ reactant matrix does not correspond to the negative entries of `V`, then it needs to be given explicitly.
- `k`: corresponds to the number of basis-functions in the GAM smoothing of the data for the estimation of $\mu(t) = (\mu_1(t), \ldots, \mu_p(t))$.

The output of the function needs to be:

- `theta`: An estimate of the kinetic parameters $\boldsymbol{\theta}$.
- `mu0`: An estimate of the initial value $\boldsymbol{\mu}_{\boldsymbol{Y}}(t_1)$.

(b) **Ozone estimation (revisited).** Consider the data presented in Exercise 7 and infer the kinetic parameters using the ODE inference algorithm presented in Section 9.4.2.

i. The true parameter settings that simulated this setting were $\boldsymbol{\theta} = (0.5, 0.25, 12, 1)$. How well did your algorithms estimate these parameters?

ii. Argue why the local linear approximation algorithm discussed in Exercise 7 is more appropriate for these data.

10

Multivariate Exploration

In some chemical experimental settings it may happen that the n by p predictions matrix $\mathbf{X}$ is horizontal in the sense of having more columns than rows, $p > n$, that is, there are more variables than observations. Then the matrix $\mathbf{X}^T\mathbf{X}$ does not have an inverse and the variance of the regression estimator in Equation (7.9) becomes undefined. It may also happen that the columns in the predictions matrix are nearly linear dependent is such a way that the inverse of $\mathbf{X}^T\mathbf{X}$ does exist, but contains extremely large diagonal values. In such a situation the estimator $\hat{\boldsymbol{\beta}} = (\mathbf{X}^T\mathbf{X})^{-1}\mathbf{X}^T\mathbf{y}$ becomes misleading.

To investigate the dependencies between a set of chemical variables requires multivariate techniques that unravel the type of dependencies in the data matrix. By matrix decompositions it can be determined whether there are a few dominating eigenvalues from which a lower dimensional subspace can be defined that summarizes the most important directions of variation in the chemical data. By a principal components analysis the variables (columns) and the observations (rows) can be projected as points in a lower dimensional space for a closer inspection. The idea of a lower dimensional space can also be applied in the context of linear regression in order to predict a chemical outcome.

In this chapter you learn:

- In Section 10.1 a technique to visualize the type of dependencies between p types of chemical measurements.
- In Section 10.2 about three important matrix decompositions to investigate the dependencies between chemical variables and how to find a lower dimensional subspace that best summarizes the most important variations in the data.
- In Section 10.3 about a visualization that projects the variables and the observations in e.g., a two-dimensional space.
- In Section 10.4 how to perform a regression in case the predictions matrix $\mathbf{X}$ has more columns than rows.

DOI: 10.1201/9781003178194-10

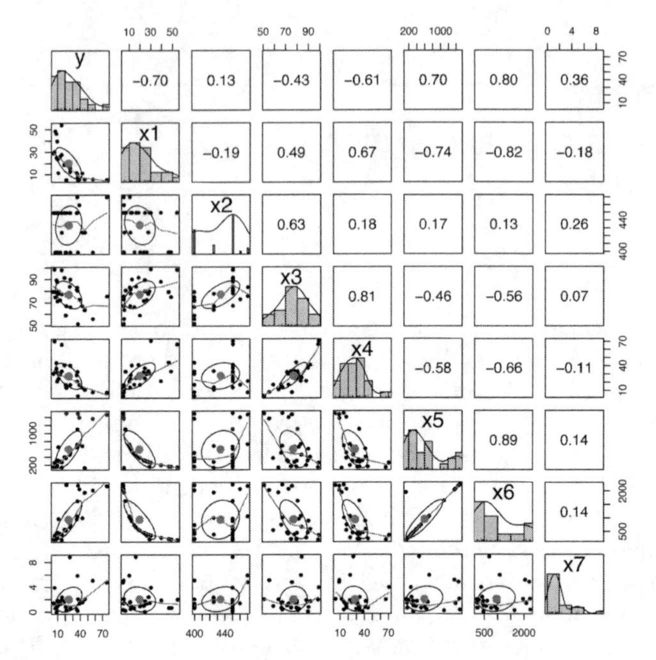

FIGURE 10.1
Pairs panel plot of the Belle Ayr Lique faction data.

10.1 Data Visualization

In a multivariate setting it often makes sense to plot the data in order to make observations about its properties. If there are p observed chemical variables, then one may construct a p by p panel with a histogram on the diagonal, a pairwise scatter plot in the cells below the main diagonal, and print pairwise correlation coefficients in the cells above the main diagonal. Such allows for a quick inspection of the pairwise dependencies between the chemical variables.

Example 10.1 *For the data from the Belle Ayr Liquefaction experiment (Cronauer et al., 1978) a pairs panel plot given in Figure 10.1 was constructed. The names of the variables for $x1$ to $x7$ are given in Table 7.5 and y is the CO2. Unfortunately the full variable names do not fit in the diagonal cells of a 8 by 8 panel plot. The splines in the pairs plot make non-linear dependencies visible between e.g., $x1$ and $x5$, and between $x1$ and $x6$. Almost all variables seem to violate the normal distribution, except possibly $x3$. Variable $x2$ attains only four different values. Variable $x5$ and $x6$ seem completely linearly dependent, except for a single outlier. The pairs plots as well as the correlation coefficients indicate weak dependencies between $x7$ and all other variables. It seems that y and $x1$ have a decreasing exponential type of dependency.*

10.2 Matrix Decomposition

Three matrix decompositions are given, each with their own properties and purposes. The so-called QR decomposition plays an important role in the computation of the linear model, in particular when the data set is large. The eigen decomposition of a correlations matrix of a set of chemical variables gives many important properties of the interdependencies between the variables. The singular value decomposition provides the basis for principal components analysis by which the variables and the observations can be projected onto a lower dimensional space.

10.2.1 QR decomposition

If the n by p matrix $\mathbf{X}$ has linearly independent columns, then there exists a decomposition

$$\mathbf{X} = \mathbf{QR}$$

where $\mathbf{Q}$ is column-wise orthonormal $\mathbf{Q}^T\mathbf{Q} = \mathbf{I}$, and $\mathbf{R}$ upper triangular. An algorithm to compute a QR decomposition is given by the **Gram-Schmidt orthogonalization** process: If we take $\mathbf{q}_1 = \mathbf{x}_1$, then we may take recursively for $k = 2, \cdots, p$, the columns of $\mathbf{Q}$ as

$$\mathbf{q}_k = \mathbf{x}_k - \sum_{j=1}^{k-1} \frac{\mathbf{q}_j^T \mathbf{x}_k}{\mathbf{q}_j^T \mathbf{q}_j} \mathbf{q}_j. \tag{10.1}$$

Using the projection matrix $\mathbf{H}_j = \mathbf{q}_j(\mathbf{q}_j^T\mathbf{q}_j)^{-1}\mathbf{q}_j^T$ we write

$$\begin{aligned}
\mathbf{q}_k &= \mathbf{x}_k - \sum_{j=1}^{k-1} \mathbf{q}_j(\mathbf{q}_j^T\mathbf{q}_j)^{-1}\mathbf{q}_j^T\mathbf{x}_k \\
&= \mathbf{x}_k - \sum_{j=1}^{k-1} \mathbf{H}_j\mathbf{x}_k \\
&= \left(\mathbf{I} - \sum_{j=1}^{k-1} \mathbf{H}_j\right) \mathbf{x}_k
\end{aligned}$$

in order to see that $\mathbf{q}_k$ are members of the space orthogonal to the space already spanned by the previous $\mathbf{x}_j$ columns ($j = 1, \cdots, k-1$). The coefficients can be collected into an upper triangular matrix $\mathbf{R}$ and the $\mathbf{q}_k$ columns into a matrix $\mathbf{Q}$ to complete the QR decomposition of the matrix $\mathbf{X}$. By placing the length of each variable in a diagonal matrix $\mathbf{D}$, we see that $\mathbf{QD}^{-1}\mathbf{DR}$ normalizes the columns of $\mathbf{Q}$ to have unit length. Then the matrix $\mathbf{QD}^{-1}$ gives the orthonormalization of the matrix $\mathbf{X}$.

Example 10.2 *In all software languages the least squares estimator of the linear model is computed by the QR decomposition. Note that we may augment the above* $\mathbf{Q}$ *matrix with orthogonal columns of unit length so that it becomes a square* n *by* n *matrix* $\mathbf{Q} = [\mathbf{Q}_1, \mathbf{Q}_2]$ *with the property* $\mathbf{Q}^T\mathbf{Q} = \mathbf{Q}\mathbf{Q}^T = \mathbf{I}$ *of column and row-wise orthonormality that characterizes a rotation matrix. Now the QR decomposition becomes*

$$\mathbf{X} = [\mathbf{Q}_1, \mathbf{Q}_2] \begin{bmatrix} \mathbf{R} \\ \mathbf{O} \end{bmatrix}.$$

We may use the fact that the Euclidian norm is invariant to a rotation matrix

$$||\mathbf{Q}^T\mathbf{y}||^2 = \mathbf{y}^T\mathbf{Q}\mathbf{Q}^T\mathbf{y} = \mathbf{y}^T\mathbf{y} = ||\mathbf{y}||^2.$$

Then we find the lower bound

$$\begin{aligned} ||\mathbf{y} - \mathbf{X}\beta||^2 &= ||\mathbf{Q}^T\mathbf{y} - \mathbf{Q}^T\mathbf{X}\beta||^2 \\ &= \left|\left| \begin{bmatrix} \mathbf{Q}_1^T\mathbf{y} \\ \mathbf{Q}_2^T\mathbf{y} \end{bmatrix} - \begin{bmatrix} \mathbf{R}\beta \\ \mathbf{O} \end{bmatrix} \right|\right|^2 \\ &= ||\mathbf{Q}_1^T\mathbf{y} - \mathbf{R}\beta||^2 + ||\mathbf{Q}_2^T\mathbf{y}||^2 \\ &\geq ||\mathbf{Q}_2^T\mathbf{y}||^2, \end{aligned}$$

which is attainable for $\widehat{\beta} = \mathbf{R}^{-1}\mathbf{Q}_1^T\mathbf{y}$. *It can be veryfied that this exaclty equals* $\widehat{\beta} = (\mathbf{X}^T\mathbf{X})^{-1}\mathbf{X}^T\mathbf{y}$.

10.2.2 Eigen decomposition

Let $\mathbf{R}$ denote a correlations matrix for p observed types of chemical measurements on n objects (sites) collected in a matrix $\mathbf{X}$. The i,jth element r_{ij} of $\mathbf{R}$ is the correlation between column i and column j, for which we have

$$r_{ij} = cor(\mathbf{x}_i, \mathbf{x}_j) = cor(\mathbf{x}_j, \mathbf{x}_i) = r_{ji}$$

so that the matrix is symmetric, a property which can be expressed as $\mathbf{R}^T = \mathbf{R}$. An eigenvalue/vector pair $(d, \mathbf{v})$ is defined by the property

$$\mathbf{R}\mathbf{v} = d\mathbf{v}. \tag{10.2}$$

In this equation we may use a scalar to normalize the eigenvector $\mathbf{v}$ to have unit length, that is $||\mathbf{v}|| = 1$. A way to search for an eigenvalue d is to write

$$(\mathbf{R} - d\mathbf{I})\mathbf{v} = \mathbf{0},$$

for the identity matrix $\mathbf{I}$ and noting that the matrix to the left is singular. Obviously such holds if and and only if its determinant zero. After properly collecting terms the determinant would give a polynomial of degree p in the variable d. From the fundamental theorem of algebra we know that there are maximally p different roots $d_1, \cdots, d_p$ to the equation $\det(\mathbf{R} - d\mathbf{I}) = 0$. For each solution d_k there is a vector $\mathbf{v}_k$ such that $(\mathbf{R} - d_k\mathbf{I})\mathbf{v}_k = \mathbf{0}$, forming an eigenvalue/vector pair $(d_k, \mathbf{v}_k)$. It is possible to analytically solve for a zero of such a polynomial up to degree 4 (Abramowitz et al., 1988), but for a larger degree this is not generally possible (Rosen, 1995).

To develop a numerical approach, we note that we are very close to a fixed point type of problem, because Equation (10.2) is equivalent to $\mathbf{v} = \mathbf{R}d^{-1}\mathbf{v}$. Also we have for positive eigenvalues that $\|\mathbf{R}\mathbf{v}\| = \|d\mathbf{v}\| = |d|\|\mathbf{v}\| = d$. After taking a starting vector $\mathbf{v}_1$ we may formulate a fixed point iterative scheme as

$$\mathbf{v}_{m+1} = \mathbf{R}\mathbf{v}_m\|\mathbf{R}\mathbf{v}_m\|^{-1}.$$

If the fixed point iterations converge, then we obtain

$$\mathbf{v}_m \to \mathbf{v} = \mathbf{R}\mathbf{v}\|\mathbf{R}\mathbf{v}\|^{-1} = \mathbf{R}\mathbf{v}d^{-1}.$$

Under weak conditions, it can be proven that the fixed point iterations do converge to the maximal eigenvalue/vector pair $(d_1, \mathbf{v}_1)$.

Recall that matrix containing pairwise correlations is symmetric matrix and therefore has real eigenvalues. Furthermore, since a correlations matrix is defined as a matrix inner product, it follows that all eigenvalues are non-negative. If we assume all eigenvalues to be distinct, then to find the second eigenvalue/vector pair, we note that we in fact seek the maximum eigenvalue of the matrix $(\mathbf{R} - d_1\mathbf{v}_1\mathbf{v}_1^T)$. That is, we seek the second pair $(d_2, \mathbf{v}_2)$ such that

$$(\mathbf{R} - d_1\mathbf{v}_1\mathbf{v}_1^T)\mathbf{v}_2 = d_2\mathbf{v}_2.$$

This pair can be found by the fixed point iterations. From the eigendecomposition in Equation (10.3) below, it can be seen that this second eigenvector $\mathbf{v}_2$ is not a member of the column space of $\mathbf{v}_1$. Using the symmetry $\mathbf{R}^T = \mathbf{R}$ we find

$$\begin{aligned} d_1\mathbf{v}_2^T\mathbf{v}_1 &= \mathbf{v}_2^T\mathbf{R}\mathbf{v}_1 = (\mathbf{v}_2^T\mathbf{R}\mathbf{v}_1)^T = \mathbf{v}_1^T\mathbf{R}\mathbf{v}_2 \\ &= d_2\mathbf{v}_1^T\mathbf{v}_2 \end{aligned}$$

Since the inner product is symmetric $\mathbf{v}_2^T\mathbf{v}_1 = (\mathbf{v}_2^T\mathbf{v}_1)^T = \mathbf{v}_1^T\mathbf{v}_2$, we have

$$0 = d_1\mathbf{v}_2^T\mathbf{v}_1 - d_2\mathbf{v}_1^T\mathbf{v}_2 = (d_1 - d_2)\mathbf{v}_1^T\mathbf{v}_2,$$

which implies $\mathbf{v}_1^T\mathbf{v}_2 = 0$, that is orthogonality of the eigenvectors. By repeating these steps we can find all eigenvectors and place these as columns next to each other in a matrix $[\mathbf{v}_1, \mathbf{v}_2, \cdots, \mathbf{v}_p] = \mathbf{V}$. The matrix $\mathbf{V}$ with eigenvectors is square and column-wise orthonormal, reason for which it is also row-wise orthonormal. That is, we have $\mathbf{V}^T\mathbf{V} = \mathbf{V}\mathbf{V}^T = \mathbf{I}$. It now follows that we have

$$\mathbf{R}\mathbf{V} = \mathbf{V}\mathbf{D},$$

from which we find, after post-multiplication by $\mathbf{V}^T$, the eigen-decomposition

$$\mathbf{R} = \mathbf{V}\mathbf{D}\mathbf{V}^T = \sum_{k=1}^{p} d_k \mathbf{v}_k \mathbf{v}_k^T. \tag{10.3}$$

The **eigendecomposition** gives the best lower rank approximation of the correlation matrix in the sense that for any matrix $\mathbf{S}$ of rank $r < p$ we have that

$$\min_{\mathbf{S}} \|\mathbf{R} - \mathbf{S}\| = \sum_{k=r+1}^{p} d_k^2,$$

where the minimum is attained by $\mathbf{S} = \sum_{k=1}^{r} d_k \mathbf{v}_k \mathbf{v}_k^T$. If the sum of the last $p - r$ eigenvalues is small, then the variations among the correlations can be summarized in a smaller dimensional space.

10.2.3 Singular value decomposition

For several purposes it is illuminating to consider the **singular value decomposition** of a n by p ($n > p$) matrix $\mathbf{X}$ with p chemical measurements on n objects. That is, there always exists a decomposition

$$\mathbf{X} = \mathbf{U}\mathbf{D}\mathbf{V}^T, \quad \text{where } \mathbf{U}^T\mathbf{U} = \mathbf{V}^T\mathbf{V} = \mathbf{V}\mathbf{V}^T = \mathbf{I}.$$

The matrix $\mathbf{U}$ contains the left singular vectors, $\mathbf{V}$ the right singular vectors of $\mathbf{X}$, and the diagonal matrix $\mathbf{D}$ contains the non-negative singular values in decreasing order $d_1 \geq d_2 \geq \cdots \geq d_p \geq 0$. The matrices are easily found from the eigen decomposition

$$\mathbf{X}^T\mathbf{X} = \mathbf{V}\mathbf{D}^2\mathbf{V}^T, \quad \text{where } \mathbf{V}^T\mathbf{V} = \mathbf{V}\mathbf{V}^T = \mathbf{I}$$

and $\mathbf{D}^2$ contains the non-negative eigen-values in decreasing order. Next, we take $\mathbf{U} = \mathbf{X}\mathbf{V}\mathbf{D}^{-1}$ to make the singular value decomposition complete.

The singular value decomposition gives the best lower rank approximation of the matrix $\mathbf{X}$ in the sense that for any matrix $\mathbf{S}$ of rank $r < p$ we have that

$$\min_{\mathbf{S}} \|\mathbf{X} - \mathbf{S}\| = \sum_{k=r+1}^{p} d_k^2,$$

where the minimum is attained by $S = \sum_{k=1}^{r} d_k \mathbf{u}_k \mathbf{v}_k^T$. If the sum of the last $p - r$ eigenvalues is small, then the variation in the data matrix $\mathbf{X}$ can be summarized well by a lower dimensional space.

10.3 Principal Components Analysis

Principal components analysis aims to find the direction of maximal variation in the data. For this purpose the columns of the data matrix $\mathbf{X}$ are standardized by column-wise centering and scaling the variances to unit length in order to make the objects (rows) and the chemical variables (columns) better comparable. To do this we compute the matrix $\mathbf{Y} = \mathbf{JXD}_n$, where the centering matrix $\mathbf{J} = \mathbf{I} - \mathbf{11}^T/n$, and the diagonal matrix $\mathbf{D}_n$ normalizes the column length. Then we have the **correlation matrix** $\mathbf{R} = \mathbf{Y}^T\mathbf{Y}$. From the singular value decomposition $\mathbf{Y} = \mathbf{UDV}^T$ we then find the matrices $\mathbf{U}, \mathbf{D}, \mathbf{V}$. The matrix $\mathbf{U}$ contains the **principal components** of the $\mathbf{Y}$. The first M columns $\mathbf{U}_M$ of the matrix $\mathbf{U}$ are the most important as these correspond to the directions of largest variation. The correlations between the principal components and the chemical variables are called the **loadings** which can be computed by

$$cor(\mathbf{X}, \mathbf{U}_M) = cor(\mathbf{Y}, \mathbf{U}_M) = \mathbf{Y}^T\mathbf{U}_M = \mathbf{V}_M\mathbf{D}_M.$$

The first M components explain $\sum_{j=1}^{M} d_j$ of the total variance p. In many cases it is reasonable to require each of the components to explain more than the variance of a single variable, which was set to one.

In case all correlations between the variables are positive, then the first eigenvector of the correlations matrix is non-negative. This implies that all the loadings are non-negative and also that the first principal component is given by a positive linear combination of the variables. In a bi-plot the variables fall in a singly quadrant.

Since the eigen values are continuously differentiable functions of the correlation matrix, the delta theorem can be applied for the construction of their 95% confidence intervals. Since these derivatives are quite complicated, instead of invoking the delta theorem we may apply the bootstrap the construct confidence intervals.

Example 10.3 *For the data from the Belle Ayr Liquefaction experiment (Cronauer et al., 1978) the eigenvalues of the correlation matrix were analyzed. The three largest eigenvalues with their 95% confidence intervals from the bootstrap are* 4.37 *95% CI [3.67, 4.83], 1.78 95% CI [1.327, 2.228], and 0.88 95% CI [0.6019, 1.1826]. Since the first two are significantly larger than one, each of these explain more variance than that of a single variable. The first explains 54.59 and the second 22.21 percent of the variance in the data.*

The first two eigenvectors and the loadings (correlations of the variables with the components) are given in Table 10.1. The variables y, x1, x3, x4, x5, and x6 have large correlations with the principal component and vary very closely in the direction of maximum variation. The variable x2 correlates largely whereas the variables x3 and x7 correlate moderately with the second component.

Example 10.4 *For the data from the Belle Ayr Liquefaction experiment (Cronauer et al., 1978) the singular value decomposition of the normalized variables was analyzed in the first two components. The values of the eigenvectors for each of the variables and the values of the runs (observations) on*

TABLE 10.1
Loadings and eigen vectors from Principal Components Analysis on the correlations of the Belle Ayr Lique faction data.

	Var	ev1	ev2	load 1	load2
CO_2	y	0.41	0.17	0.85	0.23
Space time (in min)	x1	−0.42	−0.13	−0.88	−0.17
Temperature (in degrees Celsius)	x2	0.00	0.69	0.00	0.92
Percent solvation	x3	−0.34	0.51	−0.71	0.68
Oil yield (g/100g MAF)	x4	−0.40	0.22	−0.84	0.29
Coal total	x5	0.41	0.13	0.86	0.17
Solvent total	x6	0.45	0.09	0.93	0.12
Hydrogen consumption	x7	0.11	0.39	0.23	0.52

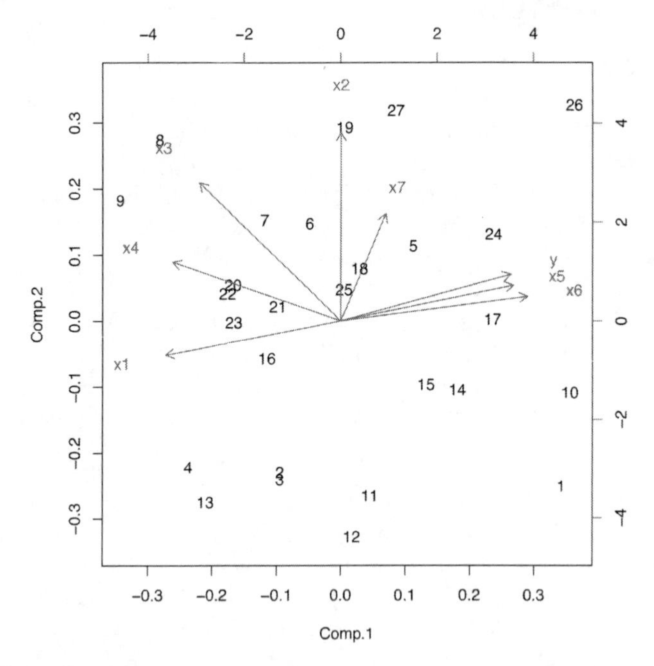

FIGURE 10.2
Pairs panel plot of the BelleAyrLiquefaction data.

the first two components are plotted simultaneously in a so-called bi-plot in Figure 10.2. Note that the values on the eigenvectors are proportional to the loadings (correlations) of the variables with the components.

The variables that stretch largely in the horizontal direction either to the right or to the left have large loadings on the first component. Variable $x2$ has a zero loading on the component 1 and a large loading on component 2. The variables y, $x5$, and $x6$ are in the positive direction of the principal component and $x1$, $x3$, $x4$ in the negative direction of the principal component.

The observations seem evenly spread in the space spanned by the first two components. The observations or runs seem evenly spread in the space spanned by the first two components.

Note that a bi-plot may reveal groups of observations in clusters of points in a lower dimensional principal component space.

Example 10.5 *Near-infrared spectra (NIR) and octane numbers of 60 gasoline samples became available from an experiment where the NIR spectra were measured using diffuse reflectance as log(1/R) from 900 nm to 1700 nm in 2 nm intervals, giving 401 wavelengths. The predictor matrix* $\mathbf{X}$ *has 60 rows and 401 columns, making it clearly horizontal. After centering its columns to have mean zero the first eigenvalue* d_1^2 *of* $\mathbf{X}^T\mathbf{X}$ *equals 2.605 and the sums of eigenvalues has a total of 3.590. The first principal component explains 72.56 percent of the variance in the predictor matrix. The second explains an additional 11.33 percent of the variance.*

10.4 Regression Using a Subspace

In case the predictor matrix $\mathbf{X}$ is horizontal, contains more columns than rows, the columns are linearly dependent. The inverse of the matrix $\mathbf{X}^T\mathbf{X}$ cannot be computed, reason for which the least squares estimator does not exit. In order to overcome this problem we may construct a new predictor matrix from collections of transformations of columns of $\mathbf{X}$.

10.4.1 Principal components regression

In order to overcome this problem of linear dependency, we collect those M linear combinations of columns from $\mathbf{X}$ into a matrix $\mathbf{Z}_M$ that have maximal variance $\mathbf{Z}_M = \mathbf{X}\mathbf{V}_M$. Next we use these as the predictor matrix and solve the minimization problem over the M-vector θ, that is

$$\widehat{\boldsymbol{\theta}} = \underset{\boldsymbol{\theta}}{\text{argmin}}\, \|\mathbf{y} - \mathbf{Z}_M\boldsymbol{\theta}\|^2 = (\mathbf{Z}_M^T\mathbf{Z}_M)^{-1}\mathbf{Z}_M^T\mathbf{y} = \mathbf{D}_M^{-2}\mathbf{Z}_M^T\mathbf{y}.$$

This technique is called Principal Components Regression which has the clear interpretation of revealing the amount of variance the components of largest variation in $\mathbf{X}$ can explain in the outcome $\mathbf{y}$. Since the columns of $\mathbf{Z}_M$ are orthogonal, the solution for its m-th element can be written as $\widehat{\theta}_m = d_m^{-2}\mathbf{z}_m^T\mathbf{y}$, for $m = 1, \cdots, M$. For a new data point $\mathbf{x}_{new}^T$ we can find the predicted value of the outcome Y_{new}. From the new $\widehat{\mathbf{z}}_{new}^T = (\mathbf{x}_{new} - \bar{\mathbf{x}})^T\mathbf{V}_M$, we find by the principal components regression vector that

$$\widehat{Y}_{new} = \hat{\mathbf{z}}_{new}^T\widehat{\boldsymbol{\theta}} = (\mathbf{x}_{new} - \bar{\mathbf{x}})^T\mathbf{V}_M\widehat{\boldsymbol{\theta}}.$$

TABLE 10.2
Cumulative percentage of variance explained by the first six components from principal components regression (PCR) and partial least squares (PLS) in the predictor matrix X and the outcome octane numbers y.

		Comp 1	Comp 2	Comp 3	Comp 4	Comp 5	Comp 6
PCR	X	72.57	83.90	90.86	95.46	96.70	97.66
	octane	18.99	19.62	46.50	97.69	97.78	97.79
PLS	X	70.97	78.56	86.15	95.40	96.12	96.97
	octane	31.90	94.66	97.71	98.01	98.68	98.93

Example 10.6 *The near-infrared spectra (NIR) measurements from Example 10.5 are used to predict the outcome octane numbers of 60 gasoline samples. The cumulative percentages of variance explained by the first six components from principal components regression (PCR) in the NIR predictor matrix* $\mathbf{X}$ *and the octane numbers are given in Table 10.2. The PCR components explain a maximum amount of variance in the NIR predictor matrix* $\mathbf{X}$*, however, the first two explain about 19 percent in the octane response, whereas component 4 explains about 50 percent of the octane response. The response octane is much closer to the fourth component than to the first.*

10.4.2 Partial least squares regression

One may be less interested in predicting from the directions of largest variation, but rather in predicting from those directions which are best associated with $\mathbf{y}$. **Partial least squares** (PLS) regression starts with computing $\widehat{\varphi}_{1j} = \mathbf{x}_j^T\mathbf{y}$, for $j = 1, \cdots p$, which are the coefficients for the regression of $\mathbf{y}$ on $\mathbf{x}_j$, where the columns $\mathbf{x}_j$ have length one. That is, the coefficients that

$$\widehat{\varphi}_{1j} = \underset{\varphi_{1j}}{\operatorname{argmin}} \|\mathbf{y} - \mathbf{x}_j\varphi_{1j}\|^2 = \mathbf{x}_j^T\mathbf{y}.$$

These are used in the summation $\mathbf{z}_1 = \sum_{j=1}^p \widehat{\varphi}_{1j}\mathbf{x}_j$ to define the first PLS direction. Next, we find the minimizing coefficients

$$\widehat{\theta}_1 = \underset{\theta}{\operatorname{argmin}} \|\mathbf{y} - \mathbf{z}_1\theta\|^2 = (\mathbf{z}_1^T\mathbf{z}_1)^{-1}\mathbf{z}_1^T\mathbf{y}.$$

The remaining are similarly computed by the following steps summarising the Algorithm for PLS:

- Step 1: Standardize each $\mathbf{x}_j$ to have mean zero and variance one. Set $\hat{\mathbf{y}}^{(0)} = \overline{y}\mathbf{1}$, and $\mathbf{x}_j^{(0)} = \mathbf{x}_j$, $j = 1, \cdots, p$.
- Step 2: For $m = 1, \cdots, p$

  1. $\mathbf{z}_m = \sum_{j=1}^p \widehat{\varphi}_{mj}\mathbf{x}_j^{(m-1)}$, where $\widehat{\varphi}_{mj} = \mathbf{y}^T\mathbf{x}_j^{(m-1)}$.

2. $\widehat{\theta}_m = (\mathbf{z}_m^T \mathbf{z}_m)^{-1} \mathbf{z}_m^T \mathbf{y}$.
3. $\hat{\mathbf{y}}^{(m)} = \hat{\mathbf{y}}^{(m-1)} + \widehat{\theta}_m \mathbf{z}_m$.
4. Orthogonalize each $\mathbf{x}_j^{(m-1)}$ with respect to $\mathbf{z}_m$ by using

$$\mathbf{x}_j^{(m)} = \mathbf{x}_j^{(m-1)} - \mathbf{z}_m (\mathbf{z}_m^T \mathbf{z}_m)^{-1} \mathbf{z}_m^T \mathbf{x}_j^{(m-1)}, \quad j = 1, \cdots, p$$

- Step 3: Output the sequence of of fitted vectors $\{\widehat{\mathbf{y}}^{(m)}\}_1^p$. Since the $\{\mathbf{z}_\ell\}_1^m$ are linear in the original $\mathbf{x}_j$, so is $\widehat{\mathbf{y}}^{(m)} = \mathbf{X}\widehat{\boldsymbol{\beta}}_{pls}^{(m)}$. These linear coefficients can be recovered from the sequence of PLS transformations.

Example 10.7 *The near-infrared spectra (NIR) measurements from Example 10.5 are used to predict the octane numbers of 60 gasoline samples. The cumulative percentages of variance explained by the first six components from PLS in the NIR predictor matrix and the octane numbers are given in 10.2. The first component explains a large amount of variance in the predictor matrix, but a relative small amount in the octane numbers. The fourth component explains less than 5 percent of the NIR variance, but about 50 percent of the octane variance. This reveals that the direction of the response octane clearly differs from that of maximum variation in the NIR space.*

The PLS components explain by definition less variance in the NIR the predictor matrix, although the size of the difference seems relatively small. The PLS components explain by definition more variance in the response octane then the PCR components, the differences are substantial (20 vs 95 percentage) if the amount explained in octane by two components are compared.

10.4.3 Determining the number of components by cross-validation

In order to prevent over- as well as under estimation we may randomly partitioning the data into a training and a test set. The PCR or PLS model can then be estimated on the basis of the training set and its predictive precision can be evaluated independently on the basis the data from the test set.

If there are n_1 objects in the training set and n_2 in the test set, then according to the root mean squared error of prediction (RMSEP) we then would have

$$RMSEP = \sqrt{\frac{1}{n_2} \sum_{i=1}^{n_2} (y_{i,Test} - \hat{y}_{i,Test})^2},$$

where i runs over the objects in the test set. Note that models which are too large would do better on the training set, but not on an independent test set. For models which are too small we could add another component in order to explain more variance. If a model performs equally well on the testing as on the training subset, then this can be interpreted as evidence for valid generalization. This approach is known as **cross validation** (Hastie et al., 2001).

TABLE 10.3
RMSEP and R^2 values for the training and the test sample from partial least squares (PLS) on the near-infrared spectra (NIR) as predictors for the octane numbers.

	(Intercept)	Comp 1	Comp 2	Comp 3	Comp 4	Comp 5
RMSEP Train	1.497	1.254	0.384	0.238	0.243	0.237
RMSEP Test	1.752	1.565	0.487	0.302	0.271	0.274
R^2 Train		0.269	0.931	0.973	0.972	0.973
R^2 Test		0.167	0.919	0.969	0.975	0.974

Example 10.8 *We continue with the near-infrared spectra (NIR) measurements from Example 10.5 to predict the octane numbers of 60 gasoline samples. 50 samples are randomly selected from the 60 to train the PLS model and the measurements from the remaining samples are used for testing the predictive power of the model.*

The RMSEP values in Table 10.3 are smaller for the training than for the test set and do almost not reduce further after taking more than three PLS dimensions. The pattern for the percentage of explained variance expressed by R^2 is similar. It seems that component three does not add much to the reduction of prediction error or to the increase of explained variance. A plot of the loadings, the correlations between the PLS components and the near-infrared spectra variables may be useful here. The loadings plot of the 401 spectra in Figure 10.3 does not give loadings from Component 3 as these are almost all very small, with a few of size 0.1. The loadings plot identifies those spectra which are of importance for Component 1 and 2 in order to predict octane numbers.

10.5 Notes and Comments

The generality of the Gram-Schmidt orthonormalization becomes clear by noting that Equation (10.1) is essentially based upon inner products. Hence, the procedure can be applied to other inner product spaces as well such as n-dimensional vectors, random variables, polynomials, periodic functions, etc. in order to construct a set of orthonormal variables or functions that span exactly the same space. Applications of QR decomposition for e.g., linear modeling are given by (Chambers and Hastie, 1992; Wood, 2017).

The fixed point method to find eigenvectors, also known as the power method, is the most basic (Horn and Johnson, 2012; Quarteroni et al., 2010). One generalization is in fact repeated QR decomposition among more advanced algorithms (Golub and Van Loan, 1996). The SVD can be very helpful

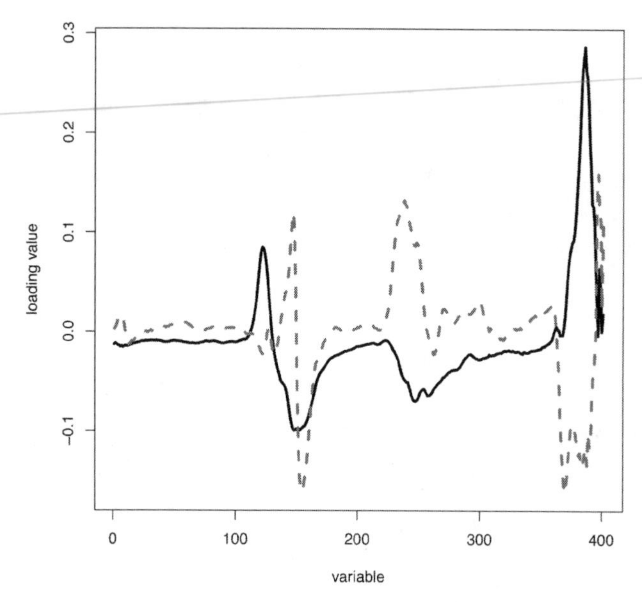

FIGURE 10.3
Plot of the loadings from partial least squares (PLS) using near-infrared spectra (NIR) as predictors for octane numbers.

in an analysis of a nearly rank deficient matrix by inspecting the smallest singular value(s). A detailed discussion measuring numerical instability by the condition number of a matrix is given by (Golub and Van Loan, 1996), see (Belsley et al., 2005) for implications for regressions analysis. The proof that the first eigen vector of the correlations matrix with positive elements is positive is given in Lancaster and Tismenetsky (1985).

A very detailed account of differential calculus on vector or matrix functions is given by Magnus and Neudecker (2019). It also covers the differentiation of eigenvalues/vectors of a symmetric matrix.

In the above we refrained from rotational indeterminacy in order not to obscure the property that each component explains a maximal amount of variance, given the previous. We concentrated in the above on the correlations as a basis for principal components analysis, which is essentially a linear approach. Properties of minimum rank approximation are given by Rao (2009), those of PCA by Jolliffe (2002) and details on non-linear principal curve analysis can be found in Hastie et al. (2001).

A very detailed state of the art discussion of prediction by PLS with several applications to Chemometrics is given by Cook and Forzani (2019).

10.6 Notes on Using R

R has functions `eigen`, `svd`, `princomp`, `pls`, and `biplot`. A matrix can easily be standardize by the `scale` function of the `MASS` library, which however uses an $n-1$ type of approach. It is easy to check the equivalence between the SVD and PCA by matrix multiplications.

By the R function `pairs.panels` from the `psych` library a pairwise panel plot of multivariate data can be constructed.

10.7 Exercises

1. **Using matrix algebra.**
   (a) Use the singular value decomposition to prove that if the matrix $\mathbf{X}$ is of full column rank, then $\mathbf{X}^T\mathbf{X}$ is positive definite. See Equation 7.8.
   (b) Show that the rank of the matrix involved in the S^2 term of the one sample t-test equals the degrees of freedom $n-1$.
2. **Column space.**
   (a) Suppose that the vector with unit elements $\mathbf{1}$ is orthogonal the column space of $\mathbf{X}$. Is this equivalent to $\mathbf{X}$ to have column mean zero?
   (b) Suppose that $\mathbf{X}$ has column mean zero. Does this imply that $\widehat{\mathbf{y}}$ from linear regression have column mean zero? Does the same hold for the principal components $\mathbf{U}$?
3. **QR decomposition for linear regression.** In example 10.2 it was stated that the $\widehat{\boldsymbol{\beta}} = \mathbf{R}^{-1}\mathbf{Q}_1^T\mathbf{y}$ equals the $\widehat{\boldsymbol{\beta}} = (\mathbf{X}^T\mathbf{X})^{-1}\mathbf{X}^T\mathbf{y}$. Show that this is indeed holds true.
4. **Cement.** In an experiment on the heat evolved in the setting of each of 13 cements. The outcome y is heat (calories per gram of cement) the predictor variable consist of percentages of weight of four constituents tricalcium aluminate x1, tricalcium silicate x2, tetracalcium alumumino ferrite x3 and dicalcium silicate x4 (Woods et al., 1932). The data are available as `cement.csv` or from the MPV library in R.
   (a) Center the data to have column-wise zero means and call the matrix $\mathbf{X}$. Compute the eigenvalues of the matrix $\mathbf{X}^T\mathbf{X}$.
   (b) Compute the singular values and check that these are the square root of the eigenvalues of $\mathbf{X}^T\mathbf{X}$.

(c) Check that the left singular vector matrix has column-wise mean zero.

(d) Check that the left singular vector matrix is column-wise orthonomal and that the right singular vector matrix is both column and row wise orthonormal.

5. **Alcohol solubility**. The solubility of alcohols in water is important in understanding alcohol transport in living organisms. The data set `alcohol` from the `robustbase` library in R collected by (Romanelli et al., 2001) contains contains 7 physicochemical characteristics of 44 aliphatic alcohols. The aim of the experiment was the prediction of the solubility on the basis of molecular descriptors.

   (a) Construct pairs panel plot such as in Figure 8.1.

   (b) Compute the eigenvalues and verify that their sum equal the number of variables 7.

   (c) Compute the bootstrap confidence intervals for the first two eigenvalue of the correlations matrix. For the second eigenvalue there are some confidence intervals with a negative left limit. Why is this unreasonable?

   (d) Check that the square root of the eigenvalues equals the standard deviation given by R library `princomp`.

   (e) Construct a biplot.

   (f) Compute the correlations between the variables and the first two principal components.

6. **NIR spectra**. In an experiment 21 NIR spectra of PET yarns were measured at 268 wavelengths for 21 corresponding densities. The 21 density values are the outcomes and the spectra the predictors. The data are available from `yarn.txt`.

   (a) Specify the order of the predictor matrix.

   (b) Perform a PCR and report the percentage of explained variance for the predictors and the outcome.

   (c) Perform a PLS and report the percentage of explained variance for the predictors and the outcome.

   (d) What is your conclusion?

7. **Olive oil.** In an experiment scores were attributed on 6 attributes from a sensory panel and measurements of 5 physico-chemical quality parameters on 16 olive oil samples. The first five oils are from Greek (G), the next five are Italian (I) and the last six are Spanish (S). The 5 chemical variables are measurements of acidity, peroxide, K232, K270, and DK. The 6 sensory variables are scores on the olive oil attributes yellow, green, brown, glossy, transp, and syrup

(Ziegel, 2000). The data `oliveoil` is available from the R library `pls`.

(a) Perform a principal component analysis on the chemical variables. Inspect the loadings matrix and whether there two dominating components present in de data.
(b) Construct a biplot and inspect whether the countries form separated clusters of olie oils.
(c) Perform a principal component analysis on the sensory variables. Inspect the loadings matrix and whether there two dominating components present in de data.
(d) Construct a biplot and inspect whether the countries form separated clusters of olie oils.
(e) Perform a PCR of the chemical variables (explanatory) on the sensory variables (outcomes).
(f) Perform a PLS regression of the chemical variables on the sensory variables.
(g) Compare the two analysis on percentage explained and cross validation.

8. **Mayonaise.** In an experiment raw NIR measurements (351 wavelengths, 1100–2500 nm in steps of 4 nm) are taken on 54 samples of mayonnaise based on six different oil types (soybean, sunflower, canola, olive, corn, and grapeseed). The resulting 54 samples were measured in triplicates, resulting in 54 x 3 = 162 different spectra (120/42 training/test) (Indahl et al., 1999). Note that the number of NIR columns is larger than the number rows which causes the function `princomp` to refuse its work. The mayonnaise data are available from the R library `pls`.

   (a) Scale the NIR data to have mean zero and unit variance.
   Hint: Simply use the `scale` function.
   (b) Apply the singular value decomposition and analyze the number of components which are the most important. How much variance explain the first and the second. Illustrate by a bar plot of the largest 25 singular values.
   (c) Compute the correlation of the first principal component with the NIR variables and plot these as a line. Do the same for the correlations with the second principal component.
   (d) Test the null hypothesis that the six oil types have equal mean on the first principal component. Repeat this for the second principal component. Illustrate by a boxplot.
   (e) Compute a biplot from the first two components with the types of mayonaise indicated. Are there clearly distinguishable clusters?

9. **Glass.** In a study of 180 archaeological glass vessels there were 13 different chemical measurements made for each (Janssens et al., 1998).

   (a) Construct a pairs plot with the correlations.
   Hint: Use the `pairs.panels` function from the `psych` library of R.

   (b) Perform a Principal Components Analysis and report the percentage of variance explained by the first two components.

   (c) Are there clusters of chemicals observable from a biplot?

Part IV

Analysis of Designed Experiments

11

Analysis of Data from Designed Experiments

Experiments are typically conducted to answer research questions about the effect size of several explanatory variables on some chemical outcome. Often, it may only be possible to set the explanatory variables at two or three levels in the experiment, for example due to practical difficulties.

There are two main experimental phases. In Phase 1 we consider potential explanatory factors of some outcome of interest. In this phase we are not so much interested in determining the effect sizes, but rather in identifying those factors, which are of real influence on the outcome of the chemical process. In Phase 2, we want to design an experiment from which we can optimize the yield of some chemical production process. In this chapter we will discuss design of experiments, product optimization, and the analysis of variance of data from an experiment.

In this chapter you learn

- In Section 11.1 which concepts are important for factorial designs.
- In Section 11.2 how to test chemical hypotheses by analysis of variance.
- In Section 11.3 how to analyze and optimize a response surface for a chemical outcome.
- In Section 11.4 how to estimate for random and fixed non-linear effects on a chemical outcome.

11.1 Concepts of Factorial Designs

Various concepts play a role in the design of experiments and their analysis. Concepts such as the sample mean and differences in mean play an important role.

DOI: 10.1201/9781003178194-11

11.1.1 Two-level one-factor design

We start with pointing out that the independent two sample t-test is a special case of a design and of regression analysis. In Section 6.4.2 the t-test was presented where the difference in mean Arsenic concentration was tested for two types of locations. Such can be seen as a one-factor design with two levels (locations) within which ten observations became available from the experiment. The t-test assuming equal variances is equivalent to linear regression with a single column, a 20 by 1 the design matrix $\mathbf{x}$ with -1 indicating the lower and 1 the higher level. Implementing this in the design matrix for linear regression, we would find the estimated coefficient

$$\widehat{\beta} = (\mathbf{x}^T\mathbf{x})^{-1}\mathbf{x}^T\mathbf{y} = \frac{1}{20}\left(\sum_{i=1}^{10} y_{i2} - \sum_{i=1}^{10} y_{i1}\right) = \frac{1}{2}(\bar{y}_2 - \bar{y}_1),$$

where $\bar{y}_1$ and $\bar{y}_2$ are the means for Location 1 and 2, respectively. Indeed, if we wish we could re-scale the column $\mathbf{x}$ by a factor $1/2$ to find an exact difference in means.

11.1.2 Two-level two-factor design

Almost always more than one factor influences the outcome. To set the scene we start with a design with two factors each having two levels. For such factors we fill the **design matrix** with orthogonal columns corresponding to effects of factor A and factor B using -1 and 1 values in a systematic manner. That is, we indicate the lower level of factor A by -1 and its higher level by 1. For the factor B we do this in a completely similar way, as summarized in Table 11.1. The two-way interaction between A and B can easily be obtained by element-wise multiplying the signs that correspond to the levels for each factor. After adding the column with the AB interaction to the design matrix, we still have column-wise orthogonality in the design matrix. We may also add a column with ones for the general mean, again keeping the orthogonality intact. This set-up is called an orthogonal 2^2 design, with the constant for the overall mean, two main effects, and an interaction effect. Next, we would conduct the experiment measuring the outcome at each combination of the levels of the factors a few times. The latter are called replicates. If this number equals n for each combination of the factors, then the design is called **balanced**. The latter would be independent if the runs of the experiment are executed in a random order for each combination of the factors and within these. That is, if the experimental conditions are set according to the factor combinations before each measurement.

The outcomes from the experiment are sometimes summarized in the totals for each combination of the levels of the factors. A common type of notation

TABLE 11.1
Yates type of matrix for a completely randomized 2^2 experiment for factor A, factor B, and, their interaction AB.

	Sum	A	B	AB
(1)	$y_{-1,-1,\bullet}$	−1	−1	1
a	$y_{1,-1,\bullet}$	1	−1	−1
b	$y_{-1,1,\bullet}$	−1	1	−1
ab	$y_{1,1,\bullet}$	1	1	1

uses a dot for the index over which the summation of the outcome runs

$$y_{ij\bullet} = \sum_{k=1}^{n} y_{i,j,k}$$

where $i \in \{-1, 1\}$, $j \in \{-1, 1\}$, and $k = 1, \cdots, n$, see Table 11.1. Another type of notation for the observed totals is to use a (1) if both factors are low, a if A is high and B low, b if A is low and B high, and ab when both A and B are high.

If we now denote the sample mean of Y at level A^+ by $\bar{Y}_{A+}$, then we get in terms of different types of notation in the literature

$$\bar{Y}_{A+} = \frac{1}{2n} Y_{1,\bullet,\bullet} = \frac{1}{2n}\left(Y_{1,1,\bullet} + Y_{1,-1,\bullet}\right) = \frac{1}{2n}\left(ab + a\right).$$

Completely similar we find the mean over all observations at the $A-$ level as

$$\bar{Y}_{A-} = \frac{1}{2n} Y_{-1,\bullet,\bullet} = \frac{1}{2n}\left(Y_{-1,1,\bullet} + Y_{-1,-1,\bullet}\right) = \frac{1}{2n}\left(b + (1)\right).$$

The main effect of factor A on the outcome is defined as the mean difference on the outcome Y between the $+$ and the $-$ level, that is the effect of A is

$$\begin{aligned} \bar{Y}_{A+} - \bar{Y}_{A-} &= \frac{1}{2n}\left(ab + a\right) - \frac{1}{2n}\left(b + (1)\right) \\ &= \frac{1}{2n}\left(-(1) + a - b + ab\right), \end{aligned}$$

where the signs exactly correspond to those in column A of Table 11.1. This is half times the regression coefficient from using the Yates matrix in linear regression. The difference in means is often visualized as the vertical lines in a main effect plot.

Completely similar we find for the main effect of factor B that

$$\begin{aligned} \bar{Y}_{B+} - \bar{Y}_{B-} &= \frac{1}{2n}\left(Y_{-1,1,\bullet} + Y_{1,1,\bullet}\right) - \frac{1}{2n}\left(Y_{-1,-1,\bullet} + Y_{1,-1,\bullet}\right) \\ &= \frac{1}{2n}\left(b + ab\right) - \frac{1}{2n}\left((1) + a\right) \\ &= \frac{1}{2n}\left(-(1) - a + b + ab\right), \end{aligned}$$

where the signs exactly correspond to those in Table 11.1.

For the interaction we need two more definitions, namely the effect of A given $B+$ and the effect of A given $B-$. The former, the effect of A given $B+$, is the difference in means between

$$\bar{Y}_{A|B+} = \bar{Y}_{A+,B+} - \bar{Y}_{A-,B+} = \frac{1}{2n}(Y_{1,1,\bullet} - Y_{-1,1,\bullet}) = \frac{1}{2n}(ab - b)$$

The latter, the effect of A given $B-$, is the difference in means between

$$\bar{Y}_{A|B-} = \bar{Y}_{A+,B-} - \bar{Y}_{A-,B-} = \frac{1}{2n}(Y_{1,-1,\bullet} - Y_{-1,-1,\bullet}) = \frac{1}{2n}(a - (-1))$$

The interaction between factor A and factor B is defined as the effect of A given $B+$ minus the effect of A given $B-$, that is

$$\begin{aligned} \bar{Y}_{A|B+} - \bar{Y}_{A|B-} &= \frac{1}{2n}(ab - b) - \frac{1}{2n}(a - (-1)) \\ &= \frac{1}{2n}((1) - a - b + ab) \end{aligned}$$

where again the signs correspond to those in column AB in Table 11.1.

The main and interaction effects can be estimated by linear regression after a proper definition of the design matrix. For the main effects, element (i, j) for the design matrix $\mathbf{X}$ equals $x_{ij} = \pm 1$ according to whether factor j is low/high in a run, see Table 11.1. In a balanced design this pattern is repeated n times for each of the replicates. The design is called orthogonal since the model matrix $\mathbf{X}$ has this property column-wise. That is, the matrix $\mathbf{X}^T\mathbf{X}$ equals a diagonal matrix $\mathbf{D}$, with the values $2 \cdot 2 \cdot n$ on its main diagonal. The estimator for the effects $\boldsymbol{\beta}$ and its distribution are

$$\hat{\boldsymbol{\beta}} = (\mathbf{X}^T\mathbf{X})^{-1}\mathbf{X}^T\mathbf{y} = \mathbf{D}^{-1}\mathbf{X}^T\mathbf{y}; \quad \hat{\boldsymbol{\beta}} \sim N(\boldsymbol{\beta}, \sigma^2\mathbf{D}^{-1}).$$

Hence, leaving out an effect from the model, comes down to leaving out the corresponding column from the design matrix and this would not change any of the other effects. That is, the effects are estimated completely independently and there is no confounding among the effects, a convenient property for making valid chemical inferences.

Example 11.1 *The data and the design of an injection molding experiment are given in Table 11.3. The first two main factors from the factorial design are orthogonal, including the constant. The estimated coefficients from these are given in Table 11.2 together with the corresponding p-values. These indicate that the null hypothesis of no effects is not to be rejected. The same conclusion follows from the general F-test. The matrix $\mathbf{X}^T\mathbf{X}$ is diagonal with elements equal to the number of runs 16. The residual standard error $\widehat{\sigma} = 4.803$, so that the standard error of the coefficients becomes $\sqrt{4.803^2/16} = 1.20075$.*

TABLE 11.2
Main and two-way interaction effects from the injection molding data.

	Estimate	Std. Error	t value	Pr(>|t|)
(Intercept)	19.75	1.20	16.45	< 0.0001
A	−0.35	1.20	−0.29	0.78
B	−0.05	1.20	−0.04	0.97
A:B	−0.30	1.20	−0.25	0.81

TABLE 11.3
2^{8-4}_{III} Fractional factorial design for the injection molding data.

run	A	B	C	D	E	F	G	H	y
1	−1	−1	−1	1	1	1	−1	1	14.0
2	1	−1	−1	−1	−1	1	1	1	16.8
3	−1	1	−1	−1	1	−1	1	1	15.0
4	1	1	−1	1	−1	−1	−1	1	15.4
5	−1	−1	1	1	−1	−1	1	1	27.6
6	1	−1	1	−1	1	−1	−1	1	24.0
7	−1	1	1	−1	−1	1	−1	1	27.4
8	1	1	1	1	1	1	1	1	22.6
9	1	1	1	−1	−1	−1	1	−1	22.3
10	−1	1	1	1	1	−1	−1	−1	17.1
11	1	−1	1	1	−1	1	−1	−1	21.5
12	−1	−1	1	−1	1	1	1	−1	17.5
13	1	1	−1	−1	1	1	−1	−1	15.9
14	−1	1	−1	1	−1	1	1	−1	21.9
15	1	−1	−1	1	1	−1	1	−1	16.7
16	−1	−1	−1	−1	−1	−1	−1	−1	20.3

11.1.3 Two-level k-factor designs

The properties above explained for two factors hold more general for all experiments with k two-level factors. In complete or full factorial k-factor two level designs all levels are combined resulting in 2^k runs. Factors with two levels can be quantitative or qualitative to model the effect of the difference between a low and a high value of a potentially determining variable of the chemical yield. To generate a full factorial design with $k = 3$ factors, we start with the word ABC consisting of the letters $\{A, B, C\}$ denoting the factors. The effects in the design correspond to all words that can be constructed from the three letters, where order is immaterial, so that $BA = AB$. The word without letters represents the constant, the single letter words $\{A, B, C\}$ represent the main effects, the words with two letters $\{AB, AC, BC\}$ represent 2-factor interactions, and the word ABC represents the 3-factor interaction. In this way we find a total of $1+3+3+1=2^3$ effects to be estimated for which we need 8 runs.

More generally, the number of effects in a 2^k design correspond to all distinct $j = 0, 1, \cdots, k$-letter words in alphabetic order. The number of j-factor interactions is given by the binomial coefficient

$$\binom{k}{j} = \frac{k!}{(k-j)! \cdot j!},$$

where $k! = k \cdot (k-1) \cdot (k-2) \cdots 1$ denotes the factorial of integer k. Hence, for $k = 5$ we have

$$\binom{5}{0} = 1, \binom{5}{1} = 5, \binom{5}{2} = 10, \binom{5}{3} = 10, \binom{5}{4} = 5, \binom{5}{5} = 1$$

for the constant, the number of main effects, 2-factor, 3-factor, 4-factor, and 5-factor interactions, respectively. In full factorial orthogonal 2^k designs all these main and interaction effects can be estimated without bias and can be statistically tested.

11.1.4 Two-level k-factor fractional designs

In industrial design it happens that $k = 20$, so that the full factorial design would require $2^{20} = 1,048,576$ runs, which obviously is not feasible to actually conduct. It is frequently argued and confirmed by literature review that interactions of order larger than 2 are relatively exceptional and small in size (Li et al., 2006). Consequently, it is fairly save to assume in industrial settings that higher order interactions are exactly zero. Under this modeling assumption, it is possible to construct designs with fewer runs for which the main and 2-order interactions can be estimated without bias. Such a design is called *regular* if it is constructed by setting defining words to be equal to the value “identity”. Next, we use the idea explicit in the Yates matrix in Table 11.1 that the square of any element in a column yields 1, implying that $I = A^2 = B^2 = C^2$. An example will make clear where this leads to.

Example 11.2 *Suppose we set identity equal to the word representing three way interaction in a design with three factors A, B, and C consisting of two levels. That is, we set $I = ABC$ which implies that the constant equals the three way interaction effect. To find the other runs of the design we use $I = A^2 = B^2 = C^2$ to compute*

$$\begin{aligned} A &= A \cdot I = A \cdot ABC = A^2 \cdot BC = BC \\ B &= B \cdot I = B \cdot ABC = B^2 \cdot AC = AC \\ C &= C \cdot I = C \cdot ABC = C^2 \cdot AB = AB \end{aligned}$$

That is, each main effect is confounded by a two-way interaction effect. These confounded or **aliased factors** *are indistinguishable, not separately estimable* *by the design. Here we have 4 aliased effects with a total of* $8 - 4 = 4$ *runs, one for the constant and one for each of the three (confounded) main effects.*

The set-up of the previous example is denoted in the literature as a 2^{3-1}_{III} design having three factors, requiring half of the runs 2^{3-1} of a full factorial 2^3 design. It is said to have resolution III as it is generated by a three letter word in the defining equality $I = ABC$. That is, the resolution equals the number of letters in the smallest word that generates the design.

Example 11.3 *Suppose we have a four factor design generated by* $I = ABCD$. *If we start writing the factors* $A, B,$ *and* C *each with two levels, then we have 8 runs. From* $I = ABCD$, *we obtain the aliases* $D = ABC$, $A = BCD$, $B = ACD$, $C = ABC$, $D = ABC$ *and* $AB = CD$, $AC = BD$, $AD = BC$. *So the main effects are only confounded by 3-factor interactions, and not by 2-factor interactions. If the assumption that the 3-factor interactions are zero is correct, then the main effects can be estimated without bias. Hence, by setting* $I = ABCD$, *we obtain* 8 *aliases or a* 2^{4-1}_{IV} *design with* 2^3 *runs with resolution* IV. *Note that this requires only half of the runs of a full* 2^4 *design.*

Similarly, a 5-factor design generated by $I = ABCDE$ gives a 2^{5-1}_{V} design requiring half of a full factorial 2^5 design. It has no aliasing among main effects or 2-factor effects.

There do exist several distinct designs with resolution IV as can be seen when we construct 2^{7-2}_{IV} designs using the generators

$$\begin{aligned} \text{Design 1}: I &= ABCF = BCDG = ADFG \\ \text{Design 2}: I &= ABCF = ADEG = BCDEFG \\ \text{Design 3}: I &= CEFG = ABCDF = ABDEG \end{aligned}$$

The number of letters in the defining words are $(4, 4, 4)$, $(4, 4, 6)$, $(4, 5, 5)$, respectively. Each has 4 as its minimum, so that the resolution for each of the designs is IV and the concept of resolution does not distinguish between these designs. Design 3, however, has only a single defining word with minimum length equal to 4 and therefore has a minimum number of main effects aliased with interactions of order 3. The latter is said to be of minimum aberration and is considered to be better as its confounding is minimal over the three designs.

Example 11.4 *A* 2^{8-4}_{III} **fractional factorial design** *for the outcome injection molding (Box et al., 2004) is given in Table 11.3. We may analyze it for all main effects and second-order interactions. The number of main effects 8 plus the number of 2-way interactions* $\binom{8}{2} = 28$ *equals 36 is much larger than the number of runs 16. This would give a horizontal design matrix in a linear* *regression. The 2-factor aliases of the model can be found to be*[1]

$$\begin{aligned} AB &= CG = DH = EF, \\ AC &= BG = DF = EH, \\ AD &= BH = CF = EG, \\ AE &= BF = CH = DG, \\ AF &= BE = CD = GH, \\ AG &= BC = DE = FH, \\ AH &= BD = CE = FG, \end{aligned}$$

This implies, from the perspective of estimation, that each of the $AB = CG = DH = EF$ *interactions are confounded and indistinguishable. The columns in Table 11.3 are orthogonal, also the main effects are not aliased with any of the 2-factor interactions. A halfnormal or Daniel plot may be of help identifying possible effects on the outcome. It gives horizontally the two times the absolute value of the effects and vertically the standard normal quantiles of the effects. In case of no effects at all all dots would occur on a straight line. The main effect of* C*, and* E *as well as the interaction* $A : E$ *seem to have the largest effect, although the latter cannot be distinguished for any of the interactions* $AE = BF = CH = DG$.

11.2 Analysis of Variance

The analysis of the effects of a factor with a number of levels is performed by testing the significance of the amount of variance explained. We go into the analysis of designs with one factor and into the analysis of designs with two factors.

11.2.1 One-way analysis of variance

In case there is one factor the with I levels, the model for the outcomes may be written as

$$Y_{ij} = \mu_i + \varepsilon_{ij}, \ \{\varepsilon_{11}, \cdots, \varepsilon_{IJ}\} \overset{IID}{\sim} N(0, \sigma^2), \tag{11.1}$$

where μ_i is the mean for level i, and ε_{ij} the error terms. To see that we can use regression analysis to estimate the model the following example is given.

[1]These can be found by the aliases function in R.

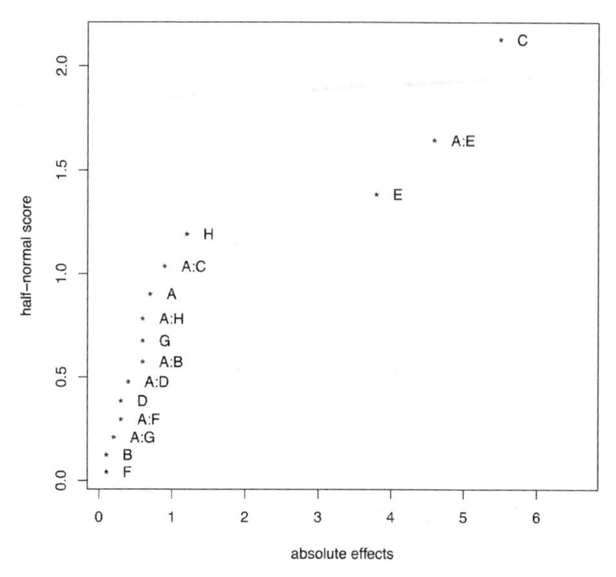

FIGURE 11.1
Halfnormal or Daniel plot from a 2^{8-4}_{III} fractional factorial design on the outcome injection molding, using : notation for two-way interactions.

Example 11.5 *Suppose that the number of levels $I = 3$ and that we have n observations of the outcome per level. Then the expected value of the outcome can be written as*

$$E[\mathbf{y}] = \begin{bmatrix} \mu_1 \mathbf{1} \\ \mu_2 \mathbf{1} \\ \mu_3 \mathbf{1} \end{bmatrix} = \begin{bmatrix} \mathbf{1} & \mathbf{0} & \mathbf{0} \\ \mathbf{0} & \mathbf{1} & \mathbf{0} \\ \mathbf{0} & \mathbf{0} & \mathbf{1} \end{bmatrix} \begin{bmatrix} \mu_1 \\ \mu_2 \\ \mu_3 \end{bmatrix} = \mathbf{X}\boldsymbol{\beta},$$

where $\mathbf{1}$ is the n-vector with ones, and $\boldsymbol{\beta} = (\mu_1, \mu_2, \mu_3)^T = \boldsymbol{\mu}^T$. The design matrix

$$\mathbf{X} = \begin{bmatrix} \mathbf{1} & \mathbf{0} & \mathbf{0} \\ \mathbf{0} & \mathbf{1} & \mathbf{0} \\ \mathbf{0} & \mathbf{0} & \mathbf{1} \end{bmatrix},$$

indicates by a one the level of the factor of an observation. When the measurements of the outcome are collected one beneath the other in a vector $\mathbf{y}$, then we obtain the estimator

$$\hat{\boldsymbol{\beta}} = (\mathbf{X}^T\mathbf{X})^{-1}\mathbf{X}^T\mathbf{y} = \begin{bmatrix} n & 0 & 0 \\ 0 & n & 0 \\ 0 & 0 & n \end{bmatrix}^{-1} \begin{bmatrix} \sum_{j=1}^{n} y_{1j} \\ \sum_{j=1}^{n} y_{2j} \\ \sum_{j=1}^{n} y_{3j} \end{bmatrix} = \begin{bmatrix} \bar{y}_1 \\ \bar{y}_2 \\ \bar{y}_3 \end{bmatrix},$$

and its distribution

$$
\begin{aligned}
\begin{bmatrix} \bar{y}_1 \\ \bar{y}_2 \\ \bar{y}_3 \end{bmatrix} = \hat{\boldsymbol{\beta}} &\sim N\left(\boldsymbol{\beta}, \sigma^2(\mathbf{X}^T\mathbf{X})^{-1}\right) \\
&= N\left(\begin{bmatrix} \mu_1 \\ \mu_2 \\ \mu_3 \end{bmatrix}, \sigma^2 \begin{bmatrix} 1/n & 0 & 0 \\ 0 & 1/n & 0 \\ 0 & 0 & 1/n \end{bmatrix}\right) = N\left(\boldsymbol{\mu}, \frac{\sigma^2}{n}\mathbf{I}\right).
\end{aligned}
$$

To investigate for the effect of the factor we wish test a larger linear model against a smaller. Note that if a smaller Model 1 with q parameters $\boldsymbol{\beta}_1$ is nested within the larger Model 2 with p parameters $\boldsymbol{\beta}_2$, then the F-statistic can be used to test Model 1 against Model 2. Moreover,

$$
\begin{aligned}
F &= \frac{\left(\text{RSS}(\widehat{\boldsymbol{\beta}}_1) - \text{RSS}(\widehat{\boldsymbol{\beta}}_2)\right)/(p-q)}{\text{RSS}(\widehat{\boldsymbol{\beta}}_2)/(n-p)} \\
&= \frac{\left(\|\mathbf{y} - \mathbf{X}_1\widehat{\boldsymbol{\beta}}_1\|^2 - \|\mathbf{y} - \mathbf{X}_2\widehat{\boldsymbol{\beta}}_2\|^2\right)/(p-q)}{\|\mathbf{y} - \mathbf{X}_2\widehat{\boldsymbol{\beta}}_2\|^2/(n-p)} \sim F_{p-q,n-p},
\end{aligned}
$$

where the right-hand side denotes the F distribution with $p-q, n-p$ degrees of freedom. Model 1 is rejected if $F > F^{(\alpha)}_{p-q,n-p}$, taking e.g., significance level $\alpha = 0.05$.

The linear model that corresponds to the null hypothesis of no effect can be written as

$$
Y_{ij} = \mu + \varepsilon_{ij}, \ \ \{\varepsilon_{11}, \cdots, \varepsilon_{IJ}\} \overset{IID}{\sim} N(0, \sigma^2), \tag{11.2}
$$

where μ is the grand mean for each level i, and ε_{ij} the normally independently distributed error terms. Putting all outcomes one beneath the other the model can be rewritten as

$$
\mathbf{y} = \mathbf{1}\mu + \varepsilon, \ \ \text{with } \varepsilon \sim N(\mathbf{0}, \sigma^2\mathbf{I}),
$$

a linear model with $\mathbf{X} = \mathbf{1}$ and $\beta = \mu$.

In order to use the F-test we have to reformulate a useful larger model with different means in such a way that it has the smaller as a sub-model. To do this we use the column invariance of the linear model, that is for all invertible matrices $\mathbf{C}$ we have $\mathbf{X}\boldsymbol{\beta} = \mathbf{X}\mathbf{C}^{-1}\mathbf{C}\boldsymbol{\beta}$. Note that regression analysis with $\mathbf{X}\mathbf{C}^{-1}$ as the predictor matrix gives exactly the same RSS as it spans the same column space. In case $I = 3$ we may rewrite the structural part of the model as

$$\begin{aligned} E[\mathbf{y}] &= \begin{bmatrix} \mathbf{1} & \mathbf{0} & \mathbf{0} \\ \mathbf{0} & \mathbf{1} & \mathbf{0} \\ \mathbf{0} & \mathbf{0} & \mathbf{1} \end{bmatrix} \begin{bmatrix} \mu_1 \\ \mu_2 \\ \mu_3 \end{bmatrix} \\ &= \begin{bmatrix} \mathbf{1} & \mathbf{0} & \mathbf{0} \\ \mathbf{0} & \mathbf{1} & \mathbf{0} \\ \mathbf{0} & \mathbf{0} & \mathbf{1} \end{bmatrix} \begin{bmatrix} 1 & 0 & 0 \\ 1 & 1 & 0 \\ 1 & 0 & 1 \end{bmatrix} \begin{bmatrix} 1 & 0 & 0 \\ -1 & 1 & 0 \\ -1 & 0 & 1 \end{bmatrix} \begin{bmatrix} \mu_1 \\ \mu_2 \\ \mu_3 \end{bmatrix} \\ &= \begin{bmatrix} \mathbf{1} & \mathbf{0} & \mathbf{0} \\ \mathbf{1} & \mathbf{1} & \mathbf{0} \\ \mathbf{1} & \mathbf{0} & \mathbf{1} \end{bmatrix} \begin{bmatrix} \mu_1 \\ \mu_2 - \mu_1 \\ \mu_3 - \mu_1 \end{bmatrix} = \mathbf{X}\boldsymbol{\beta}, \end{aligned} \tag{11.3}$$

where $\boldsymbol{\beta} = (\mu_1, \mu_2 - \mu_1, \mu_3 - \mu_1)^T$. The first column of this design matrix $\mathbf{X}$ equals the vector with ones that corresponds to the model with equals means. The matrix

$$\mathbf{C} = \begin{bmatrix} 1 & 0 & 0 \\ -1 & 1 & 0 \\ -1 & 0 & 1 \end{bmatrix}.$$

is called **contrast matrix** as it aranges for the differences in means. Using the fact that linear combinations of normally distributed random variables are normally distributed, we find

$$\mathbf{C}\hat{\boldsymbol{\beta}} \sim N\left(\mathbf{C}\boldsymbol{\beta}, \sigma^2\mathbf{C}(\mathbf{X}^T\mathbf{X})^{-1}\mathbf{C}^T\right),$$

and more specifically that

$$\mathbf{C}\hat{\boldsymbol{\beta}} = \begin{bmatrix} \bar{y}_1 \\ \bar{y}_2 - \bar{y}_1 \\ \bar{y}_3 - \bar{y}_1 \end{bmatrix} \sim N\left(\begin{bmatrix} \mu_1 \\ \mu_2 - \mu_1 \\ \mu_3 - \mu_1 \end{bmatrix}, \sigma^2/n \begin{bmatrix} 1 & -1 & -1 \\ -1 & 2 & 1 \\ -1 & 1 & 2 \end{bmatrix} \right).$$

For practical purposes the σ^2 on the right-hand side is replaced by its estimator $\hat{\sigma}^2$. This is what all sofware systems give as output when a single factor is used in the model notation to indicate the level of each run.

Example 11.6 *The interply bond strength of a thermoplastic composite was measured in an experiment under the influence of a laser with power at 40, 50 or 60W (Mazumdar and Hoa, 1995). The factor laser has three levels denoted by 40W, 50W, and 60W. The model with the differences in means is routinely applied software systems by a factor which indicates the level to which an observation belongs. The three measurements for each level of the factor are given in Table 11.4 together with their means and standard deviations. The values of the measurements together with the mean and standard deviations per level of the factor laser are given in Figure 11.2 and the accompanying table. We have* $n = 9$ *observations divided over* $p = 3$ *levels of the factor, and the smaller model with the intercept only has* $q = 1$*. From the summary of the analysis of variance in Table 11.5, it can be observed that the residual sum of squares for*

TABLE 11.4
Interply bond strength measurements for three laser levels, repeated across three runs.

Run	40W	50W	60W
1	25.66	29.15	35.73
2	28.00	35.09	39.56
3	20.65	29.79	35.66
mean	24.77	31.34	36.98
SD	3.75	3.26	2.23

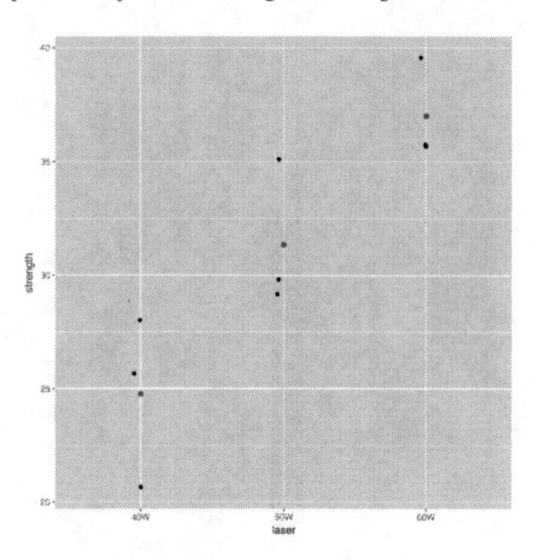

FIGURE 11.2
The interply bond strength measurements of a thermoplastic composite depending is tested in an experiment on depending on the power of a laser with levels 40, 50 or 60W. The thicker dots represent the mean per level.

TABLE 11.5
F-testing in analysis of variance for null hypothesis of no differences in mean.

	Df	Sum Sq	Mean Sq	F value	Pr(>F)
Laser	2	224.18	112.09	11.32	0.0092
Residuals	6	59.42	9.90		

the smaller model is 283.6056 and for the larger model 59.42173, leading to a reduction in residual sum of squares of 283.6056-59.42173=224.1838. The F-value

$$\frac{(283.6056 - 59.42173)/(3-1)}{283.6056/(9-3)} = 11.31827$$

is relatively large, resulting in a p-value smaller than 0.01, *indicating to reject the model with equal means corresponding to the null-hypothesis of no effect. The first estimated beta coefficient from regression analysis in Table 11.6 equals the means of the outcomes for the 40W reference level 1 the second and the third are the difference in means with respect to the 40W reference level 1. The p-values for marginal testing of the differences in mean reveal that both higher laser power levels have a positive effect, with that of 60W being twice as large as that of 50W relative to the reference 40W.*

When the null hypothesis of equal means is rejected one may wonder which of the means are different. Tukey (1949) considered all possible differences

TABLE 11.6
Coefficients from linear regression analysis of composite measurements giving the mean as the intercept and the differences in means as effects.

	Estimate	Std. Error	t value	Pr(>|t|)
(Intercept)	24.7700	1.8169	13.63	0.0000
laser 50W - 40W	6.5733	2.5695	2.56	0.0430
laser 60W - 40W	12.2133	2.5695	4.75	0.0031

TABLE 11.7
All pairwise honest significant differences in means, with adjusted p-values according to Tukey's procedure.

	diff	lwr	upr	p adj
50W-40W	6.5733	−1.3106	14.4573	0.0948
60W-40W	12.2133	4.3294	20.0973	0.0075
60W-50W	5.6400	−2.2440	13.5240	0.1507

$(\bar{Y}_i - \bar{Y}_j)$ in mean and showed that

$$(\bar{Y}_i - \bar{Y}_j) \pm q^* \sqrt{\frac{1}{2} s^2 \left(\frac{1}{n_i} + \frac{1}{n_j} \right)},$$

covers the true differences with an overall confidence level $1 - \alpha$, where q^* is the quantile for studentized range distribution and s^2 the sample variance. The procedure keeps the significance level α fixed at e.g., 5% as all pairwise differences are compared. In experiments consisting of factors with many levels where also all differences in means are relevant, Tukey's method may be of great help.

Example 11.7 *For the data of the previous example all pairwise differences between the means are computed together with the confidence intervals according to Tukey's HSD method. From Table 11.7 it can be observed that only the difference between 60 and 40W is significant.*

The model for one-way analysis of variance is also written as

$$Y_{ij} = \mu_1 + \alpha_i + \varepsilon_{ij}, \ \{\varepsilon_{11}, \cdots, \varepsilon_{IJ}\} \overset{IID}{\sim} N(0, \sigma^2), \tag{11.4}$$

where $\alpha_1 = 0$, and $\alpha_i = \mu_i - \mu_1$ for $i = 2, \cdots, I$.

11.2.2 Two-way analysis of variance

We generalize our testing of equality of means to a setting where two factors are of influence on the measurements. To determine the size of the influence we need a model

$$Y_{ijk} = \mu + \alpha_i + \beta_j + (\alpha\beta)_{ij} + \varepsilon_{ijk}, \quad \{\varepsilon_{111}, \cdots, \varepsilon_{IJK}\} \overset{IID}{\sim} N(0, \sigma^2) \quad (11.5)$$

where μ is the grand mean, α_i the i-th main effect of factor α, β_j the j-th effect of factor β, $(\alpha\beta)_{ij}$ interaction effect, and ε_{ijk} the error terms. To make the parameters of Model (11.5) estimable suitable constraints need to be imposed. A suitable possibilities is to fix the first level of each a factor to be zero $\alpha_1 = \beta_1 = 0$ and to use contrasts for the main effects. For the interaction effects we may use $(\alpha\beta)_{i1} = 0$ for all i and $(\alpha\beta)_{1j}$ for all j and estimate the remaining. The corresponding design matrix for two-way analysis of variance with two factors each with three levels equals

$$\mathbf{X} = \begin{bmatrix} 1 & 0 & 0 & 0 & 0 & 0 & 0 & 0 & 0 \\ 1 & 1 & 0 & 0 & 0 & 0 & 0 & 0 & 0 \\ 1 & 0 & 1 & 0 & 0 & 0 & 0 & 0 & 0 \\ 1 & 0 & 0 & 1 & 0 & 0 & 0 & 0 & 0 \\ 1 & 1 & 0 & 1 & 0 & 1 & 0 & 0 & 0 \\ 1 & 0 & 1 & 1 & 0 & 0 & 1 & 0 & 0 \\ 1 & 0 & 0 & 0 & 1 & 0 & 0 & 0 & 0 \\ 1 & 1 & 0 & 0 & 1 & 0 & 0 & 1 & 0 \\ 1 & 0 & 1 & 0 & 1 & 0 & 0 & 0 & 1 \end{bmatrix},$$

where the second column pertains to mains effect α_2, the third to α_3, the fourth to β_2, the fifth to β_3, the sixth interaction effect $(\alpha\beta)_{22}$, the seventh to $(\alpha\beta)_{32}$, the eight to $(\alpha\beta)_{23}$, the ninth to $(\alpha\beta)_{33}$. The interaction effects follow from the element-wise multiplication of the corresponding columns for the main effects.

Model (11.5) is the complete model with the maximum number of parameters. Chemical engineering hypotheses can be formulated in terms of submodels which are tested against larger models. For this we have the model without interaction effects:

$$Y_{ijk} = \mu + \alpha_i + \beta_j + \varepsilon_{ijk}, \quad (11.6)$$

without β effects:

$$Y_{ijk} = \mu + \alpha_i + \varepsilon_{ijk}, \quad (11.7)$$

without α effects:

$$Y_{ijk} = \mu + \beta_j + \varepsilon_{ijk}, \quad (11.8)$$

and without main or interaction effects:

$$Y_{ijk} = \mu + \varepsilon_{ijk}. \quad (11.9)$$

This gives a hierarchy of models each of which corresponds to a chemical hypothesis on effects on the outcome. The design matrices easily follow from leaving out the columns in the design matrix corresponding to the effects assumed to be zero. These hypotheses can be tested by comparing the model fit with an F-test.

TABLE 11.8
The effect of material type and temperature on the life of battery times.

	Temperature	15			70			125		
Replication	**Material**	1	2	3	1	2	3	1	2	3
1		130	150	138	34	136	174	20	25	96
2		74	159	168	80	106	150	82	58	82
3		155	188	110	40	122	120	70	70	104
4		180	126	160	75	115	139	58	45	60

TABLE 11.9
Two-way analysis of variance table for effect of material type (M) and temperature (T) on the life of battery.

	Df	**Sum Sq**	**Mean Sq**	**F value**	**Pr(>F)**
M	2	10683.72	5341.86	7.91	0.0020
T	2	39118.72	19559.36	28.97	0.0000
Interaction M:T	4	9613.78	2403.44	3.56	0.0186
Residuals	27	18230.75	675.21		

Example 11.8 *A chemical engineer has no control on the temperature variations once a device leaves the factory. The task of the engineer is to investigate two major problems: (i) The effect of material type and temperature on the life of the device, and (ii) Finding the type of material which has the least variation among the varying temperature levels. By an experiment the dependence of the life of a battery is investigated under the temperature conditions 15, 70, and, 125 Fahrenheit, using three types of plate material. For each combination of Temperature and Material, 4 replications of the life of battery are tested, which resulted in the measurements given in Table 11.8.*

To perform the analysis of variance by linear regression the design matrix **X** *for the two-factor model with interaction effects is constructed, see Equation 11.6, where* **1** *denotes a vector with 4 ones and* **0** *a vector with four zeros. Note that there is no column for Material 1 and for Temperature 15, which corresponds to taking $\alpha_1 = \beta_1 = 0$. The columns for the for the interaction effects are the element wise products of the corresponding columns for the main effects for Material and Temperature. The analysis of variance Table 11.9 gives the testing results for the presence of effects on the basis of the residual sums of squares. It can be concluded that there is a main effect for Temperature, a main effect of Material, and an interaction effect. Note that these are overall effects of factors of which differences in means may not be significant.*

The estimates of effects, the corresponding standard errors, t-values, and p-values are given in Table 11.10. The null-hypothesis of zero Material effects is not rejected, that of zero Temperature effects is rejected for both contrasts, and that of zero interaction effects are not rejected except for the M3 by T70

TABLE 11.10
Effect estimates for two-way analysis of variance model.

	Estimate	Std. Error	t value	Pr(>|t|)
(Intercept)	134.7500	12.9924	10.3714	0.0000
M2 - M1	21.00	18.3741	1.1429	0.2631
M3 - M1	9.25	18.3741	0.5034	0.6187
T70 - T15	−77.50	18.3741	−4.2179	0.0002
T125 - T15	−77.25	18.3741	−4.2043	0.0003
M2:T70	41.50	25.9849	1.5971	0.1219
M3:T70	79.25	25.9849	3.0499	0.0051
M2:T125	−29.00	25.9849	−1.1160	0.2742
M3:T125	18.75	25.9849	0.7216	0.4768

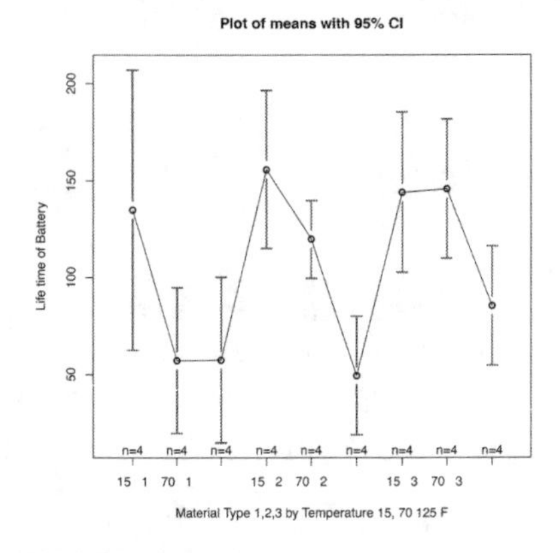

FIGURE 11.3
Means with 95% confidence lines of battery life times depending on Material and Temperature.

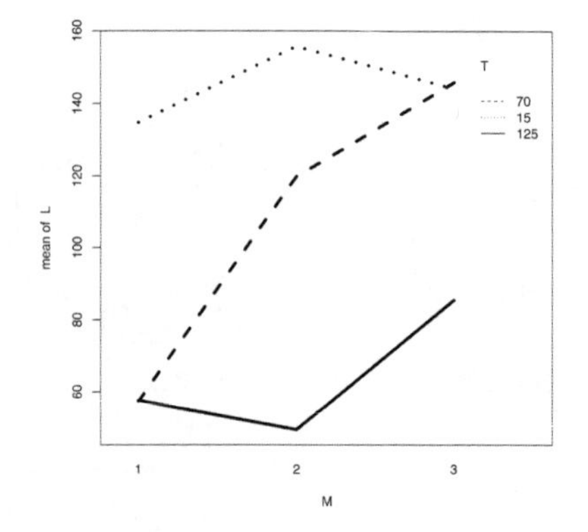

FIGURE 11.4
Interactions plot for battery life depending upon three types of material and three levels of Temperature.

effect. There is not sufficient evidence for specific difference in means effect of material.

The data do not provide sufficient evidence for Material 2 and 3 to be better than Material 1 with respect to life time of battery. There is no clear direct difference between Material 2 and 3, however, Material 3 has a positive interaction effect at the 70 F condition on life time of batteries. Diagnostic plots gave no clear evidence for influential observations in the data. High temperature has a decreasing effect on life time for all three materials, but there is a differential decreasing effect for material 3 at 70 Fahrenheit.

Note that the F-test indicates an overall effect of a complete factor, but that each of the level effects may not be sufficiently large for the p-values to become smaller than the significance level.

Example 11.9 *To give more meaning to the model in Equation (11.5) and to connect different notations in the literature, we compute the expected value of effect A, using the ± values in the design matrix. Then we obtain*

$$\begin{aligned}
E\left[\bar{Y}_{A+} - \bar{Y}_{A-}\right] &= \frac{1}{2}E\left[-\bar{Y}_{-1,1} + \bar{Y}_{1,-1} - \bar{Y}_{-1,1} + \bar{Y}_{1,1}\right] \\
&= \frac{1}{2}(-\mu - \alpha_{-1} - \beta_{-1} - (\alpha\beta)_{-1,-1} + \mu + \alpha_1 + \beta_{-1} \\
&\quad + (\alpha\beta)_{1,-1} - \mu - \alpha_{-1} - \beta_1 \\
&\quad - (\alpha\beta)_{-1,1} + \mu + \alpha_1 + \beta_1 + (\alpha\beta)_{1,1}) \\
&= \frac{1}{2}2(\alpha_1 - \alpha_{-1}) - \sum_{j=-1}^{1}(\alpha\beta)_{-1j} + \sum_{j=-1}^{1}(\alpha\beta)_{1j} \\
&= \alpha_1 - \alpha_{-1}.
\end{aligned}$$

Similarly the expected effect of the B factor equals $\beta_1 - \beta_{-1}$.

11.2.3 Blocking factors

In many situations is impossible to conduct all runs in the design under strictly homogeneous conditions. Then it makes often sense to add an additional factor to the design which is indicating that the runs are conducted in a certain order, that is within certain blocks. The factor corresponding to this is called a blocking factor. The purpose of the blocking factor is to reduce residual variance so that testing of the effects improves. It may be the case that no attempt is made to actually interpret the effect of the blocking factor.

Example 11.10 *The chemical reaction data (Myers et al., 2009) with yield as outcome in a central composite design with 2 factors in 2 blocks. The factor Time has levels 80, 85, and 90 and the factor Temperature has the levels 170, 175, and 180. The experiment was conducted in two blocks (B1 and B2) each consisting of 7 runs out of a total of 14. The data are given in Table 11.14.*

A succinct way to summarize the analysis of variance is in terms of first order (FO), pure quadratic (PQ), and two-way interaction (TWI) effects. For the chemical reaction data these effects are given in Table 11.15 without blocking effects and in Table 11.16 with blocking effects.

Example 11.11 *The chemical reaction data (Myers et al., 2009) with yield as outcome in a central composite design with 2 factors in 2 blocks. The factor Time has roughly has levels 80, 85, and 90 and the factor Temperature roughly has the levels 170, 175, and 180, both with some deviations from these levels. The experiment was conducted in two blocks, B1 and B2, each consisting of 7 runs out of a total of 14. The data given in Table 11.14 have a mean yield of 82.8 for B1 and 78.4 for B2, which are different according to the two-sample t-test (p-value = 0.00013). A succinct way to summarize the analysis*

TABLE 11.11
Chemical reaction data Effect estimates for two-way analysis of variance model.

	Time	Temp	Block	Yield
1	80.00	170.00	B1	80.50
2	80.00	180.00	B1	81.50
3	90.00	170.00	B1	82.00
4	90.00	180.00	B1	83.50
5	85.00	175.00	B1	83.90
6	85.00	175.00	B1	84.30
7	85.00	175.00	B1	84.00
8	85.00	175.00	B2	79.70
9	85.00	175.00	B2	79.80
10	85.00	175.00	B2	79.50
11	92.07	175.00	B2	78.40
12	77.93	175.00	B2	75.60
13	85.00	182.07	B2	78.50
14	85.00	167.93	B2	77.00

TABLE 11.12
Analysis of variance summary table without blocking effects.

	Df	Sum Sq	Mean Sq	F value	Pr(>F)
FO(x1, x2)	2	9.63	4.81	0.55	0.5962
TWI(x1, x2)	1	0.06	0.06	0.01	0.9346
PQ(x1, x2)	2	17.78	8.89	1.02	0.4031
Residuals	8	69.73	8.72		

TABLE 11.13
Analysis of variance summary table with blocking effects.

	Df	Sum Sq	Mean Sq	F value	Pr(>F)
Block	1	69.53	69.53	2611.10	0.0000
FO(x1, x2)	2	9.63	4.81	180.73	0.0000
TWI(x1, x2)	1	0.06	0.06	2.35	0.1694
PQ(x1, x2)	2	17.79	8.90	334.05	0.0000
Residuals	7	0.19	0.03		

of variance is in terms of first order (FO), pure quadratic (PQ), and two-way interaction (TWI) effects. The analysis of variance in Table 11.15 without blocking effects reveals not to reject the null hypothesis of zero FO, PQ, and TWI effects. Note that the residual sum of squares as well as mean sum of squares are rather large compared to those associated with the FO, PQ, and TWI effects. The analysis of variance Table 11.16 with blocking effects reveals a relatively large mean sum of squares for the blocking effect, reason for which

TABLE 11.14
Chemical reaction data with 14 run with two factors Time and Temperature in two blocks B1 and B2.

Run	Time	Temp	Block	Yield	Run	Time	Temp	Block	Yield
1	80.00	170.00	B1	80.50	7	85.00	175.00	B2	79.70
2	80.00	180.00	B1	81.50	8	85.00	175.00	B2	79.80
3	90.00	170.00	B1	82.00	9	85.00	175.00	B2	79.50
4	90.00	180.00	B1	83.50	10	92.07	175.00	B2	78.40
5	85.00	175.00	B1	83.90	11	77.93	175.00	B2	75.60
6	85.00	175.00	B1	84.30	12	85.00	182.07	B2	78.50
7	85.00	175.00	B1	84.00	13	85.00	167.93	B2	77.00

TABLE 11.15
Analysis of variance summary table without blocking effects.

	Df	Sum Sq	Mean Sq	F value	Pr(>F)
FO(x1, x2)	2	9.63	4.81	0.55	0.5962
TWI(x1, x2)	1	0.06	0.06	0.01	0.9346
PQ(x1, x2)	2	17.78	8.89	1.02	0.4031
Residuals	8	69.73	8.72		

TABLE 11.16
Analysis of variance summary table with blocking effects.

	Df	Sum Sq	Mean Sq	F value	Pr(>F)
FO(x1, x2)	2	9.63	4.81	180.73	0.0000
TWI(x1, x2)	1	0.06	0.06	2.35	0.1694
PQ(x1, x2)	2	17.79	8.90	334.05	0.0000
Block	1	69.53	69.53	2611.10	0.0000
Residuals	7	0.19	0.03		

the residual mean sum of squares is drastically smaller. The p-values from the F-tests reveals that the FO and PQ effects are significant, whereas that of TWI is not.

This example illustrates that the conclusions and decisions with respect to the effects of the other factors on the yield may be completely different after incorporating the blocking effect into the model.

11.3 Analysis of the Response Surface

If the most influential factors are identified, then we may use the numerical values of the factors in a linear model with main, two-way interaction, and quadratic effects written as

$$\begin{aligned} Y_i &= \beta_0 + \beta_1 x_{1i} + \beta_2 x_{2i} + \beta_{12} x_{1i} x_{2i} + \beta_{11} x_{1i}^2 + \beta_{22} x_{2i}^2 + \varepsilon_i \\ &= \beta_0 + (\beta_1, \beta_2) \begin{pmatrix} x_{1i} \\ x_{2i} \end{pmatrix} + (x_{1i}, x_{2i}) \begin{bmatrix} \beta_{11} & \beta_{12}/2 \\ \beta_{12}/2 & \beta_{22} \end{bmatrix} \begin{bmatrix} x_{1i} \\ x_{2i} \end{bmatrix} + \varepsilon_i \\ &= \beta_0 + \mathbf{b}^T \mathbf{x}_i + \mathbf{x}_i^T \mathbf{B} \mathbf{x}_i + \varepsilon_i, \text{ where } \varepsilon_i \overset{IID}{\sim} N(0, \sigma^2). \end{aligned}$$

The structural part of the model is a scalar valued vector function $y(\mathbf{x})$ for any value of the predictive variables $\mathbf{x}^T = (x_1, x_2)$. Note that the matrix $\mathbf{B}$ is symmetric by definition. After estimating the beta coefficients from the data by least squares, computing $\widehat{\boldsymbol{\beta}} = (\mathbf{X}^T\mathbf{X})^{-1}\mathbf{X}^T\mathbf{y}$, it is straightforward to compute $\widehat{\mathbf{b}}$ and $\widehat{\mathbf{B}}$ as functions of the estimated coefficients. The closer the estimates are to the true values in the process, the better the estimated **response surface**

$$\widehat{y}(\mathbf{x}) = \hat{\beta}_0 + \widehat{\mathbf{b}}^T \mathbf{x} + \mathbf{x}^T \widehat{\mathbf{B}} \mathbf{x}$$

approaches to true surface $y(\mathbf{x})$. Such occurs if the residual error variance $\widehat{\sigma}$ is small, the number of replicates is sufficiently large, and the most important effects are in the model.

In applications of **industrial design** we often wish to optimize the response surface $y(\mathbf{x})$ over all possible values of the predictor values $\mathbf{x}$. For this purpose the first and the second-order derivatives

$$\frac{\partial \widehat{y}}{\partial \mathbf{x}} = \widehat{\mathbf{b}} + 2\widehat{\mathbf{B}}\mathbf{x}, \quad \frac{\partial^2 y}{\partial \mathbf{x} \partial \mathbf{x}^T} = 2\widehat{\mathbf{B}}$$

are important. Setting the first order derivative to zero and solving for $\mathbf{x}$ gives the **stationary point** as a unique solution

$$\mathbf{x}_s = -\frac{1}{2}\widehat{\mathbf{B}}^{-1}\widehat{\mathbf{b}}$$

if and only if the inverse of $\widehat{\mathbf{B}}$ exists. The type of stationary point $\mathbf{x}_s$ of the surface function is characterized by the eigenvalues of the matrix $\widehat{\mathbf{B}}$. Since the latter is by definition symmetric, all its eigenvalues and eigenvectors consist of real numbers. The eigen decomposition $\widehat{\mathbf{B}} = \mathbf{V}\mathbf{D}\mathbf{V}^T$ exists with an orthonormal **eigenvector matrix** $\mathbf{V}$, that is $\mathbf{V}^T\mathbf{V} = \mathbf{V}\mathbf{V}^T = \mathbf{I}$, and a diagonal matrix $\mathbf{D}$ containing the eigenvalues on its diagonal. If all eigenvalues d_j, $j = 1, \cdots, J$, of $\mathbf{B}$ are positive, then, taking $\mathbf{V}^T\mathbf{x} = \mathbf{a}$, gives

$$\mathbf{x}^T\widehat{\mathbf{B}}\mathbf{x} = \mathbf{x}^T\mathbf{V}\mathbf{D}\mathbf{V}^T\mathbf{x} = \mathbf{a}^T\mathbf{D}\mathbf{a} = \sum_{j=1}^{J} a_j^2 d_{jj} > 0$$

for all non-zero $\mathbf{x}$, suggesting a minimum for the stationary point. If all eigenvalues are negative, then $\mathbf{x}^T\widehat{\mathbf{B}}\mathbf{x} < 0$ for all non-zero $\mathbf{x}$, suggesting a maximum for the stationary point. If there are positive as well as negative eigenvalues,

TABLE 11.17
Regression coefficients for the main, interaction, and quadratic effects of time and temperature on a chemical reaction.

	Estimate	**Std. Error**	**t value**	**Pr(>|t|)**
(Intercept)	84.10	0.08	1056.07	< 0.0001
BlockB2	−4.46	0.09	−51.10	< 0.0001
x1	0.93	0.06	16.16	< 0.0001
x2	0.58	0.06	10.01	< 0.0001
x1:x2	0.13	0.08	1.53	0.1694
x1^2	−1.31	0.06	−21.79	< 0.0001
x2^2	−0.93	0.06	−15.54	< 0.0001

then the stationary point $\mathbf{x}_s$ is a saddle point. When the coefficients in the matrix $\widehat{\mathbf{B}}$ are large, then the eigenvalues can become large so that the surface function is steep near the stationary point. In such a case small changes from $\mathbf{x}_s$ may have a large impact on the level of the surface.

Example 11.12 *Data became available from the execution of a central composite design with the factors Time having levels 80, 85, and 90, and Temperature having levels 170, 175, and 180 in 2 blocks over 14 runs measuring the outcome of a chemical reaction (Lenth, 2010). The blocks, main, and interaction effects are orthogonal, but the quadratic effects are confounded somewhat with the blocks and the intercept. The estimated regression coefficients for the main, interaction and quadratic effects of Time and Temperature on chemical reaction are given in Table 11.17. All effects are significant from the marginal t-tests except for the interaction. The matrix* $\mathbf{B}$ *has rows* $-1.31, 0.06$ *and* $0.06, -0.93$ *with negative eigenvalues* $-0.92, -1.32$ *characterizing the stationary point* $86.86, 176.67$ *for Time and Temperature, respectively, as a maximum. The latter is illustrated by the contour and the perspective plots in Figure 11.5 and 11.6. According to the adjusted R-squared the percentage of explained variance is 99.64, indicating an excellent fit.*

11.4 Mixed Effects Models

Experiments may involve entities such as operators, laboratory technicians, or, animals which are repeatedly involved in the measurements. Such type of effects rests in randomly varying objects essentially different from effects that are fixed in nature. Modeling for the random effects may greatly reduce the error variance and the standard error for the estimated fixed effects. In exceptional research situations it may happen that all effects are random,

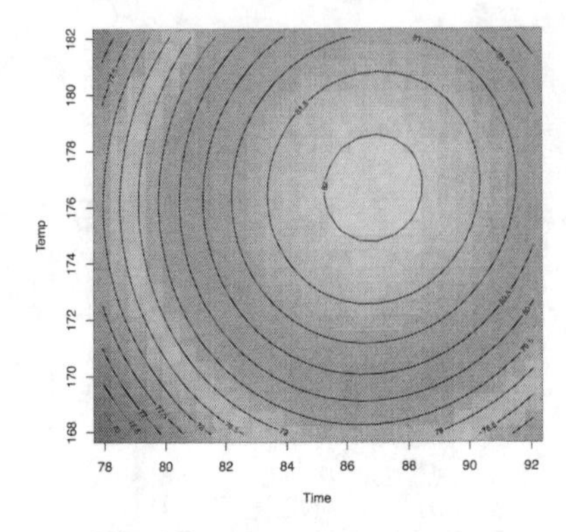

FIGURE 11.5
Contour plot of the surface of chemical reaction depending on Temperature and Time.

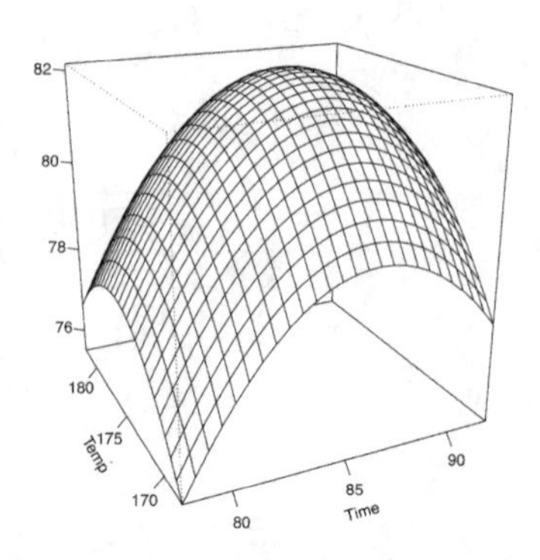

FIGURE 11.6
Perspective plot of the surface of chemical reaction depending on Temperature and Time.

more commonly is the setting where part of the effects are random and part are fixed. The latter are called mixed effects models which are also available for modeling non-linear processes.

11.4.1 Linear random effects models

Experiments may involve entities such as operators, laboratory technicians, or, animals which are repeatedly involved in the measurements. Such type of effects rests in randomly varying objects essentially different from effects that rest in nature. To illustrate the basic idea we consider a mixed model having a fixed constant μ and a random intercept a_i for each operator

$$Y_{ij} = \mu + b_i + \varepsilon_{ij}, i = 1 \cdots, n_i, \;\; j = 1 \cdots, n_j; \;\; \varepsilon_{ij} \overset{IID}{\sim} N(0, \sigma^2), \qquad (11.10)$$

where the number of operators is b, the number of measurements n_i for each operator, and μ the mean. We make the reasonable assumption that the measurements within operators are dependent and those between operators are independent. The operator effects are given by the **random parameters** $b_i \overset{IID}{\sim} N(0, \sigma_a^2)$. We also assume the b_i to be independent to the error terms ε_{ij} for all i and j. For μ to be estimable we need to make the identification type of assumption $\sum_{i=1}^{b} b_i = 0$. In this model we have $E[Y_{ij}] = \mu$ and the decomposition of variances $\mathrm{Var}[Y_{ij}] = \mathrm{Var}[b_i] + \mathrm{Var}[\varepsilon_{ij}] = \sigma_b^2 + \sigma^2$. Note that if we would increase the number of operators and the number of measurements per operator, then we would expect $\hat{\mu}$ to converge to the fixed overall mean μ. For this reason this parameter is called a **fixed effect**. The $\hat{b}_1, \cdots, \hat{b}_{n_i}$ estimates are called random effects, not generalizable to an effect resting in nature. These models are mixed in having fixed as well as random effects, the latter not being reproducible.

Example 11.13 *In an experiment the brightness of pulp was measured by a reflectance meter, repeatedly during a shift of four different operators (Sheldon, 1960). For each of the four operators 5 determinations of brightness became available. The means for Operator 1 to 4 are 60.24, 60.06, 60.62 and 60.68. The intercept* $\hat{\mu}$ *from the random effects model was 60.40 and the random effects for Operator 1 through 4 are -0.122, -0.259, 0.168, and 0.213. The Operator effects seem relatively small compared to the size of the intercept. The estimated error variance was* $\widehat{\sigma} = 0.2609$ *and that for the random effects of the operators* $\widehat{\sigma}_b = 0.3260$.

It may happen that there are two types of random effects where one varies within the other. A model with such type of interaction effects is

$$Y_{ijk} = \mu + b_i + (bc)_{ij} + \varepsilon_{ijk}, \tag{11.11}$$

where the effects $b_i \overset{IID}{\sim} N(0, \sigma_b^2)$, $(bc)_{ij} \overset{IID}{\sim} N(0, \sigma_{bc}^2)$, $\varepsilon_{ijk} \overset{IID}{\sim} N(0, \sigma^2)$ are independent and random in the sense that these vary over units of measurement in such a way that these are not systematic in representing a fixed population type of effect. We use, as before, the conditions

$$\sum_{i=1}^{r} b_i = \sum_{i=1}^{r} \sum_{j=1}^{c} (bc)_{ij} = 0.$$

In the model we have $E[Y_{ij}] = \mu$ and

$$\text{Var}[Y_{ij}] = \text{Var}[b_i] + \text{Var}[(bc)_{ij}] + \text{Var}[\varepsilon_{ij}] = \sigma_b^2 + \sigma_{bc}^2 + \sigma^2.$$

Example 11.14 *Seven specimen were sent to 6 laboratories in the form of 3 separate batches and each was analyzed for the concentration of an unknown analyte in (g/kg) (Committee et al., 1987). The laboratories and the batches are considered to be random in an experiment which was conducted in such a way that the batches varied within the laboratories. For illustrative purposes the analysis was conducted on the concentrations of the analyte (g/kg) for Specimen 5, which are given in duplicates in Figure 11.7. From the linear mixed model the intercept from the fixed effect was* $\hat{\mu} = 7.7614$. *For the Laboratories we have* $\hat{\sigma}_a = 0.6665$ *with 95% CI* $[0.1916; 1.3132]$, *for the interactions of batches within Laboratories we have* $\hat{\sigma}_{ab} = 0.5018$ *with 95% CI* $[0.3042; 0.8411]$ *and for the residuals we have* $\hat{\sigma} = 0.3110$ *with 95% CI* $[0.2316, 0.4486]$. *The standard deviation due to the laboratories and due to the batches within laboratories have comparable size, being twice as large as the error standard deviation. The confidence intervals are sufficiently small and positive suggesting the conclusion that the random effects are present.*

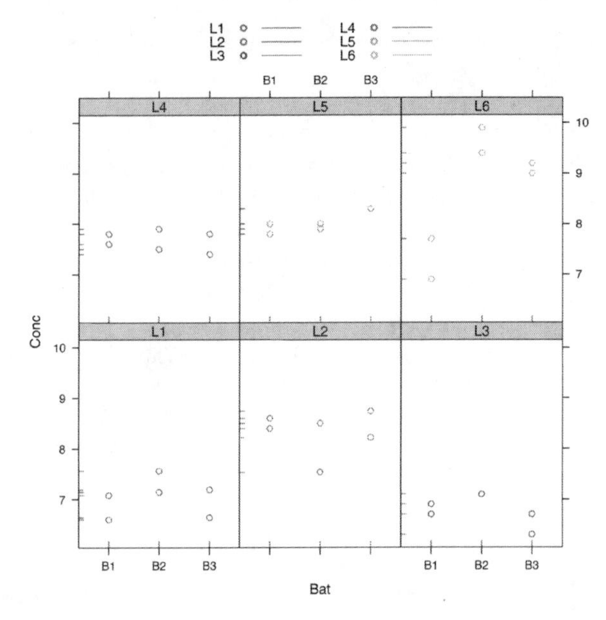

FIGURE 11.7
Concentrations of five specimens over six laboratories and three batches.

TABLE 11.18
Brightness of pulp measurements for specimen 5, random laboratory effects, random batch within laboratory interaction random effects.

	Measurements			Laboratory	Batch within lab		
	B1	**B2**	**B3**	**Random effects**	**B1**	**B2**	**B3**
L1	6.84	7.35	6.92	−0.59	−0.28	0.15	−0.21
L2	8.50	8.02	8.48	0.47	0.23	−0.17	0.21
L3	6.80	7.10	6.50	−0.78	−0.15	0.10	−0.40
L4	7.70	7.70	7.60	−0.08	0.01	0.01	−0.07
L5	7.90	7.95	8.30	0.24	−0.08	−0.04	0.25
L6	7.30	9.65	9.10	0.75	−1.02	0.95	0.49

From the random effects in Table 11.18 it can observed that the random effects of Laboratory 1 and 3 are negative and that of Laboratory 6 is positive. Note that Laboratory 6 has a large negative random Batch 1 effect and large positive effects for Batch 2 and 3. The concentrations for Laboratory 4 are constant over the Batches and all very close to $\hat{\mu} = 7.7614$*, reason for which the corresponding random effects are close to zero. These descriptive type of evaluations are line with observations from Figure 11.7.*

11.4.2 Linear mixed effects models

It may be noted the previous random effect models (11.10) and (11.11) can be extended to have an expected values that linearly depend on several explanatory variables as well as on several random effects. Such gives the more general model

$$\begin{aligned} Y_{ij} &= x_{ij1}\beta_1 + \cdots + x_{ijp}\beta_p + z_{ij1}b_1 + \cdots + z_{ijq}b_q + \varepsilon_{ij}, \\ & i = 1\cdots, M, \;\; j = 1\cdots, n_i; \;\; \varepsilon_{ij} \overset{IID}{\sim} N(0,\sigma^2). \end{aligned} \tag{11.12}$$

After collecting the linear combinations for each outcome j in vectors and putting these one beneath the other, we obtain the linear mixed model as

$$\mathbf{y}_i = \mathbf{X}_i\boldsymbol{\beta} + \mathbf{Z}_i\mathbf{b}_i + \boldsymbol{\varepsilon}_i,$$

where $\mathbf{X}_i$ is $n_i \times p$, $\mathbf{Z}_i$ is $n_i \times q$, the error $\boldsymbol{\varepsilon}_i \sim N(\mathbf{0}, \sigma^2\mathbf{I})$ independent to the random effects $\mathbf{b}_i \sim N(\mathbf{0}, \boldsymbol{\Psi})$, and $\sum_{i=1}^{M} n_i = N$. The columns of $\mathbf{X}_i$ and $\mathbf{Z}_i$ may contain co-variates as well as zeros and ones that corresponding to e.g., one-way or two-way analysis of variance models, such as in (11.3) or (11.6). There also may be identical columns in these matrices accounting for fixed as well as random effects, where the latter are constrained to have average zero in the estimation procedure. In such models is may be reasonable to assume a population mean as well as random deviations from this due to operators or laboratories. Similar remarks hold for slopes, which may be taken fixed as well as random for models in certain settings. In this way a great variety of models are subsumed by the setup.

Since the error and the random effects are independent the distribution of the outcome from the mixed model is now given by

$$\mathbf{y}_i \sim N\left(\mathbf{X}_i\boldsymbol{\beta}, \sigma^2(\mathbf{I} + \mathbf{Z}_i\boldsymbol{\Psi}\mathbf{Z}_i^T)\right), \tag{11.13}$$

for $i = 1, \cdots, M$. Note that the random effects are represented only by their variance matrix $\boldsymbol{\Psi}$. To see how the random effects depend on the outcome variables we derive their covariance

$$\text{Cov}[\mathbf{b}_i, \mathbf{y}_i] = \text{Cov}[\mathbf{b}_i, \mathbf{X}_i\boldsymbol{\beta} + \mathbf{Z}_i\mathbf{b}_i + \boldsymbol{\varepsilon}_i] = \text{Cov}[\mathbf{b}_i, \mathbf{b}_i]\mathbf{Z}_i^T = \sigma^2\boldsymbol{\Psi}\mathbf{Z}_i^T.$$

From this we find the joint distribution of the outcomes and the random effects as

$$\begin{bmatrix} \mathbf{b}_i \\ \mathbf{y}_i \end{bmatrix} \sim N\left(\begin{bmatrix} \mathbf{0} \\ \mathbf{X}_i\boldsymbol{\beta} \end{bmatrix}, \begin{bmatrix} \sigma^2\boldsymbol{\Psi} & \sigma^2\boldsymbol{\Psi}\mathbf{Z}_i^T \\ \sigma^2\mathbf{Z}_i\boldsymbol{\Psi} & \sigma^2(\mathbf{I} + \mathbf{Z}_i\boldsymbol{\Psi}\mathbf{Z}_i^T) \end{bmatrix}\right)$$

From this we may derive the conditional expectation of the random effects given the outcome variables

$$E[\mathbf{b}_i|\mathbf{y}_i] = \text{Cov}[\mathbf{b}_i, \mathbf{y}_i]\text{Var}[\mathbf{y}_i]^{-1}(\mathbf{y}_i - E\mathbf{y}_i]) = \boldsymbol{\Psi}\mathbf{Z}_i^T(\mathbf{I} + \mathbf{Z}_i\boldsymbol{\Psi}\mathbf{Z}_i^T)^{-1}(\mathbf{y}_i - \mathbf{X}_i\boldsymbol{\beta})$$

This would suggest to take as the estimated random effects

$$\widehat{\mathbf{b}}_i = \widehat{\boldsymbol{\Psi}}\mathbf{Z}_i^T(\mathbf{I} + \mathbf{Z}_i\widehat{\boldsymbol{\Psi}}\mathbf{Z}_i^T)^{-1}(\mathbf{y}_i - \mathbf{X}_i\widehat{\boldsymbol{\beta}})$$

To derive the maximum likelihood estimator we use equation (11.13) with the definition $\mathbf{\Sigma}_i = \mathbf{I} + \mathbf{Z}_i \mathbf{\Psi} \mathbf{Z}_i^T$. Then

$$
\begin{aligned}
L(\boldsymbol{\beta}, \boldsymbol{\theta}, \sigma^2 | \mathbf{y}) &= \prod_{i=1}^{M} (2\pi\sigma^2)^{-n_i/2} |\mathbf{\Sigma}_i|^{-1/2} \\
&\quad \times \exp\left\{ -(\mathbf{y}_i - \mathbf{X}_i \boldsymbol{\beta})^T \mathbf{\Sigma}_i^{-1} (\mathbf{y}_i - \mathbf{X}_i \boldsymbol{\beta}) / 2\sigma^2 \right\} \\
&= (2\pi\sigma^2)^{-N/2} \left(\prod_{i=1}^{M} |\mathbf{\Sigma}_i|^{-1/2} \right) \\
&\quad \times \exp\left\{ - \left\| \begin{bmatrix} \mathbf{\Sigma}_1^{-1/2} \mathbf{y}_1 \\ \vdots \\ \mathbf{\Sigma}_M^{-1/2} \mathbf{y}_M \end{bmatrix} - \begin{bmatrix} \mathbf{\Sigma}_1^{-1/2} \mathbf{X}_1 \\ \vdots \\ \mathbf{\Sigma}_M^{-1/2} \mathbf{X}_M \end{bmatrix} \boldsymbol{\beta} \right\|^2 /2\sigma^2) \right\} \\
&= (2\pi\sigma^2)^{-N/2} \left(\prod_{i=1}^{M} |\mathbf{\Sigma}_i|^{-1/2} \right) \exp\left\{ -\|\tilde{\mathbf{y}} - \tilde{\mathbf{X}} \boldsymbol{\beta}\|^2 / 2\sigma^2) \right\}.
\end{aligned}
$$

Comparing with Equation (7.7) we see that this function is maximal over the regression parameters for

$$\widehat{\boldsymbol{\beta}}(\boldsymbol{\theta}) = (\tilde{\mathbf{X}}^T \tilde{\mathbf{X}})^{-1} \tilde{\mathbf{X}}^T \tilde{\mathbf{y}}$$

and maximal over the error variance if

$$\widehat{\sigma}^2(\boldsymbol{\theta}) = \|\tilde{\mathbf{y}} - \tilde{\mathbf{X}} \boldsymbol{\beta}\|^2 / N.$$

Substitution of these into the likelihood function gives

$$L(\widehat{\boldsymbol{\beta}}(\boldsymbol{\theta}), \boldsymbol{\beta}, \widehat{\sigma}^2(\boldsymbol{\theta})) = \left(\prod_{i=1}^{M} |\mathbf{I} + \mathbf{Z}_i \mathbf{\Psi}(\boldsymbol{\theta}) \mathbf{Z}_i^T|^{-1/2} \right) \left[2\pi \widehat{\sigma}^2(\boldsymbol{\theta}) \right]^{-N/2} \cdot \exp(-N/2),$$

the so-called profiled likelihood, which is to be maximized over $(\boldsymbol{\theta})$. In case the random effects variance matrix $\mathbf{\Psi}$ is assumed to be diagonal or ¨proportional to the identity matrix, that is $\mathbf{\Psi} = \sigma_b^2 \mathbf{I}$, maximization of the profiled likelihood simplifies considerably (Pinheiro and Bates, 2006). An approach using the QR decomposition is given in the Notes and Comments to this chapter.

Example 11.15 *In an experiment three brands of machines were compared for an industrial process (Milliken and Johnson, 2017). Six workers were randomly chosen from the employees of a factory to operate each machine three times. The response is an overall productivity score taking into account the number and quality of components produced. From the interaction plot of productivity scores by the workers depending on the machines in Figure 11.8 it can be seen that workers have different levels of productivity depending on the type of machine. As a first model for the productivity data we may take*

$$Y_{ijk} = \beta_j + b_i + \varepsilon_{ijk}, \;\; b_i \overset{IID}{\sim} N(0, \sigma_b^2), \;\; \varepsilon_{ijk} \overset{IID}{\sim} N(0, \sigma^2),$$

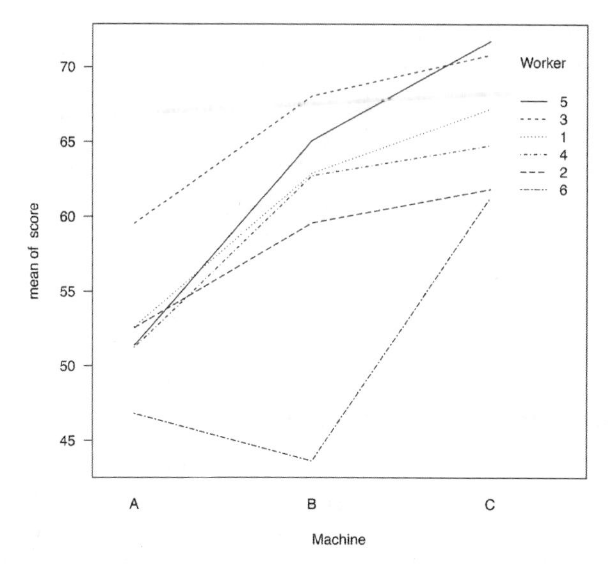

FIGURE 11.8
Machine by Worker interaction plot of productivity scores.

TABLE 11.19
Likelihood ratio testing of Model 1 without against Model 2 with Machine/Worker interaction effects.

Model	df	AIC	BIC	logLik	Test	L.Ratio	p-value
1	5	296.8782	306.5373	−143.4391			
2	6	227.6876	239.2785	−107.8438	1 vs 2	71.1906	0.0000

for Worker $i = 1\cdots,6$, Machine $j = 1\cdots,3$, and repeat $k = 1\cdots,3$. This model does take the different worker levels into account, but not the clear Worker 6 by Machine B interaction among possibly several others. A second model for the data is

$$Y_{ijk} = \beta_j + b_i + b_{ij} + \varepsilon_{ijk},\ b_i \overset{IID}{\sim} N(0,\sigma_b^2),\ b_{ij} \overset{IID}{\sim} N(0,\sigma_2^2),\ \varepsilon_{ijk} \overset{IID}{\sim} N(0,\sigma^2),$$

This model takes Machine within Worker interaction effects as well as Worker level effects into account.

Model 1 is clearly rejected; the data do provide evidence for Worker/Machine interaction effects. The residual standard error $\hat{\sigma} = 3.16$ for Model 1 is considerably larger than that of Model 2 (0.96, 95% CI [0.76, 1.21]). This is obviously due to the Worker/Machine interaction effects. The Worker $\hat{\sigma}_1 = 4.78$ with 95% CI [2.24; 10.16] and the Machine $\hat{\sigma}_2 = 3.72$ with 95% CI [2.38, 5.84] are both clearly larger than zero. From the marginal testing of the fixed effects in Table 11.20 it can be concluded that Machine B performs about 8 points

TABLE 11.20
Summary of fixed machine effects.

	Value	Std.Error	DF	t-value	p-value
(Intercept)	52.3556	2.4858	36.0000	21.0616	0.0000
MachineB	7.9667	2.1770	10.0000	3.6595	0.0044
MachineC	13.9167	2.1770	10.0000	6.3927	0.0001

better, and Machine C performs about 14 points better than Machine A. The analysis indicates Machine C to have the largest overall productivity.

11.4.3 Nonlinear mixed effects models

For nonlinear functions measured on units in an experiment we may want to estimate general fixed effects in addition to random effects. The fixed effects hold for all units measured, while the random effects depend on the units and cannot be generalized to the effect that rests in nature. Element wise we obtain a function f that depends on fixed parameters $\boldsymbol{\beta}$ as well as on random parameters $\mathbf{b}_i$ as follows

$$y_{ij} = f(x_{ij}, \boldsymbol{\beta}, \mathbf{b}_i) + \varepsilon_{ij}.$$

By writing outcomes one underneath the other we obtain the model in terms of a vector-valued function

$$\mathbf{y}_i = \mathbf{f}(\mathbf{x}_i, \boldsymbol{\beta}, \mathbf{b}_i) + \boldsymbol{\varepsilon}_i,$$

where the error terms $\boldsymbol{\varepsilon}_i \sim N(\mathbf{0}, \sigma^2\mathbf{I})$ are independent to the random effects $\mathbf{b}_i \sim N(\mathbf{0}, \boldsymbol{\Psi})$, for $i = 1, \cdots, M$. Usually the random effects are added to the fixed effects to obtain a flexible method for fitting curves to the chemical measurements. Note that to make the model identified we have to constrain the random parameters e.g., by $\sum_{i=1}^{M} \mathbf{f}_i = \mathbf{0}$. Writing $\mathbf{f}_i(\boldsymbol{\beta}, \mathbf{b}_i)$ for $\mathbf{f}(\mathbf{x}_i, \boldsymbol{\beta}, \mathbf{b}_i)$, we note that a way to estimate the parameters is by minimizing the penalized nonlinear least squares objective function

$$\sum_{i=1}^{M} \left[\|\mathbf{y}_i - \mathbf{f}_i(\boldsymbol{\beta}, \mathbf{b}_i)\|^2 + \mathbf{b}_i^T \boldsymbol{\Psi}^{-1} \mathbf{b}_i \right].$$

A closely related approach to linear mixed modeling is obtained by a first-order approximation. That is, by defining the Jacobian matrices

$$\widehat{\mathbf{X}}_i^{(w)} = \left.\frac{\partial \mathbf{f}_i}{\partial \boldsymbol{\beta}^T}\right|_{\widehat{\boldsymbol{\beta}}^{(w)}, \widehat{\mathbf{b}}_i^{(w)}}, \quad \widehat{\mathbf{Z}}_i^{(w)} = \left.\frac{\partial \mathbf{f}_i}{\partial \mathbf{b}_i^T}\right|_{\widehat{\boldsymbol{\beta}}^{(w)}, \widehat{\mathbf{b}}_i^{(w)}},$$

and the response vector

$$\widehat{\mathbf{w}}_i^{(w)} = \mathbf{y}_i - \mathbf{f}_i(\widehat{\boldsymbol{\beta}}^{(w)}, \widehat{\mathbf{b}}_i^{(w)}) + \widehat{\mathbf{X}}_i^{(w)} \widehat{\boldsymbol{\beta}}^{(w)} + \widehat{\mathbf{Z}}_i^{(w)} \widehat{\mathbf{b}}_i^{(w)}.$$

Using $\sigma^2\boldsymbol{\Psi}^{-1} = \boldsymbol{\Delta}^T\boldsymbol{\Delta}$ and $\boldsymbol{\Sigma}_i(\boldsymbol{\Delta}) = \mathbf{I} + \mathbf{Z}_i\boldsymbol{\Delta}^{-1}\boldsymbol{\Delta}^{-T}\mathbf{Z}_i^{(w)T}$, the approximate log-likelihood function

$$\ell(\boldsymbol{\beta}, \sigma^2, \boldsymbol{\Delta}|\mathbf{y}) = -\frac{N}{2}\log(2\pi\sigma^2) - \frac{1}{2}\left\{\log|\boldsymbol{\Sigma}_i(\boldsymbol{\Delta})| + \sigma^{-2}\left[\widehat{\mathbf{w}}_i^{(w)} - \widehat{\mathbf{X}}_i^{(w)}\boldsymbol{\beta}\right]^T \right.$$
$$\left. \times\, \boldsymbol{\Sigma}_i(\boldsymbol{\Delta})^{-1}\left[\widehat{\mathbf{w}}_i^{(w)} - \widehat{\mathbf{X}}_i^{(w)}\boldsymbol{\beta}\right]\right\}.$$

This is similar to the log-likelihood for the linear mixed model with the response vector $\widehat{\mathbf{w}}_i^{(w)}$ and the design matrices $\widehat{\mathbf{X}}_i^{(w)}$ and $\widehat{\mathbf{Z}}_i^{(w)}$ for $i = 1, \cdots, M$. Accordingly, we may alternate between updating the vectors $\widehat{\boldsymbol{\beta}}^{(w)}$ and $\widehat{\mathbf{b}}_i^{(w)}$ for all i, and updating $\boldsymbol{\Psi}$ via $\boldsymbol{\Delta}$ by a profiled likelihood approach.

Example 11.16 *An experiment was conducted to investigate the influence of ambient CO2 concentration on the CO2 uptake depending on the type of plant and night temperature (Potvin et al., 1991). The CO2 uptake of six plants from Quebec and six from Mississippi was measured at several levels of ambient CO2 concentration. Half of the plants of each type were chilled overnight before the measurements were collected. The experiment consists of the factors Type and Temperature each crossed over their two levels with three measurements per cell. The measurements presented as dots in Figure 11.9 indicate a monotonically increasing process toward an upper bound (up to measurement error). A class of functions that represents such a process is*

$$f(x) = f(x, \beta_1, \beta_2, \beta_3) = \beta_1 + (\beta_2 - \beta_1)\exp[-x\exp(\beta_3)],$$

where x is CO2 concentration, $\beta_2 - \beta_1 < 0$ and $\beta_1 > 0$. The process starts for $x = 0$ at $f(0) = \beta_2$ and monotonically increases to the horizontal asymptote $\lim_{x\to\infty} f(x) = \beta_1$, since $0 < e^{\beta_3}$. The random effects b_{1i}, b_{2i}, b_{3i} can be build into the model by

$$\begin{aligned} y_{ij} &= f(x, \beta_1 + b_{1i}, \beta_2 + b_{2i}, \beta_3 + b_{3i}) + \varepsilon_i \\ &= (\beta_1 + b_{1i}) + (\beta_2 + b_{2i} - (\beta_1 + b_{1i}))\exp[-x_{ij}\exp(\beta_3 + b_{3i})] + \varepsilon_i \end{aligned} \tag{11.14}$$

The random effects make the structural part of the model flexible to fit the plant specific rate of CO2 uptake y depending on CO2 concentration x. By a likelihood ratio type of test or comparison of AIC values, it can be evaluated which random effects are actually necessary to better fit the model to the data. We estimate the random effects for the three parameters and subject these to a further analysis of variance to investigate the effects of the two experimental factors. Such obviates the specification of factor effects in starting values for the Newton type of algorithm, which is non-trivially difficult in case interaction effects are to be tested. Three asymptotic regression models we estimated with a different set of random parameters in addition to the three fixed β parameters. Model 1 has random parameters b_{1i}, b_{2i}, b_{3i}, Model 2 has b_{1i}, b_{2i}, and

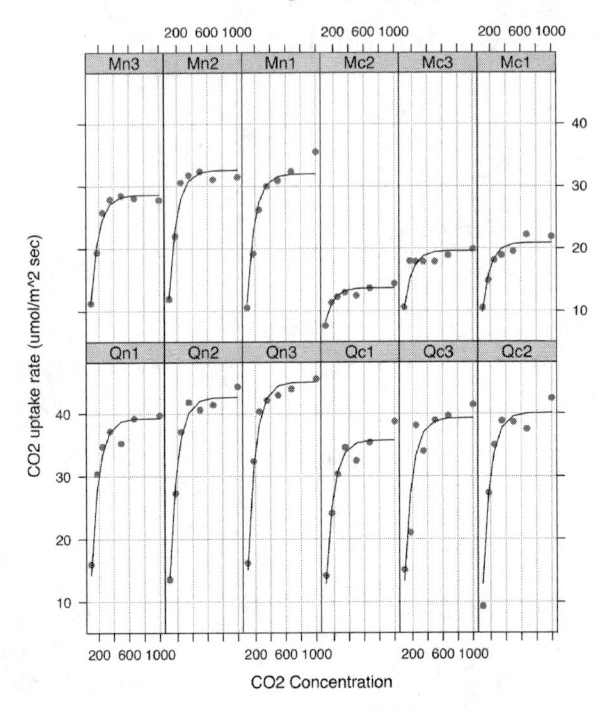

FIGURE 11.9
The CO2 uptake of plants from Quebec (Q) and Mississippi (M) under a chilled (c) and unchilled (n) condition depending on ambient CO2 concentration (horizontally).

TABLE 11.21
Likelihood ratio testing of three models for CO2 uptake.

Model	df	AIC	BIC	logLik	Test	L.Ratio	p-value
1	10	420.8776	445.1857	-200.4388			
2	7	417.0045	434.0202	-201.5023	1 vs 2	2.1270	0.5465
3	5	445.4093	457.5634	-217.7046	2 vs 3	32.4048	0.0000

Model 3 has only b_{1i} as random parameters, for $i = 1, \cdots, 12$. The asymptotic regression models were tested against one another to examen the systematic differences in plant behavior, see Table 11.21.

The p-value 0.5465 suggests not to reject Model 2 against the larger Model 1, that is the CO2 uptake data do not provide evidence for random b_{3i} effects. The p-value < 0.0001 from the likelihood ratio test of Model 3 against Model 2 indicates to reject Model 3. That is, the data do provide evidence for random b_{1i} and random b_{2i} effects in CO2 uptake. From the fixed effects of Model 2 in Table 11.22 it can be observed that the estimated horizontal asymptote $\widehat{\beta}_1 = 32.45$ ($\mu mol/m^2$ sec) fits well to the non-chilled Mississippi plants, but

TABLE 11.22
Fixed effects for three β parameters for the plant CO2 uptake data.

Fixed parameter	Value	Std.Error	DF	t-value	p-value
$\widehat{\beta}_1$	32.45	2.82	70.00	11.49	< 0.0001
$\widehat{\beta}_2$	−17.95	3.97	70.00	−4.52	< 0.0001
$\widehat{\beta}_3$	−4.65	0.05	70.00	−95.58	< 0.0001

TABLE 11.23
Addition of fixed and random plant specific effects for the CO2 uptake.

	$\widehat{\beta}_1+b_{i1}$	$\widehat{\beta}_2+b_{i2}$	$\widehat{\beta}_3+b_{i3}$	Treatment	Type
Qn1	39.26	−22.63	−4.65	Quebec	Non-Chilled
Qn2	42.60	−28.63	−4.65	Quebec	Non-Chilled
Qn3	45.01	−29.10	−4.65	Quebec	Non-Chilled
Qc1	35.75	−20.77	−4.65	Quebec	Chilled
Qc3	39.26	−24.97	−4.65	Quebec	Chilled
Qc2	40.06	−27.63	−4.65	Quebec	Chilled
Mn3	28.62	−14.29	−4.65	Mississippi	Non-Chilled
Mn2	32.61	−18.11	−4.65	Mississippi	Non-Chilled
Mn1	32.02	−19.20	−4.65	Mississippi	Non-Chilled
Mc2	13.70	0.34	−4.65	Mississippi	Chilled
Mc3	19.61	−3.97	−4.65	Mississippi	Chilled
Mc1	20.90	−6.41	−4.65	Mississippi	Chilled

TABLE 11.24
Two-way analysis of variance on the $\widehat{\beta}_1 + f_1$ horizontal asymptote Model 2 parameters for the CO2 uptake.

	Df	Sum Sq	Mean Sq	F value	Pr(>F)
Treatment	1	215.4448	215.4448	26.1305	0.0009
Type	1	743.8235	743.8235	90.2157	0.0000
Treatment:Type	1	61.8515	61.8515	7.5018	0.0255
Residuals	8	65.9596	8.2449		

not for several others. The estimated standard error $\hat{\sigma} = 1.84$ with 95% CI [1.54, 2.20] is relatively small indicating a good overall fit of the curves from Model 2 to the plant CO2 uptake measurements.

The plant specific effects given by the addition of the fixed β and corresponding b_{ij} effects are given in Table 11.23. It can be observed from Figure 11.9 that, apart from measurement error, the horizontal asymptote values in the first column correspond with the estimated end levels of CO2 uptake. A two-way analysis of variance on the asymptotic plant levels gives significant main as well as an interaction effect for both plant Type and Temperature condition, see Table 11.24. From the estimated effects in Table 11.25 lead the following interpretations. The Quebec Type in the Non-Chilled condition has

TABLE 11.25
Estimated analysis of variance effects of the plant specific horizontal asymptote $\widehat{\beta}_1 + f_1$ parameters for the CO2 uptake.

$\widehat{\beta}_1+f_1$	Estimate	Std. Error	t value	Pr(>|t|)
(Intercept)	42.2898	1.6578	25.5095	0.0000
TreatmentChilled	−3.9338	2.3445	−1.6779	0.1319
TypeMississippi	−11.2055	2.3445	−4.7795	0.0014
TreatmentChilled:TypeMississippi	−9.0812	3.3156	−2.7389	0.0255

on estimated asymptote of 42.3 ($\mu mol/m^2$ sec) rate of CO2 uptake. The effect of the Chilled condition is a decrease of the asymptote by 3.9 ($\mu mol/m^2$ sec), the Mississippi Type has a decreasing effect of 11.2 ($\mu mol/m^2$ sec) and the interaction between the Chilled condition and the Mississippi Type leads to a decrease in asymptote of 9.1 ($\mu mol/m^2$ sec) rate of CO2 uptake. The Adjusted R-squared of 0.92 for the two-way analysis of variance model indicates a good fit.

Note that the post hoc analysis of the random effects by analysis of variance makes it possible to make statistical inferences about a specific parameter in of a fitted non-linear mixed model. Obviously, the strict upper bound interpretation of the horizontal asymptote is obscured by the measurement error, however, the estimated size of the error terms by the residual standard deviation is smaller than the size of the estimated effects from the two-way analysis of variance.

11.5 Notes and Comments

Using the QR decomposition linear mixed model estimation.

In case the model matrices $\mathbf{X}_i$ and $\mathbf{Z}_i$ in linear mixed modeling are relatively large, an approach for linear mixed model estimation using the QR decomposition is useful. To detail this maximum likelihood estimation procedure we start with some preparations. Let's define $\sigma^2\mathbf{\Psi}^{-1} = \mathbf{\Delta}^T\mathbf{\Delta}$ e.g., as a Choleski factorization. The matrix $\mathbf{\Psi}$ is a function of $\boldsymbol{\theta}$, so that we also have $\mathbf{\Delta} = \mathbf{\Delta}(\boldsymbol{\theta})$. If we define the determinant of $\mathbf{\Psi}$ by absolute value signs, then we find

$$|\mathbf{\Psi}|^{-1/2} = \sigma^{-2q}|\sigma^2\mathbf{\Psi}^{-1}|^{1/2} = \sigma^{-2q}|\mathbf{\Delta}^T\mathbf{\Delta}|^{1/2} = \sigma^{-2q}|\mathbf{\Delta}|,$$

where the latter denotes the absolute value of the determinant. Furthermore we have

$$\mathbf{b}_i^T\mathbf{\Psi}^{-1}\mathbf{b}_i \cdot \frac{\sigma^2}{\sigma^2} = \sigma^{-2}\mathbf{b}_i^T\mathbf{\Delta}^T\mathbf{\Delta}\mathbf{b}_i = \sigma^{-2}\|\mathbf{\Delta}\mathbf{b}_i\|^2.$$

Now the marginal density function for the random effects is

$$p(\mathbf{b}_i|\boldsymbol{\theta},\sigma^2) = (2\pi)^{-q/2}|\boldsymbol{\Psi}|^{-1/2}\exp-(\mathbf{b}_i^T\boldsymbol{\Psi}^{-1}\mathbf{b}_i/2) \tag{11.15}$$

$$= (2\pi\sigma^2)^{-q/2}|\boldsymbol{\Delta}|\exp\left(-\|\boldsymbol{\Delta}\mathbf{b}_i\|^2/2\sigma^2\right) \tag{11.16}$$

From the independence of the error terms and the random effects, we obtain the conditional density

$$p(\mathbf{y}_i|\mathbf{b}_i,\boldsymbol{\beta},\sigma^2) = (2\pi\sigma^2)^{-n_i/2}\exp(-\|\mathbf{y}_i - \mathbf{X}_i\boldsymbol{\beta} - \mathbf{Z}_i\mathbf{b}_i\|^2/2\sigma^2). \tag{11.17}$$

Note that the random effects have expected value zero we must integrate the conditional density given the random effects in order to find a meaningful likelihood function. We start with conditioning, apply a QR decompostion twice and use a change of variable to derive the profiled likelihood as follows

$$L(\boldsymbol{\beta},\boldsymbol{\theta},\sigma^2|\mathbf{y}) = \prod_{i=1}^{M} p(\mathbf{y}_i|\boldsymbol{\beta},\boldsymbol{\theta},\sigma^2) = \prod_{i=1}^{M}\int p(\mathbf{y}_i|\mathbf{b}_i,\boldsymbol{\beta},\sigma^2)\cdot p(\mathbf{b}_i|,\boldsymbol{\theta},\sigma^2)d\mathbf{b}_i$$

$$= \prod_{i=1}^{M}(2\pi\sigma^2)^{-n_i/2}|\boldsymbol{\Delta}|\int(2\pi\sigma^2)^{-q/2}$$

$$\times\ \exp\left\{-(\|\mathbf{y}_i - \mathbf{X}_i\boldsymbol{\beta} - \mathbf{Z}_i\mathbf{b}_i\|^2 + \|\boldsymbol{\Delta}\mathbf{b}_i\|^2)/2\sigma^2)\right\}d\mathbf{b}_i$$

$$= \prod_{i=1}^{M}(2\pi\sigma^2)^{-n_i/2}|\boldsymbol{\Delta}|\int(2\pi\sigma^2)^{-q/2}$$

$$\times\ \exp\left\{-\left\|\mathbf{Q}_i^T\left(\begin{bmatrix}\mathbf{y}_i\\ \mathbf{0}\end{bmatrix} - \begin{bmatrix}\mathbf{X}_i\\ \mathbf{0}\end{bmatrix}\boldsymbol{\beta} - \begin{bmatrix}\mathbf{Z}_i\\ \boldsymbol{\Delta}\end{bmatrix}\mathbf{b}_i\right)\right\|^2/2\sigma^2)\right\}d\mathbf{b}_i$$

Now using the QR decomposition

$$\begin{bmatrix}\mathbf{Z}_i & \mathbf{X}_i & \mathbf{y}_i\\ \boldsymbol{\Delta} & \mathbf{0} & \mathbf{0}\end{bmatrix} = \mathbf{Q}_i\begin{bmatrix}\mathbf{R}_{11i} & \mathbf{R}_{10i} & \mathbf{c}_{1i}\\ \mathbf{0} & \mathbf{R}_{00i} & \mathbf{c}_{2i}\end{bmatrix},$$

and the change of variable $\boldsymbol{\phi}_i = (\mathbf{c}_{1i} - \mathbf{R}_{10i}\boldsymbol{\beta} - \mathbf{R}_{11i}\mathbf{b}_i)/\sigma$ with $d\boldsymbol{\phi}_i = \sigma^{-q}|\mathbf{R}_{11i}|d\mathbf{b}_i$, we obtain

$$= \prod_{i=1}^{M}(2\pi\sigma^2)^{-n_i/2}|\boldsymbol{\Delta}|\exp\left\{-\|\mathbf{c}_{0i} - \mathbf{R}_{00i}\boldsymbol{\beta}\|^2/2\sigma^2\right\}|\mathbf{R}_{11i}|^{-1}$$

$$\cdot\int(2\pi)^{-q/2}\exp\left\{-\|(\mathbf{c}_{1i} - \mathbf{R}_{10i}\boldsymbol{\beta} - \mathbf{R}_{11i}\mathbf{b}_i)/\sigma\|^2/2\right\}\sigma^{-q}|\mathbf{R}_{11i}|d\mathbf{b}_i$$

$$= \prod_{i=1}^{M}(2\pi\sigma^2)^{-n_i/2}\frac{|\boldsymbol{\Delta}|}{|\mathbf{R}_{11i}|}\exp\left\{-\|\mathbf{c}_{0i} - \mathbf{R}_{00i}\boldsymbol{\beta}\|^2/2\sigma^2\right\}$$

$$\cdot\int(2\pi)^{-q/2}\exp\left\{-\|\boldsymbol{\phi}_i\|^2/2\right\}d\boldsymbol{\phi}_i.$$

For the function to be integrated on the right-hand side we recognize the multivariate standard normal density, so that we conclude that it equals one. Writing the terms in the exponent one beneath the other using a second QR decomposition

$$\begin{bmatrix} \mathbf{R}_{001} & \mathbf{c}_{01} \\ \vdots & \vdots \\ \mathbf{R}_{00M} & \mathbf{c}_{0M} \end{bmatrix} = \mathbf{Q} \begin{bmatrix} \mathbf{R}_{00} & \mathbf{c}_0 \\ \mathbf{0} & \mathbf{c}_1 \end{bmatrix},$$

were arrive at

$$(2\pi\sigma^2)^{-N/2} \cdot \exp\left\{ - \left\| \mathbf{Q}^T \left(\begin{bmatrix} \mathbf{c}_{01} \\ \vdots \\ \mathbf{c}_{0M} \end{bmatrix} - \begin{bmatrix} \mathbf{R}_{001} \\ \vdots \\ \mathbf{R}_{00M} \end{bmatrix} \boldsymbol{\beta} \right) \right\|^2 /2\sigma^2 \right\} \cdot \prod_{i=1}^{M} \frac{|\boldsymbol{\Delta}|}{|\mathbf{R}_{11i}|}$$

$$= (2\pi\sigma^2)^{-N/2} \cdot \exp\left\{ - \left(\|\mathbf{c}_1\|^2 + \|\mathbf{c}_0 - \mathbf{R}_{00}\boldsymbol{\beta}\|^2 \right) /2\sigma^2 \right\} \cdot \prod_{i=1}^{M} \frac{|\boldsymbol{\Delta}|}{|\mathbf{R}_{11i}|}. \tag{11.18}$$

To maximize the right-hand side of Equation (11.18) one may use least squares theory to find

$$\widehat{\boldsymbol{\beta}}(\boldsymbol{\theta}) = \mathbf{R}_{00}^{-1}\mathbf{c}_0, \text{ and } \hat{\sigma}^2(\boldsymbol{\theta}) = \frac{\|\mathbf{c}_1\|^2}{N}. \tag{11.19}$$

Substitution of these into Equation (11.18)gives the profiled likelihood

$$L(\boldsymbol{\theta}|\mathbf{y})) = L(\widehat{\boldsymbol{\beta}}(\boldsymbol{\theta}), \boldsymbol{\theta}|\mathbf{y}, \hat{\sigma}^2(\boldsymbol{\theta}) = \left(\frac{N}{2\pi \|\mathbf{c}_1\|^2} \right)^{N/2} \cdot \exp(-N/2) \cdot \prod_{i=1}^{M} \frac{|\boldsymbol{\Delta}|}{|\mathbf{R}_{11i}|}$$

so that the profiled log-likelihood becomes

$$\ell(\boldsymbol{\theta}|\mathbf{y}) = \frac{N}{2}(\log N - \log(2\pi) - 1) - N \log \|\mathbf{c}_1\| + \sum_{i=1}^{M} \log\left(\frac{|\boldsymbol{\Delta}|}{|\mathbf{R}_{11i}|} \right).$$

Maximization of the profiled log-likelihood yields the $\widehat{\boldsymbol{\theta}}$ from which the $\widehat{\boldsymbol{\beta}}$ and $\hat{\sigma}^2$ are found using (11.19).

The optimization of the profiled log-likelihood for (non)-linear mixed models is usually accomplished by (Quasi-) Newton-Raphson iterations, but several other algorithms are also available (Pinheiro and Bates, 2006; Thisted, 2017; Demidenko, 2013).

Additional notes and remarks.

An excellent book length treatment of experimental design is Montgomery (2013). For a deep theoretical approach to the subject we refer to Mukerjee and Wu (2007).

The fact that a negative definite Hessian matrix implies that a stationary point is locally maximal is given e.g., in a standard calculus book such as Apostol (2007b).

It is possible to use mixed generalized linear modeling in an experimental setting where counts are observed within regions over time (Venables and Ripley, 2000), or to use a generalized additive approach with random effects in spline functions (Wood, 2017).

11.6 Notes on Using R

References to many R libraries are given by CRAN its Experimental Design page, from which further access to the literature is easily found. More specifically, properties of a design such as the generalized word length or the generators can be investigated by

```
library(FrF2)
design  <- FrF2(2^(7-2), 7, generators = c("ABC","BCD"))
GWLP(design)
generators(design)
```

Obviously, one could simplify the notation 2^{7-2}, it, however, kindly reminds us that the design has 7 factors in 32 runs. To construct a smallest design with 6 factors

```
FrF2(nfactors = 6, resolution =4, randomize=FALSE)
FrF2(design="6-2.1", randomize=FALSE)
```

Alternatively, to construct a smallest design keeping certain effects estimable one may use for instance

```
FrF2(estimable=formula("~(A+B+C)^2+D+E"))
```

To construct a 3^2 design in 3 blocks one uses

```
fac.design(nlevels=c(3,3),blocks=3,randomize=FALSE)
```

A 3^5 design in 27 blocks

```
fac.design(nlevels=c(3,3,3,3,3),blocks=27,randomize=FALSE)
```

Binomial coefficients to compute the number of effects can conveniently be computed from R its `choose` function. The output of the basic `lm` function can be used with the generic functions `summary` and `anova`. Analysis of variance by the `aov` function which allows for correlated errors in repeated measurements post hoc analysis by `TukeyHSD` or Fisher's method. An `interaction.plot` often facilitates a correct interpretation of main and interaction effects.

Linear mixed models can be estimated by the `lme` or the `lmer` functions. The likelihood ratio testing can be performed by the generic `anova` function. The random effects can easily be extracted by the `ranef` function, or the `coef` function, whereas the confidence interval are generated by the `intervals` function. Non linear mixed effects model can be estimated by the `nlme` function, generalized linear mixed models by the `geeglm` or `glmm`, and generalized additive models by the `gamm` function.

11.7 Exercises

1. **Inverse of contrast matrix.** The contrast matrix

$$\mathbf{C} = \begin{bmatrix} 1 & 0 & 0 \\ -1 & 1 & 0 \\ -1 & 0 & 1 \end{bmatrix}.$$

has inverse, as claimed in the above,

$$\mathbf{C}^{-1} = \begin{bmatrix} 1 & 0 & 0 \\ 1 & 1 & 0 \\ 1 & 1 & 1 \end{bmatrix}.$$

   (a) Show that this holds true.
   (b) Use R to find the inverse.
   (c) Use R to verify that

$$\mathbf{C}\mathbf{C}^T = \begin{bmatrix} 1 & -1 & -1 \\ -1 & 2 & 1 \\ -1 & 1 & 2 \end{bmatrix}$$

2. **Constructing and analysing designs.**
   (a) Construct a full factorial design 2^5 and compute its generalized words of lengths. What do you conclude?
   Hint: Use the library `DoE.base`.
   (b) Construct a full factorial design 2^5 with 4 blocks and compute its generalized words of lengths. What do you conclude?
   (c) Construct a 2^{10-5} Design A with generators =
   c("ABCD","ABCE","ABDE","ACDE","BCDE")
   and a 2^{10-6} Design B with generators =
   c("ABC","BCD","ACD","ABD","ABCD","AB")
   and give their properties. If estimation of main effect is deemed import, which of the would you prefer?
   Hint: Use e.g., the `FrF2` function of the `FrF2` library.

Hint: The summary of a constructed design gives detailed information on its properties.

(d) Construct a 64 runs design with all (A+B+C)*(D+E+F+G+H) interactions estimable.

3. **Geraniol production**. Use the `VSGFS` data (Vasilev et al., 2014) from the library `DoE.base` with the data from a Tageugi design.

   (a) Compute the generalized words of lengths 3 and 4 indicate and explain what these mean.

   (b) Construct mosaic plots of the 2 factor interactions. What do you conclude from these?

   (c) Construct a correlation plot of the design. Are the main effects confounded with the 2-way interaction effects?
   Hint Use R its function `corrPlot`.

   (d) Use an analysis of variance to select a model with main and 2-factor inactions for the outcome yield. Find the model with significant main and 2-factor interaction effect.

   (e) Use an analysis of variance to select a model with main and 2-factor inactions for the outcome yield. Find the model with significant main and 2-factor interaction effect. Give a brief interpretation of the effects. Report its Adjusted R-squared.

   (f) Compute the predicted values of the yield with their corresponding prediction intervals, add these to the data. Order the data set according to the maximum predicted yield. What is your conclusion?

4. **Bottling.** In a three factorial experiment bottling is investigated. The height of the fills in the soft drink bottle is required to be as consistent as possible. The height is controlled through three factors:

   - the percent carbonation of the drink,
   - the operating pressure in the filler, and
   - the line speed which is the number of bottles filled per minute.

   The first factor variable gives the percentage of carbonation at the levels 10, 12, and 14, the operating pressure is available at the levels 25 and 30 psi units, while the line speed is available at the levels 200 and 250 bottles per minute. Two replicates are available for each combination of the levels of the three factors (Montgomery, 2013). In this experiment, the deviation from the required height level is measured. A data frame with 24 observations contains the following 4 variables.
   Deviation: Deviation from required height level
   Carbonation: the percent carbonation of the drink

Pressure: the operating pressure in the filler
Speed: the number of bottles filled per minute
The data are available from the R library `ACSWR` under the name `bottling`.

(a) Formulate the research question corresponding to the situation at hand.
(b) Perform an informal analysis to explore the research question. That is, give statistics and plots that summarize the data.
(c) Perform a formal analysis testing a relevant hypothesis.
(d) Give the conclusion including the answer on the research question.
(e) Provide one point of discussion about critical aspects of the research.

5. **PVC.** Each of three Operators used eight different Resin rail cars to produce PVC. The outcome is the particle size called "psize" (Morris and Watson, 1998). Assume that the particle size measurements are all independent. The data are available as `pvc` from the R library `faraway`.

   (a) Compute the operator by resin interactions plot.
   (b) Test for an interaction effect between Operators and Resin.
   (c) Test differences between Operators and levels of Resin. Report the fit of the main effects model to the data.
   (d) What is the main conclusion from the analysis? What would you add to make the analysis more complete?

6. **Pastry dough**. In a pastry dough experiment (Lawson, 2014) three factors were investigated on the yield using a blocking factor with seven levels. The data are available from the R library `daewr` under the name `pastry`, or from `pastry.csv`.

   (a) Test the model with blocking factor using only main effects. Is there a blocking effect? Briefly check the diagnostics by visualization.
   (b) Estimate a second-order model with the two significant factors. What is the conclusion?
   Hint: use the R its `rsm` function from the library `rsm`.
   (c) Investigate whether the response surface contains an optimum. Illustrate your findings with a contour plot.

7. **Maillard reaction in milk powder**. In an experiment with production of milk powder the effect of water activity and temperature on the formation of maillard reaction products was investigated. There were 9 treatment combinations of the two factors and three replicates (blocks) of the experiment giving a total of 27 productions. The factors and levels were: water activity (approx. 0.15, 0.25,

and 0.10, coded as 1, 2, 3 in the data set), and temperature (100° C, 110° C, 120° C, 140° C). The 27 samples were stored and measurements were made after 4, 6, and 8 weeks. The measurements (response variables) were: concentration of maillard reaction products (which may give a bad taste), and sensory evaluation of taste (high = good taste). The data are available from file `maillaird.csv`.

(a) Arrange for factors for the predictive variables of maillard and test for the main and interaction affects leaving out repeat from the model.

(b) Add repeats in the repeated measurement model and compare with the previous. What effects do exist?
Hint: Use aov with rep in the Error term.

(c) Do there exist effects on taste?

8. **Olive oil.** In an experiment scores were attributed on 6 attributes from a sensory panel and measurements of 5 physico-chemical quality parameters on 16 olive oil samples. The first five oils are from Greek (G), the next five are Italian (I) and the last six are Spanish (S). The 5 chemical variables are measurements of acidity, peroxide, K232, K270, and DK. In the above analysis from one of the exercises in Chapter 10 the biplot suggested differences between countries of olive oil production on the first principal component. The 6 sensory variables are scores on the olive oil attributes "yellow," "green," "brown," "glossy," "transp," and "syrup" (Ziegel, 2000). The data `oliveoil` is available from the library `pls`.

   (a) Perform a principal component analysis on the chemical variables and construct the biplot from the first two components.

   (b) Test the null hypothesis that the countries have equal means on the first principal component.

   (c) A boxplot suggests that the variances are unequal. Redo the testing dealing with heterogeneity.

9. **Carbachol.** In an experiment (Douglas et al., 2004) the effect of Carbachol on nucleotide activation (guanine nucleotide bonding) was investigated in three different brain regions among five adult male rats. Carbachol, also known as carbamylcholine and sold under the brand name Miostat, is a cholinomimetic drug that binds and activates acetylcholine receptors. The brain regions that studied were abbreviated as "BST," "LS," and "VDB" and treatment factor had the levels Basal and Carbachol. The continuous dependent variable Nucleotide activation is called "activate." The data are available from the R library `WWGbook` (West et al., 2006) under the name `rat.brain`.

(a) Construct a factor treatment with the labels "Basal," "Carbachol" and a factor region with the labels "BST," "LS," "VDB." Construct a line plot of the data by e.g., `xyplot` and give your observations with respect to a linear mixed model. Are there clear random rat effects, main treatment and region effects observable? How about interaction effects?

(b) Estimate two models with a random intercept for rat. The first model has main region and treatment effects and the second has these as well as interaction affects. Test the first against the second.

(c) Extract the random effects and relate these to the ordering in the plot.

(d) Interpret the main and interaction effects.

(e) Construct confidence intervals for the main and interaction effects as well as for the error standard deviations.

10. **Simulation Study on non-linear mixed model**. Here we undertake a simulation study close to the above CO2 example with model formula in Equation (11.14). Suppose we have a 2 by 2 cross over design on the factors Type and Treatment each with two levels with 5 plants per condition, where we focus on the parameter β_1 determining the horizontal asymptote of y, the CO2 uptake. We take for the beta parameters $\beta_1 = 40; \beta_2 = -15; \beta_3 = -5.5$ with random effects added to β_1 from the normal distribution $N(0, 0.5^2)$ and no random effects on the other two beta parameters. Suppose further that on the β_1 parameter we have for Type 2 a decreasing main effect of -5, for the Treatment 2 condition a decreasing main effect of -10 and for the interaction of these conditions an interaction -7 effect. Suppose that the independent error terms come from the normal distribution $N(0, 2^2)$. Take the x CO2 concentration from 100 to 1000 with steps of 100.

    (a) Construct the data according to model. The following steps are handy.
    Hint 1: Construct a design matrix with zeros and ones according to the conditions of the two factors.
    Hint 2: Add to the matrix the parameters with the fixed for β_2, and β_3 and the fixed plus added for β_1.
    Hint 3: Use a for loop to construct the structural part of the model $f(x, \beta_1 + b_{1i}, \beta_2, \beta_3)$, the random error from the normal distribution, and the outcome y by addition.
    Hint 4: Construct the data part necessary for non-linear mixed modeling in a matrix and store these one underneath the other for each plant.

    (b) Perform the non-linear mixed modeling, gives its model summary, and add the coefficients to the design matrix. Perform a two-way analysis of variance on the $\beta_1 + b_{1i}$ parameters and compare the estimated effects with the true effects from the design. Are the true effects within their 95% confidence intervals? Hint: Use starting values e.g., c(f1 $= 30$, f2 $= -25$, f3 $= -5$).

(c) What is the effect of increasing the number of measurements x CO2 concentration between 100 and 1000?

12

Robust Analysis of Models

Above, we often assumed normally distributed and IID errors. This type of assumption was frequently made for estimation and testing chemical hypotheses while fitting a (non-)linear model or an analysis of variance. This assumption simplifies maximum likelihood estimation and results in many explicit formula for estimators, test-statistics and their distributions.

In chemical research, especially outside the laboratory, it frequently happens that several measurements failed or were otherwise recorded far away from the bulk of the data. Based on the residuals from estimated models we have defined various diagnostics to evaluate the validity of these assumptions and to detect unusual observations. Although these procedures do give useful information in many situations, sometimes, however, procedures such as Cook's distance may fail to detect all outlying observations as some may be masked (Maronna et al., 2019). Since all estimators are continuous functions of the data, a small fraction of extreme outlying observations could have a large influence on the estimator making valid inferences more difficult. An obvious example is the mean, where each observation is weighted equally by $1/n$, so that a single extreme observation can have a large influence on the value of the estimator $\widehat{\mu} = \sum_{i=1}^{n} y_i/n$. It seems clear that manual outlier rejection is both undesirable as well as time consuming, especially when several are present in the data. Rather, it is desirable to have estimation procedures which are robust against outlying data points, relatively unbiased, consistent, as well as near efficient. This can be accomplished by generalizing maximum likelihood estimation with specifically designed objective functions. The basic ideas of robust estimation can nowadays easily be applied to almost all methods for analyzing chemical data.

In this chapter you learn

- In Section 12.1 how to test for outliers and how large the effect of an outlying data point can be on the estimated curve.
- In Section 12.2 how to estimate a location (target) or scale (variability) in a robust manner, when there exist outliers in the data.

DOI: 10.1201/9781003178194-12

- In Sections 12.3 and 12.4 how to estimate the coefficients of a linear or non-linear model in the presence of outlying data points.

12.1 Outlying Data Points

A classical test for outlying data points is discussed and illustrated. Next the drastic effect of an outlying data point on the estimated line or curve is illustrated.

12.1.1 A classical test for detecting an outlier

A classical way for detecting outliers is by the Grubbs test (Grubbs, 1969), which is based on the assumption that the measurements have a normal distribution. Specifically, the null hypothesis H_0: *There are no outliers in the data set* is tested against the alternative H_A: *There is at least one outlier in the data set.* The test statistic is the largest absolute deviation from the sample mean relative to the sample standard deviation. That is,

$$G = \frac{\max_{i=1,\cdots,n} |\overline{Y} - Y_i|}{s_y}.$$

The null hypothesis of no outliers is rejected at significance level α if

$$G > \frac{n-1}{n} \sqrt{\frac{t^2_{\alpha/(2n),n-2}}{n-2+t^2_{\alpha/(2n),n-2}}}$$

where $t_{\alpha/(2n),n-2}$ is the upper critical value of the t-distribution with $n-2$ degrees of freedom and significance level $\alpha/(2n)$.

Example 12.1 *24 measurements in ($\mu g g^{-1}$) of copper in wholemeal flour (Committee, 1989) were subjected to the Grubbs test. The measurements obtained are (Venables and Ripley, 2002)*

```
2.90  3.10  3.40  3.40  3.70  3.70  2.80  2.50  2.40  2.40  2.70  2.20
5.28  3.37  3.03  3.03 28.95  3.77  3.40  2.20  3.50  3.60  3.70  3.70
```

The data yield the value $G = 4.6569$, which corresponds to a p-value $2.2. \cdot 10^{-16}$ from the Grubbs test. The data provide strong evidence to reject the null hypothesis of no outliers in the data. It can be concluded that the highest value 28.95 is an outlier.

There are several general or omnibus normality tests available for the null hypothesis that the data come from an unknown normal distribution. A test with good asymptotic properties is the Shapiro-Wilk normality test.

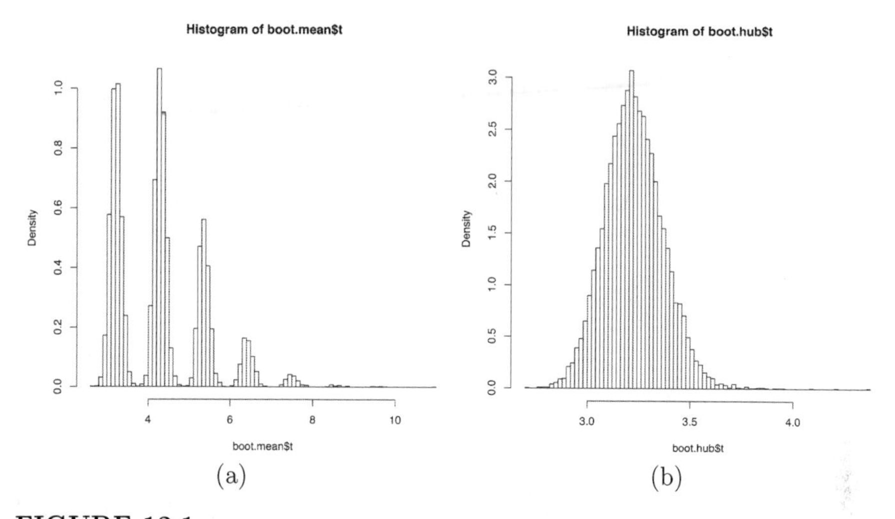

FIGURE 12.1
Empirical distribution of the mean (a) and Huber's location (b) from 10,000 bootstrap replicates of the copper in wholemeal data.

Example 12.2 *Testing the null-hypothesis that the 24 measurements in* (μgg^{-1}) *of copper in wholemeal come from an unknown normal distribution by the Shapiro-Wilk normality test gives the p-value* $1.283 \cdot 10^{-9}$, *providing strong evidence to reject normality.*

12.1.2 The effect of an outlier on the estimated curve

It the above the sample mean $\overline{Y}$ was used as an estimator of the center of location of the distribution of measurements. If the sample of chemical measurements are symmetrically and normally distributed around the target (expected value), then the sample mean is an unbiased and efficient estimator of location. If such is the case, then the empirical distribution of the mean will be symmetric and bell-shaped according to a normal distribution.

Example 12.3 *To illustrate the problem of an extreme outlying measurement in the data the empirical distribution of the mean of the 24 measurements of copper* (μgg^{-1}) *in wholemeal was computed on the basis of 10,000 bootstrap samples, with replacement. The histogram of the empirical distribution of the mean in Figure 12.1(a) is multimodal, typically revealing a local maximum for those samples containing the extreme outlier several times. It is clear that the empirical distribution of the mean does not resemble a normal distribution making the construction of a confidence interval problematic. Furthermore, since the mean 4.28 differs largely from the median 3.385* (μgg^{-1}), *due to the extreme outlier the position of the true location is uncertain.*

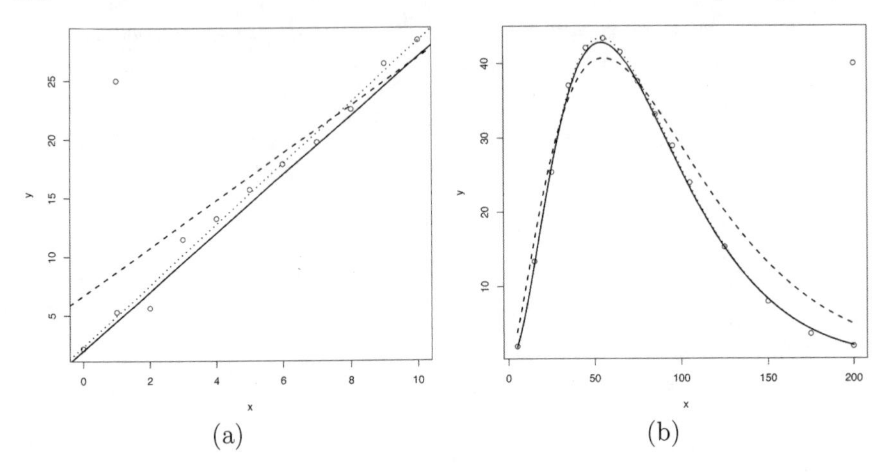

FIGURE 12.2
Linear (a) and non-linear regression (b) with a clear outlier in the data, where the unbroken curve or line represents the true, the broken the non-robust least squares, and the dotted the robust approximation from the data.

The outlying measurement makes the position of the target location uncertain and the distribution of the estimator non-normal complicating the construction of a confidence interval.

Example 12.4 *To illustrate the influence of outlying data points on linear a simple linear model was simulated adding a few clearly outlying points. Suppose that* $y_i = \beta_0 + \beta_1 x_i + \varepsilon_i$, *with* $\beta_0 = 2$, $\beta_1 = 2.5$, $\varepsilon_i \overset{IID}{\sim} N(0,1)$, *where* y_i *is the Absorption and* x_i *the Concentration, say, for* $x_1 = 0$ *to* $x_{11} = 10$, *with steps size equal to one and the error terms generated from the standard normal distribution. There is one extreme outlying data point added to the data, which is completely unrelated to the linear model. The estimated maximum likelihood coefficients* $\hat{\beta}_0 = 6.69$ *and* $\hat{\beta}_1 = 2.02$ *do not approximate the true parameters well. The data, the true model* $y = 2 + 2.5x$ *(unbroken) and the estimated model (broken) are given in Figure 12.2.*

Example 12.5 *To illustrate the influence of outlying data points on non-linear model, data were simulated and an outlier was added. Suppose that we have the model* $y_i = \beta_1 t_i^{\beta_2} e^{-\beta_3 \cdot x_i} + \varepsilon_i$, *with* $\varepsilon_1, \cdots, \varepsilon_n\} \overset{IID}{\sim} N(0, 0.5^2)$. *taking* $\beta_1 = 0.07$, $\beta_2 = 2.15$, *and* $\beta_3 = 0.04$. *The estimated coefficients from maximum likelihood were* $\beta_1 = 0.34$, $\beta_2 = 1.58$, *and* $\beta_3 = 0.028$, *resulting in the estimated broken curve in Figure 12.2(b), which does not approximate the true unbroken curve very well.*

Of course the outlying data point can also be discovered by diagnostic plots, outlier testing, or bootstrapping, but this need not always be the case if there are several weakly outlying data points.

12.2 Robust Estimation

12.2.1 Robust estimation a location parameter

Recall the location model $y_i = \mu + \varepsilon_i$, for $i = 1, \cdots, n$ with target μ, and measurement error terms, ε_i, which are independent random variables that come from a symmetric distribution around zero $F(y) = F_0(y - \mu)$, e.g., the normal. The likelihood is defined as the function

$$L(y_1, \cdots, y_n; \mu) = \prod_{i=1}^{n} f_0(y_i - \mu)$$

over the target μ, with density $f_0 = F_0'$ equal to the derivative of the distribution function F_0. The maximum likelihood estimator is defined as

$$\widehat{\mu} = \underset{\mu}{\operatorname{argmax}}\, L(y_1, \cdots, y_n; \mu). \tag{12.1}$$

In the context of robust estimation the terminology for the chemical target is generalized somewhat into "location". The idea of robust estimation hinges upon generalizing maximum likelihood in such a way that the estimator becomes robust against outlying data points and at the same time keeping the favorable properties of maximum likelihood estimators intact. To do so we consider the class of M-estimators.

Definition 12.1 *An* **M-estimator** *is defined as the argument that minimizes the additive estimation function ρ, that is*

$$\widehat{\mu} = \underset{\mu}{\operatorname{argmin}} \sum_{i=1}^{n} \rho(y_i - \mu). \tag{12.2}$$

If we take $\rho(y) = -\log f_0$, then we obtain the maximum likelihood estimator of location as a special case.

Example 12.6 *Suppose that F_0 is the standard normal distribution with density $f_0(y) = e^{-y^2/2}/\sqrt{2\pi}$. Then maximizing the log-likelihood is equivalent to taking $\rho(y) = y^2$, so that*

$$\widehat{\mu} = \underset{\mu}{\operatorname{argmin}} \sum_{i=1}^{n} (y_i - \mu)^2. \tag{12.3}$$

This gives the arithmetic mean of the measurements as an estimator $\widehat{\mu} = \bar{y} = \sum_{i=1}^{n} y_i/n$ of the location μ.

Example 12.7 *If F_0 is the double exponential distribution with $f_0(y) = e^{-|y|}/2$, then taking $\rho(y) = -\log f_0$ leads to*

$$\widehat{\mu} = \underset{\mu}{\operatorname{argmin}} \sum_{i=1}^{n} |y_i - \mu|. \tag{12.4}$$

The solution to this minimization problem gives the median $\widehat{\mu} =$ Median$(y_1, \cdots, y_n)$ *as the estimator of location.*

Several estimators can conveniently be defined after ordering the measurements from the smallest to the largest

$$y_{(1)} \leq y_{(2)} \leq \cdots \leq y_{(n)}.$$

If n is odd, then $n = 2m - 1$ for an integer m, and we may write the median a scalar valued vector function $\text{Med}(\mathbf{y}) = y_{(m)}$. If n is even, then it is custom to write

$$\text{Med}(\mathbf{y}) = \frac{y_{(m)} + y_{(m+1)}}{2}.$$

The median is the 50th percentile of the measurements, separating the smallest half from the largest.

It can be shown that the median is very robust in the sense that it remains stable even when there are 50% outlying measurements, which is known in the literature under the notion that the median has breakdown point 1/2 (Huber, 1981; Maronna et al., 2019). If, however, the measurements follow a normal distribution, then the standard error of the median is large, so that it is not an efficient estimator of location. Since the mean is a maximum likelihood estimator, its standard error is the smallest possible making it an efficient estimator of location. If, however, the largest measurement $y_{(n)}$ increases linearly with n, then this would also hold for the mean $\bar{y}_n$, showing that the mean is a non-robust estimator of the location μ. The idea is to use keenly defined functions from which the estimator for location is highly robust, efficient, and asymptotically normal.

If ρ is differentiable, then, taking $\psi = \rho'$, a necessary condition for the solution of the estimator in Equation (12.2) is

$$\sum_{i=1}^{n} \psi(y_i - \widehat{\mu}) = 0. \tag{12.5}$$

To see that M-estimators of location are in fact a weighted sum of the measurements, we define the weight function

$$W(y) = \begin{cases} \psi(y)/y & \text{for} \quad y \neq 0 \\ \psi'(0) & \text{for} \quad y = 0 \end{cases}$$

assuming that $\psi'(0) > 0$. For non-zero y we have $\psi(y) = yW(y)$, so that we may rewrite the necessary condition in Equation (12.5) as

$$\sum_{i=1}^{n} W(y_i - \widehat{\mu})(y_i - \widehat{\mu}) = 0.$$

Now taking $w_i = W(y_i - \widehat{\mu})$ we obtain the location estimator $\widehat{\mu}$ as a weighted sum of the measurements

$$\widehat{\mu} = \frac{\sum_{i=1}^{n} w_i y_i}{\sum_{i=1}^{n} w_i}.$$

Now we can define Huber's (Huber, 1981) robust estimator of location by taking

$$\rho_k(y) = \begin{cases} y^2 & \text{for} \quad |y| \le k \\ 2k|y| - k^2 & \text{for} \quad |y| > k, \end{cases} \tag{12.6}$$

with derivative $2\psi_k(y)$, where

$$\psi_k(y) = \begin{cases} y & \text{for} \quad |y| \le k \\ k \cdot \text{sign}(y) & \text{for} \quad |y| > k. \end{cases}$$

Note that if k is equal to infinity, then $\rho_\infty(y) = y^2$, which gives the efficient but non-robust $\widehat{\mu} = \bar{y}$ and if we take $k = 0$, then $\psi_0(y) = \text{sign}(y)$ which gives the $\widehat{\mu} = \text{Med}(\mathbf{y})$ which is not efficient, but maximally robust. The weight function now is

$$W_k(y) = \min\left\{1, \frac{k}{|y|}\right\}.$$

Since $k/|y| \to 0$, as $|y| \to \infty$, the latter illustrates the general property that M-estimators tend to downweigh the contribution of extremely large measurements. M-estimators are asymptotically normal with mean μ and variance v/n, where the asymptotic variance v is estimated by

$$\widehat{v} = \widehat{\sigma}^2 \frac{\frac{1}{n}\sum_{i=1}^n [\psi(y_i - \widehat{\mu})/\widehat{\sigma}]^2}{[\frac{1}{n}\sum_{i=1}^n \psi'(y_i - \widehat{\mu})/\widehat{\sigma}]^2}. \tag{12.7}$$

The asymptotic normality can be used to construct a confidence interval for the location μ by taking $\widehat{\mu} \pm z_{1-\alpha/2}\sqrt{\widehat{v}/n}$. It is recommended to take $k = 1.5$, but in applications its value can be changed by the investigator, if desired. Large values of k will make the Huber estimator similar to the mean and values of k close to zero will make it similar to the median. M-estimators have many favorable properties such a large breakdown point 1/2, good large sample behavior for symmetric distributions with heavy tail and it behaves like a maximum likelihood estimator in being consistent and efficient.

Example 12.8 *To study the empirical distribution of Huber's estimator of location 10,000 bootstrap samples from the 24 measurements ($\mu g g^{-1}$) of copper in wholemeal were computed. The empirical distribution of Huber's estimator of location, using $k = 1.5$, given in Figure 12.1(b) is symmetric around 3.2067. The 95% CI from the bootstrap $[2.91, 3.47]$ resembles the 95% CI $[2.93, 3.48]$ taken from asymptotic normal distribution with SE 0.1416 from Equation (12.7).*

Example 12.9 *A study revealed 13 different chemical measurements for each of 180 archaeological glass vessels (Janssens et al., 1998). The mean, median, and Huber's M location of five of these are given in Table 12.1 together with a those from a simulated normally distributed variable with mean $\mu = 10$ and standard deviation $\sigma = 2$, added as a reference.*

TABLE 12.1
Location estimates for the normal distribution and five chemical measurements on glass vessels.

	Normal	Na2O	P2O5	Fe2O3	BaO	PbO
mean	10.16	12.55	0.50	0.40	0.07	0.38
median	9.98	14.17	0.28	0.37	0.04	0.14
Hubers M	10.13	13.75	0.31	0.38	0.06	0.15

TABLE 12.2
Location, standard error (SE) and 95% CI for 180 normal deviates $N(10, 2)$ and PbO glass measurements.

normal	$\widehat{\mu}$	SE	CI		PbO	$\widehat{\mu}$	SE	CI	
mean	10.05	0.15	9.74	10.37	mean	0.380	0.069	0.244	0.517
median	9.97	0.21	9.64	10.37	median	0.145	0.006	0.130	0.158
Huber's M	10.11	0.15	9.81	10.41	Huber's M	0.158	0.008	0.143	0.174

For the simulated normal measurements the estimates of location have the expected behavior of being nearly equal. The mean differs largely from the median and Huber's M location estimates for the chemicals P2O5 and PbO, nicely illustrating the sensitivity of the mean for outlying measurements. The three estimates of location are fairly equal for Na2O and Fe2O3.

Example 12.10 *Table 12.2 gives various estimates of location, their standard error (SE), and 95% confidence intervals for 180 simulated normal measurements and the PbO measurements in the 180 glass vessels from the previous example. The asymptotic variance of the median depends on the assumed distribution, which we currently assume to be unknown. The bootstrap was used to compute the SE and the CI for the median. For the normal deviates the SE is the smallest, implying the same for the width of the CI, however, for practical purposes the differences in SE are clearly relatively small. For PbO due to outliers at the right-hand side, the SE for the mean is the largest, but the latter is more than ten times larger than that of the median and Huber's M. This illustrates once more the sensitivity of the mean for outlying measurements and that in such situations inferences about location can better be based upon the confidence interval for the median or Huber's M.*

12.2.2 Robust estimation of scale

For an estimate of the spread in the data we need to analyze the size of the scale σ in a multiplicative model $y_i = \sigma u_i$, where $u_1, \cdots, u_n$ are independently and identically distributed (IID). A classical robust way, after ordering $y_{(1)} \leq \cdots \leq y_{(n)}$, to do this is by the **inter-quartile range**. The latter is defined as

$$IQR(\mathbf{y}) = y_{(\lfloor \frac{3}{4}n \rfloor)} - y_{(\lfloor \frac{1}{4}n \rfloor)},$$

the distance between the 75th and the 25th percentile, where the floor function $\lfloor x \rfloor$ rounds downwards to the nearest integer. When the measurements have a distribution F, then $y_{(\lfloor \frac{3}{4}n) \rfloor}$ will converge to the 75th percentile $F^{-1}(3/4)$ of the distribution. If the measurements come from a normal distribution $N(\mu, \sigma^2)$, then $IQR(\mathbf{y})/1.349$ is an estimator converging almost surely to σ. The normalization constant $1.349 = 2 \cdot \Phi^{-1}(3/4)$ comes from the 0.75 quantile of the standard normal distribution.

Suppose that the multiplicative model $y_i = \sigma u_i$, where $u_1, \cdots, u_n$ are IID with density f_0. The distribution of y_i generates a scale family with density $\frac{1}{\sigma} f_0(\frac{y}{\sigma})$, so that the **maximum likelihood** estimator is

$$\widehat{\sigma} = \underset{\sigma}{\operatorname{argmax}} \frac{1}{\sigma^n} \prod_{i=1}^{n} f_0\left(\frac{y_i}{\sigma}\right). \tag{12.8}$$

If we now take the log and set the derivative to zero, then we obtain

$$\frac{\partial}{\partial \sigma} \log\left(\frac{1}{\sigma^n} \prod_{i=1}^{n} f_0\left(\frac{y_i}{\sigma}\right)\right) = -\frac{n}{\sigma} - \frac{1}{\sigma} \sum_{i=1}^{n} \frac{y_i}{\sigma} \cdot \frac{f_0'(y_i/\sigma)}{f_0(y_i/\sigma)} = 0.$$

If we now take $\rho(t) = t\psi(t)$, with $\psi = -f_0'/f_0$, we have the equivalent expression

$$\frac{1}{n} \sum_{i=1}^{n} \rho\left(\frac{y_i}{\widehat{\sigma}}\right) = \delta, \tag{12.9}$$

which defines an M-estimator of scale for a positive constant δ, e.g., taken equal to 1. It is illuminating to consider a few choices for ρ.

Example 12.11 *If F_0 is the standard normal distribution with $f_0(x) = e^{-y^2/2}/\sqrt{2\pi}$, then $f_0'(y) = -y f_0(y)$, so that $\psi(y) = y$ and $\rho(y) = y\psi(y) = y^2$. Taking $\delta = 1$ in Equation (12.9) gives*

$$\frac{1}{n} \sum_{i=1}^{n} \rho\left(\frac{y_i}{\widehat{\sigma}}\right) = \frac{1}{n} \sum_{i=1}^{n} \left(\frac{y_i}{\widehat{\sigma}}\right)^2 = 1,$$

from which we obtain the root mean square estimator $\widehat{\sigma} = \sqrt{\sum_{i=1}^{n} y_i^2/n}$. If we now consider the deviances with respect to the mean of a sample of chemical measurements, then we obtain $\widehat{\sigma} = \sqrt{\sum_{i=1}^{n}(y_i - \bar{y})^2/n}$ the standard deviation (SD).

Example 12.12 *If F_0 is the* **double exponential distribution** *with $f_0(y) = e^{-|y|}/2$, then $f_0'(y) = f_0(y) \cdot -\text{sgn}(y)$ and $\psi(y) = \text{sgn}(y)$, so that $\rho(y) = y\psi(y) = y \cdot \text{sgn}(y) = |y|$ and Equation (12.9) with $\delta = 1$ reads*

$$\frac{1}{n} \sum_{i=1}^{n} \rho\left(\frac{y_i}{\widehat{\sigma}}\right) = \frac{1}{n} \sum_{i=1}^{n} \frac{|y_i|}{\widehat{\sigma}} = 1.$$

This yields the mean of the absolute values $\widehat{\sigma} = \sum_{i=1}^{n} |y_i|/n$ *as an estimator for scale.*

Example 12.13 *By taking* $\delta = 1/2$ *and* $\rho(y) = I(|y| > c)$*, we obtain* $\widehat{\sigma} = \mathrm{Med}(|\mathbf{y}|)/c$*. If we now consider the deviances with respect to the median of a sample of chemical measurements, then we get the* **median absolute deviation** *(MAD) defined as*

$$\mathrm{MAD}(\mathbf{y}) = \underset{1 \leq i \leq n}{\mathrm{Med}} \left(|y_i - \mathrm{Med}(\mathbf{y})| \right).$$

We may normalize the MAD, by dividing it by $\Phi^{-1}(3/4) = 0.675$*, in order to obtain*

$$\mathrm{MADN}(\mathbf{y}) = \mathrm{MAD}(\mathbf{y})/0.675$$

which is an estimator that converges (Maronna et al., 2019) to the scale σ*, if the data* $y_1, \cdots, y_n$ *are normally distributed* $N(\mu, \sigma^2)$*.*

When ρ is quadratic near zero in the sense of $\rho'(0) = 0$ and $\rho''(0) > 0$, then the M-scale estimator can be written in terms of a root mean square. That is, if we define the weight function

$$W(y) = \begin{cases} \rho(y)/y^2 & \text{for} \quad y \neq 0 \\ \rho''(0) & \text{for} \quad y = 0 \end{cases}$$

then the defining Equation (12.9) becomes equivalent to

$$\widehat{\sigma}^2 = \frac{1}{n\delta} \sum_{i=1}^{n} W\left(\frac{y_i}{\widehat{\sigma}}\right) y_i^2,$$

a weighted root mean square estimate. If ρ is bounded, then the weights become smaller for larger values y_i, so that these contribute less to the value of $\widehat{\sigma}^2$.

Example 12.14 *In a study 13 different measurements for 180 archaeological glass vessels were obtained (Janssens et al., 1998). A pseudo random variable from the normal distribution with mean* $\mu = 25$ *and standard deviation* $\sigma = 2$ *is added as a reference.*

The scale estimates for five chemicals are given in Table 12.3. The so-called scale τ_2 *is a robust and efficient scale M-estimator to be defined in Section 12.6, the IQR and the MADN are robust non-efficient scale estimators. The SD, MADN and IQR are very similar for the normal deviates and close to the true value* $\sigma = 2$*. For the chemicals P2O5 and PbO a very large difference in scale estimate occurs between the non-robust SD and the robust. Observe that the SD can even be more than ten times larger compared to the robust. This illustrates the sensitivity of the standard deviation for outliers.*

TABLE 12.3
Various types of scale estimators $\hat{\sigma}$ for values from the normal distribution and from five amounts of chemicals in 180 glass vessels. We used the normalized IQR for the normalized deviate and not for the glass measurements.

	Normal	Na2O	P2O5	Fe2O3	BaO	PbO
SD	2.1218	4.3351	0.6717	0.1473	0.0917	0.9310
MADN	2.0275	2.0667	0.1290	0.1001	0.0482	0.0756
IQR	2.0489	3.4565	0.2048	0.1330	0.0830	0.1052
scaleTau2	2.1256	2.3649	0.1536	0.1097	0.0530	0.0914

12.3 Robust Linear Regression

In Section 7.3.3 various diagnostics for detecting outlying observations were given. Methods such as Cook's distance are certainly useful in relatively simple situations, but these may fail in more complex situations with more than one predictive variable, because each of $\hat{\sigma}^2_{(i)}$, h_{ii} and r_i^2 may be largely influenced by an outlier. If there are several influential data points caused by either large values of y_i, extreme rows $\mathbf{x}_i^T$, or combinations of these, then a leave-one-out approach may obviously fail to properly detect outliers. The same holds for the phenomenon of outliers which are masked in the data. Furthermore, if an outlier is detected and the subjective decision is taken to leave it out of the analysis, new outlying data points may emerge. All this may lead to time consuming procedures hardly enhancing scientific scrutiny.

It seems clear that robust methods provide a way out by a single analysis. To define methods we write the residuals as $r_i(\boldsymbol{\beta}) = Y_i - \mathbf{x}_i^T\boldsymbol{\beta}$. The **M-estimator** is defined as the parameter $\boldsymbol{\beta}$ that minimizes the function

$$S_n(\boldsymbol{\beta}) = \sum_{i=1}^{n} \rho_B\left(\frac{r_i(\boldsymbol{\beta})}{\widehat{\sigma}_n}\right). \tag{12.10}$$

Equivalently, if the derivative of ρ_B with respect to $\boldsymbol{\beta}$ exists, the minimizing parameter vector solves the first order equation

$$\sum_{i=1}^{n} \psi_B\left(\frac{r_i(\boldsymbol{\beta})}{\widehat{\sigma}_n}\right)\mathbf{x}_i = \mathbf{0}.$$

An important choice for ρ is the bisquare ρ function

$$\rho(x) = \begin{cases} 1 - \left[1 - \left(\frac{x}{k}\right)^2\right]^3 & \text{for} \quad |x| \le k \\ 1 & \text{for} \quad |x| > k \end{cases}, \tag{12.11}$$

which is bounded, differentiable, even, nondecreasing in $|x|$, and attains its

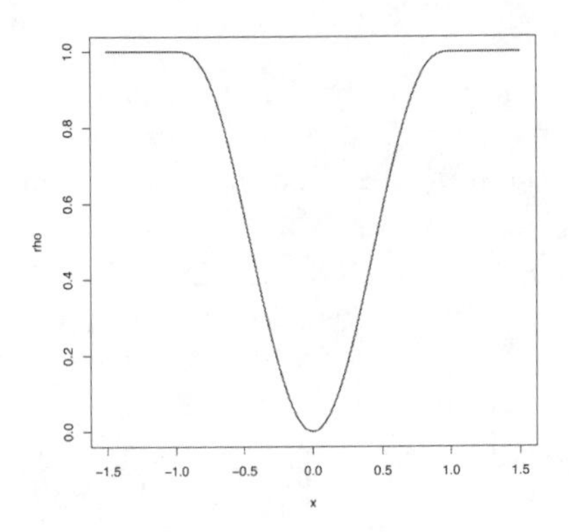

FIGURE 12.3
Bisquare ρ function with $k = 1$.

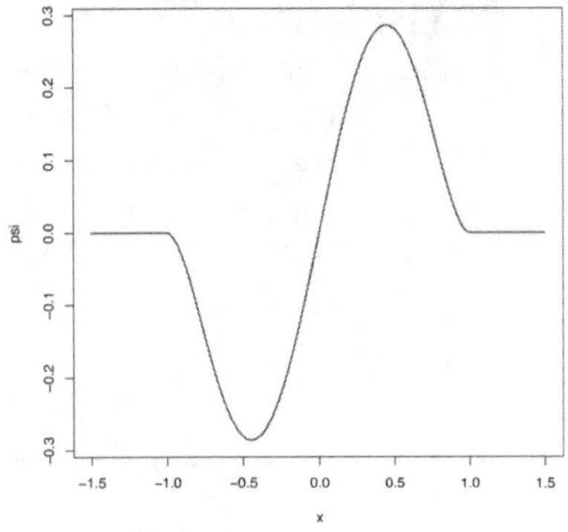

FIGURE 12.4
Bisquare ψ function with $k = 1$.

minimum zero at $x = 0$. To illustrate these properties the graph of the ρ function with $k = 1$ is given in Figure 12.3. Its first order derivative $\rho'(x) = 6\psi(x)/k^2$ equals the so-called **bisquare psi-function**, defined as

$$\psi(x) = x\left[1 - \left(\frac{x}{k}\right)^2\right]^2 \cdot I_{[-k,k]}(x),$$

with $I_{[-k,k]}(x)$ the indicator function attaining one for x inside the interval $[-k, k]$ and zero outside. The graph of the bisquare ψ function with $k = 1$ in Figure 12.4 is descending to its minimum, increasing to a maximum, and then redescending to zero as x goes from $-k$ to k.

The minimization of object function in Equation (12.10) is started with a fast and highly robust $\widehat{\boldsymbol{\beta}}_0$, from which the robust scale estimator $\widehat{\sigma}_n$ is computed by e.g., the normalized median of absolute values. This results in the so-called **MM-regression**. It is an example of an MM-estimator that attempts to combine robustness with efficiency. MM-regression is computed by an iterative weighted least squares algorithm that converges to the minimizing parameter. The MM-estimator is asymptotically normal with

$$\sqrt{n}(\widehat{\boldsymbol{\beta}} - \boldsymbol{\beta}) \to N(0, v\boldsymbol{V}^{-1})$$

where $\boldsymbol{V} = \frac{1}{n}\boldsymbol{X}^T\boldsymbol{X}$ and v is estimated by

$$\widehat{v} = \widehat{\sigma}^2 \frac{n}{n-p} \frac{\frac{1}{n}\sum_{i=1}^n \psi(r_i/\widehat{\sigma})^2}{[\frac{1}{n}\sum_{i=1}^n \psi'(r_i/\widehat{\sigma})]^2}.$$

The MM-regression fits the linear regression model to the data resistant against outliers and provides an efficient estimator having the smallest possible standard error.

TABLE 12.4
Regression of outcome city miles per US gallon with several explanatory variables and estimated by MM.

	Estimate	**Std. Error**	**t value**	**Pr(>|t|)**
(Intercept)	31.1661	7.5600	4.1225	0.0001
Fuel.tank.capacity	−0.4591	0.2071	−2.2172	0.0292
Passengers	−0.1204	0.4092	−0.2943	0.7692
Wheelbase	0.2299	0.1064	2.1608	0.0334
Weight	−0.0080	0.0015	−5.1594	0.0000
Robust	**Estimate**	**Std. Error**	**t value**	**Pr(>|t|)**
(Intercept)	40.8357	6.0714	6.7259	0.0000
Fuel.tank.capacity	−0.4336	0.1769	−2.4517	0.0162
Passengers	0.3914	0.2933	1.3346	0.1854
Wheelbase	0.0218	0.0779	0.2793	0.7807
Weight	−0.0052	0.0011	−4.5548	0.0000

Example 12.15 *To illustrate robust linear regression the data in Figure 12.2 we analyzed, where the estimated line by MM-regression is given as the dotted line. The robust parameters obtained by MM-regression are* $\hat{\beta}_0 = 2.34$ *and* $\hat{\beta}_1 = 2.60$ *are much closer to the true* $\beta_0 = 2$, $\beta_1 = 2.5$ *than the previously obtained by maximum likelihood. The 95% CI from MM-regression for the intercept* $[1.19; 3.50]$ *and the slope* $[2.46; 2.75]$ *contain the true parameters. The 95% CI for the intercept* $[0.13; 13.24]$ *and the slope [0.87; 3.18] do contain the true parameters, but are quite large in size. The estimated error standard deviation from MM-estimation is* $\hat{\sigma} = 0.97$ *is quite close to the true value* $\sigma = 1$*, whereas that from maximum likelihood estimation is* $\hat{\sigma} = 5.79$ *is much larger.*

Example 12.16 *To illustrate the usefulness of robust MM-regression in a situation where several observations are influential we analyze a sample of 93 cars that were selected at random from a total of 1993 listed in both the Consumer Reports issue and the PACE Buying Guide (Lock, 1993). We concentrate on a linear model for the outcome variable City miles per US gallon (City MPG) with explanatory variables Fuel Tank Capacity, Number of Passengers, Wheelbase, and Weight of the car.*

The linear model has two observations with studentized residuals larger than 4.7 both significant after Bonferroni correction for multiple testing. The largest Cooks distance is 0.41, the largest leverage hat value 0.31 and several DFBETA values are larger than $2/\sqrt{n}$*. The adjusted R-squared values indicate that the linear model explains 73.3% and the robust model 79.7% of the variance in City MPG. The regression coefficients in Table 12.4 reveal drastic differences due to reducing the influence of outlying observations using robust MM-estimation. According to linear regression there is an effect of Wheelbase on City MPG, however, this effect is not confirmed by robust MM regression.*

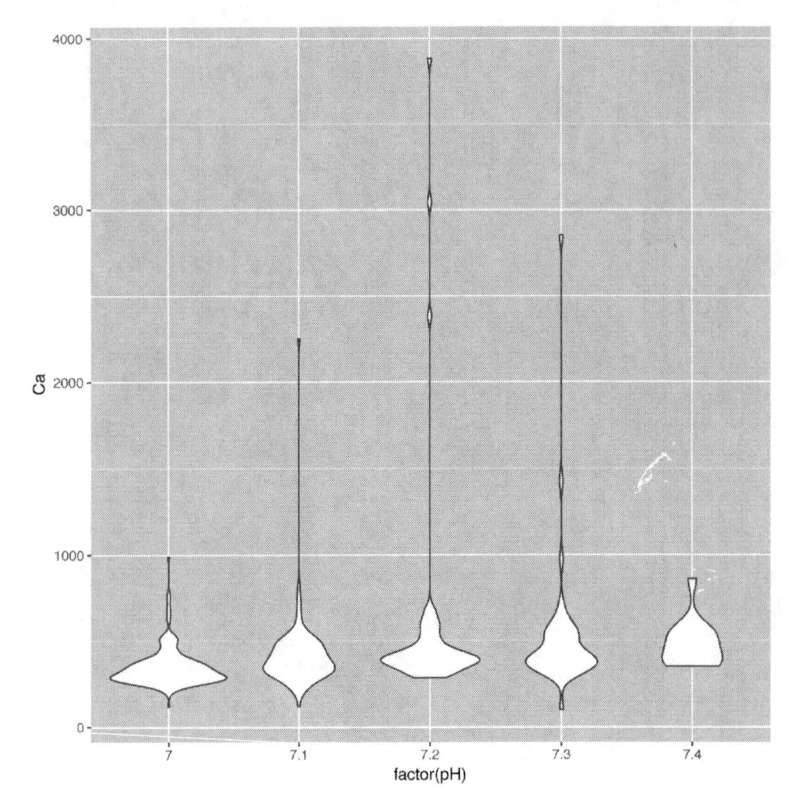

FIGURE 12.5
Violin plot of Ca amount in soil depending on pH.

According to both analysis there is an effect of car Weight on City MPG, however, the size of the effect from robust MM regression is smaller.

12.3.1 Robust one-way analysis of variance

If we take for $\boldsymbol{X}$ the design matrix corresponding to one-way analysis of variance, then MM-regression can be used to test the null hypotheses that the locations parameters are equal.

Example 12.17 *In order to investigate whether Calcium content in soil samples depends on soil pH, a total 428 samples were collected in different communities of the Condroz region in Belgium. The soil pH ranges between 7.0 and 7.5 with steps of size 0.1. The violin plot in Figure 12.5 clearly indicates that at all pH levels the Ca distributions are skewed to the right with several extremely outlying data points. The pH values are taken as levels of a factor in order to test for differences with respect to the reference pH 7.0.*

TABLE 12.5
Estimates from the linear model compared with those from MM based on Ca amount measurement in soil depending on levels of pH.

	Estimate	SE	t value	Pr (>|t|)	MM-Est	SE	t value	Pr (>|t|)
(Intercept)	347.55	20.69	16.79	0.0000	327.34	5.62	58.1832	0.0000
factor(pH)7.1	68.09	35.23	1.932	0.0539	56.72	11.58	4.8956	0.0000
factor(pH)7.2	218.51	44.67	4.890	0.0000	78.15	13.35	5.8504	0.0000
factor(pH)7.3	178.64	50.69	3.524	0.0005	94.61	19.12	4.9486	0.0000
factor(pH)7.4	130.07	81.85	1.589	0.1128	118.67	27.90	4.2527	0.0000

The results from classical and robust analysis of variance are given in Table 12.5. The estimates for pH 7.1 are quite similar, but the other reveal some large differences. The non-robust estimated effect of pH 7.2 is almost three times larger than that by the robust MM-estimator. The non-robust standard error is three times larger than that of the robust. The non-robust adjusted R-square 0.06071 is much smaller than the 0.172 from robust MM-estimation.

12.3.2 Robust two-way analysis of variance

If we take for $\boldsymbol{X}$ the design matrix corresponding to two-way analysis of variance, then MM-regression can be used to test for the main and interaction effects of the two factors.

Example 12.18 *The strength of a paper depends on the factors: (i) the percentage of hardwood concentration in the raw pulp, (ii) the vat pressure, and (iii) the cooking time of the pulp (Montgomery, 2013). In an experiment the hardwood concentration is tested at three levels of 2, 4, and 8 percentage, three of vat pressure at 400, 500, and 650, and cooking time at the levels 3 and 4 hours. For each combination of these three-factor variables, 2 observations are available, and thus a total of 3.3.2.2 = 36 observations. The goal of the study is to investigate the size of the effects of the three factor variables on the strength of the paper, and the presence of interaction effect, if any. A three way analysis a variance was conducted with an interactions between Hardwood and Pressure. There was however one significant outlier found possibly increasing the error variance.*

The adjusted R-squared was 0.74 and 0.64, respectively for from robust MM and non-robust regression, indicating that the robust model fits the data better. The results from robust MM-regression and regression are summarized in Table 12.6. The Hardwood main effects by regression are not confirmed by MM-regression. The effect of Pressure 500 relative to the reference Pressure 400 found by regression is not confirmed by robust MM-regression. The effect of Pressure 650 relative to Pressure 400 increases Strength of paper by 2.9 according to robust MM-regression. The Cooking Time effect of 4 relative to 3 hours increases the effect on Hardwood Strength by 1.4 according

TABLE 12.6
Estimated coefficients from linear regression to left and robust linear MM-regression to the right from outcome Strength of Paper from experiment factors hardwood (H) with levels percentage 2, 4, and 8, levels vat pressure (P) 400, 500, and 650, and cooking time (CT) 3 and 4 hours.

	Est lm	SE	t value	Pr (>|t|)	MM-Est	SE	t value	Pr (>|t|)
(Intercept)	195.761	0.471	415.416	0.000	196.496	0.306	642.328	0.000
H4-H2	1.425	0.632	2.254	0.033	0.538	0.650	0.827	0.416
H8-H2	1.125	0.632	1.779	0.087	0.311	0.425	0.731	0.471
P500-P400	2.025	0.632	3.203	0.004	1.385	0.731	1.895	0.069
P650-P400	3.775	0.632	5.971	0.000	2.963	0.349	8.493	0.000
CT4-CT3	1.278	0.298	4.287	0.000	1.431	0.212	6.750	0.000
H4:P500	−2.450	0.894	−2.740	0.011	−1.730	0.950	−1.822	0.080
H8:P500	−2.900	0.894	−3.243	0.003	−2.258	0.801	−2.818	0.009
H4:P650	−2.950	0.894	−3.299	0.003	−2.063	0.683	−3.021	0.006
H8:P650	−2.850	0.894	−3.188	0.004	−2.034	0.546	−3.726	0.001

MM-regression. The data provide evidence for three decreasing interaction effects on the Strength of Paper according to robust MM-regression. It is interesting to stress that the standard errors for some of the effects by MM-regression are considerably smaller compared to those from regression.

According to these findings it is reasonable to expect the largest predicted Strength of Paper $\hat{y}$ for the combination percentage of Hardwood 2, Pressure 650, and CT 4. The predicted Paper Strength value from the robust MM-regression for this combination is 200.8899 with 95% Prediction Interval (199.26, 202.51). From ordering all predicted values it can easily be checked that this is indeed the optimal combination of factors for Paper Strength.

12.4 Robust Nonlinear Regression

For non-linear regression we define the residual as

$$r_i(\boldsymbol{\beta}) = Y_i - f(\mathbf{x}_i, \boldsymbol{\beta}),$$

in order to conform to the definitions to robust estimation. Then the above robust MM-estimation method can be generalized to non-linear regression using the above bi-square ρ and ψ functions.

Example 12.19 *To illustrate the robust non-linear MM-regression the data in Figure 12.2(b) were analyzed. The estimated parameters and their standard errors are given in Table 12.7. It can be seen that the MM-estimated parameters are much closer to the true* $\beta_1 = 0.07$*,* $\beta_2 = 2.15$*, and* $\beta_3 = 0.04$*, than the estimated parameters from maximum likelihood. The dotted curve*

TABLE 12.7
Estimated coefficients for the non-linear model and their standard errors from maximum likelihood and robust MM-estimation.

Maximum likelihood	Estimate	Std. Error	t value	Pr(>|t|)
$\hat{\beta}_1$	0.3434	0.6090	0.5639	0.5824
$\hat{\beta}_2$	1.5851	0.5722	2.7701	0.0159
$\hat{\beta}_3$	0.0287	0.0093	3.0903	0.0086
MM-estimation	**Estimate**	**Std. Error**	**t value**	**Pr(>|t|)**
$\hat{\beta}_1$	0.0677	0.0091	7.4171	0.0000
$\hat{\beta}_2$	2.1650	0.0441	49.0877	0.0000
$\hat{\beta}_3$	0.0402	0.0007	53.8335	0.0000

TABLE 12.8
Parameter estimates for Model 2 and their standard errors, t- and p-values for non-linear least squares and robust non-linear least squares by M-estimation.

	Estimate	Std. Error	t value	Pr(>|t|)
$\hat{\beta}_1$	0.1666	0.0383	4.3489	0.0001
$\hat{\beta}_2$	0.0052	0.0007	7.7532	< 0.0001
$\hat{\beta}_3$	0.0122	0.0015	7.9390	< 0.0001
MM	**Estimate**	**Std. Error**	**t value**	**Pr(>|t|)**
$\hat{\beta}_1$	0.1439	0.0291	4.9418	< 0.0001
$\hat{\beta}_2$	0.0048	0.0006	8.3287	< 0.0001
$\hat{\beta}_3$	0.0134	0.0013	10.2511	< 0.0001

from the MM-estimated parameters is much closer to the unbroken true than the broken curve one from maximum likelihood. Indeed, the curve from MM-regression is hardly be distinguished from the true. The standard errors from MM-estimation are much smaller compared to the maximum likelihood parameters. In the presence of a clearly outlying data point the MM-estimation is less biased and much more precise compared to maximum likelihood.

Example 12.20 *In the above non-linear regression analysis of the data from a NIST study (Chwirut, 1979) involving ultrasonic calibration, it was found that*

$$\text{Model 2: } f_2(\beta_1, \beta_2, \beta_3, x) = \frac{e^{-\beta_1 x}}{\beta_2 + \beta_3 x}.$$

fitted the data best. From the parameters from least squares and robust MM estimation in Table 12.8 it can be observed that these are fairly equal, confirming the previous conclusions drawn. However, the residual standard error $\hat{\sigma}$ *from non-linear least squares was* $= 3.172$, *which is considerably larger than the* 1.707. *obtained from robust non-linear least squares.*

TABLE 12.9
HC3 and HC4 correction for heterogeneity of variances on analysis of variance of City miles per US gallon usage of three types of cars.

	Estimate	**Std. Error**	**t value**	**Pr(>|t|)**
(Intercept)	18.3636	1.2270	14.9666	0.0000
TypeMidsize	1.1818	1.5027	0.7864	0.4352
TypeSmall	11.4935	1.5146	7.5884	0.0000
HC3	**Estimate**	**Std. Error**	**t value**	**Pr(>|t|)**
(Intercept)	18.3636	0.4748	38.6749	0.0000
TypeMidsize	1.1818	0.6297	1.8767	0.0663
TypeSmall	11.4935	1.4463	7.9466	0.0000
HC4	**Estimate**	**Std. Error**	**t value**	**Pr(>|t|)**
(Intercept)	18.3636	0.4667	39.3509	0.0000
TypeMidsize	1.1818	0.6162	1.9179	0.0607
TypeSmall	11.4935	1.4082	8.1619	0.0000

12.5 Dealing with Heterogeneity

It may happen that we wish to test the hypothesis of equal means with normally distributed measurements, but that the assumption of equal variance between groups is violated. Data contain **heterogeneity** in the measurement if not all observations have equal variance σ^2. For such data the estimated variance of the regression coefficients needs to be adjusted in order to make valid inferences. The seminal publication by White (1980) gives, amongst others, two methods based upon what is known as HC3 : $\omega_i = \varepsilon_i^2/(1 - h_{ii})$ and HC4 : $\omega_i = \varepsilon_i^2/(1 - h_{ii})^{\delta_i}$, where $\delta_i = \min(4, h_{ii}/\bar{h})$ and the h_{ii} terms come from the hat matrix. Upon defining $\widehat{\boldsymbol{\Omega}} = \text{Var}(\omega_i)$ the variance matrix becomes

$$\text{Var}(\hat{\boldsymbol{\beta}}) = (\mathbf{X}^T\mathbf{X})^{-1}\mathbf{X}^T\widehat{\boldsymbol{\Omega}}\mathbf{X}(\mathbf{X}^T\mathbf{X})^{-1}$$

The corrected variances imply different standard errors eventually leading to different conclusions from marginal t-testing.

Example 12.21 *We use a the data from a random sample of 93 cars out of a total of 1993 listed in the Consumer Reports issue or the PACE Buying Guide. Pickup trucks and Sport/Utility vehicles were eliminated and duplicated models were listed at most once (Lock, 1993). We concentrate on the null hypotheses that the types "Large," "Midsize," and "Small" have an equal mean City miles per US gallon by EPA rating (MPG). The standard deviations for these 1.50, 1.89, and 6.10, respectively for "Large," "Midsize," and "Small" show large differences.*

The standard errors in Table 12.9 are quite similar for the HC3 and HC4 corrections, with some large differences compared with standard linear regression. The HC3 and HC4 standard errors for contrast in means between Large and Midsize cars is considerably smaller. The same holds for the p-values, although, in the current example, this does not lead to different conclusions.

12.6 Appendix: Scale Tau Estimator

For the constants the default values $c_1 = 4.5$ and $c_2 = 3.0$ are taken (Maronna and Zamar, 2002). First, $s_0 = \text{MAD}(\mathbf{y})$ is computed. Next, the weights using the function $w_{c_1}(u) = \max(0, (1 - u^2/c_1^2)^2)$ in

$$w_i = w_{c_1}((y_i - \text{Med}(\mathbf{y}))/s_0),$$

From this the robust location measure is taken as the weighted sum $\hat{\mu} = (\sum_i w_i y_i)/(\sum_i w_i)$, from which the robust tau-estimate is computed using

$$s(\mathbf{y})^2 = s_0^2 \cdot \frac{1}{n} \sum_{i=1}^{n} \rho_{c_2}((y_i - \hat{\mu})/s_0),$$

where $\rho_{c_2}(u) = \min(c_2^2, u^2)$.

12.7 Notes and Comments

It seems clear that in the presence of outlying data points, chemical inferences from robust analysis are scientifically more sound than those based upon non-robust model estimation.

If a large part of the data deviates largely from the bulk, then MM-regression will tend to estimate a model close to the bulk of the data. It may in such a situation very well be the case that there are in facto two separable models pertaining to distinct processes in de data.

It seems clear that robust estimation in many cases safeguards against a superficial non-robust analysis and that a comparison often makes sense for drawing valid conclusions from an analysis of the chemical data.

Robust estimation and modeling has a much wider scope of applications than rank based testing. In addition, the attempt by robust procedures to estimate the size of effects in the unit of measurement of the chemical outcome is a great advantage.

The search for object functions that behave asymptotically robust against outliers has a long history starting with rank type of estimation, using quantiles, influence functions (Hampel et al., 2011), M-, L-, S-estimators

(Huber, 2009; Jureckova and Picek, 2005). There is a modern book length treatment of robust estimation (Maronna et al., 2019) which is fully integrated with comments on using R. MM robust estimation procedures are developed for non-linear least squares and to generalized least squares (Cantoni and Ronchetti, 2001).

12.8 Notes on Using R

There are several other functions available for fitting regression models in a way resistant to the influence of extreme outliers in the data. The function `lqs` in the library `MASS` is based on quantiles and provides algorithms for robust linear model estimation to the data. Less resistant, but statistically more efficient, M-estimators are available from as the `rlm` function in the library `MASS`.

MM-estimation for robust linear regression is implemented by `lmrob` and for robust non-linear regression by the function `nlrob`. Predicted values collected in a matrix can be ordered by the basic `order` function for finding a maximum/minimum of the yield. 95 percent prediction intervals can be investigated by the generic `predict` function using the newdata option as in

```
predict(mod,interval="prediction",newdata=data.frame(H="2",
 P="650", CT="4"))
```

Non-normality can be investigated by R its `skewness` and `kurtosis` functions. That is, the empirical distribution of PbO in the glass vessel example is estimated by the bootstrap using

```
> x <- glass[,13]
> boot.out<-boot(x, function(x,i){skewness(x[i])}, R=2000)
```

A convenient way to apply the bootstrap to linear regression by the `Boot` function from the `car` library. Many robust estimation procedures are available from R its Task View Robust.

12.9 Exercises

1. **Minimization**. Suppose we have the data $y = (1, 3, 11)$. Then the median is 3 and the mean equals 5.

   (a) Show that minimizing the function $\rho(x) = \sum_{i=1}^{3}(y_i - x)^2$, yields the mean.

(b) The paper and pen minimization the function in equation 12.4 leads to somewhat unusual mathematics, however, we may approximate this by numerical minimization of the object function

$$\rho(x) = \sum_{i=1}^{3} |y_i - x|^p,$$

with running from $p = 2$, $p = 1.9$, to $p = 1.1, , p = 1.0$. Minimize the function numerically and print the power p and the minimizing point to the screen. What do you conclude?
Hint: Use a for loop and e.g., the R function `optimize`.

2. **Properties of the rho functions.**

(a) Show that the $\rho_k(y)$ function in Equation (12.6) indeed has derivative $2\psi_k(y)$ for the $2\psi_k(y)$ function given.

(b) Compute the weight function and verify that it conforms the claimed expression $W_k(y) = \min\left\{1, \frac{k}{|y|}\right\}$.

(c) Verify by calculus methods that the ρ function in (12.11) has the claimed properties of being bounded, differentiable, even, nondecreasing in $|x|$, and attains its minimum zero at $x = 0$, with first order derivative $\rho'(x) = 6\psi(x)/k^2$ for the bisquare psi-function

$$\psi(x) = x\left[1 - \left(\frac{x}{k}\right)^2\right]^2 \cdot I_{[-k,k]}(x).$$

3. **Glass.** In a study of 180 archaeological glass vessels there were 13 different chemical measurements made for each (Janssens et al., 1998). The glass data are available from the R library `chemometrics`.

(a) Perform a simple linear regression on outcome BaO with explanatory variable PbO. Test for outliers.

(b) Perform a robust simple linear regression of PbO on BaO.

(c) Plot the both estimated lines and comment on your findings.

4. **Solar thermal energy**. From an experiment solar thermal energy test data became available, see `table.b2` from the R library `MPV`. The meaning of the variables stored as columns in the table are
y : Total heat flux (kwatts)
x1 : Insolation (watts/m^2)
x2 : Position of focal point in east direction (inches)
x3 : Position of focal point in south direction (inches)
x4 : Position of focal point in north direction (inches)
x5 : Time of day

(a) Fit a linear model of y by x5 and compare with a robust MM-estimation. What do you conclude?

(b) Fit a linear model of y by x1 + x2 + x3 + x4 and compare with a robust MM-estimation. What do you conclude?

(c) What is the fit of the model that you propose?

5. **Wine quality**. From a testing experiment 38 observations became available on the quality of Pinot Noir wine. We found by by minimum AIC in the above that the measure of wine quality was predictable from Flavor, Oakiness, and the factor(Region). The data are available from the R library `MPV` under the name `table.b11`.

   (a) Investigate the diagnostics of the model using Quality as outcome and Flavor, Oakiness, and Region (factor) as explanatory variables.

   (b) Compare the linear model from least squares with that from MM-estimation.

6. **Kinematic viscosity**. In a study it was investigated whether the outcome Kinematic viscosity $10^{-6}m^2/s$ depends upon (x1) the ratio 2-methoxyethanol to 1,2-dimethoyethane (dimensionless) and (x2) Temperature ^oC. The data are available from the R `MPV` library under the name `table.10`.

   (a) Investigate the diagnostics of the least squares fit. Are there outliers present.

   (b) Compare the estimated coefficients with those from robust MM-estimation.

7. **Damage on carrots.** In an experiment the damage on carrots was investigated by a soil experiment with three blocks and eight levels of insecticide. The data consists of 24 observations on the success as the number of carrots with insect damage, total as the total number of carrots per experimental unit, a numeric logdose vector giving log(dose) values, and block the factor indicating the levels B1, B2, and B3 (McCullagh and Nelder, 1989). The data carrots are available from the R library `robustbase`.

   (a) Fit the binomial generalized least squares model to the data using the blocking factor.

   (b) Test for outliers.
   Hint: Simply use the outlier test from the `car` library.

   (c) Fit the model to the data by robust generalized least squares.
   Hint: Use the `glmrob` function form the `robustbase` library.

8. **Phosphorus content.** In an experiment the effect of inorganic and organic Phosphorus in the soil was investigated upon the phosphorus content of the corn grown in this soil. The variable plant is the

outcome with the Phosphorus content, inorg gives the Inorganic soil Phosphorus and organic the Organic soil Phosphorus. The phosphor data are available from the R library `robustbase`.

(a) Plot the plant phosphorus content against organic soil Phosphorus, and against Inorganic soil Phosphorus. What do you observe?

(b) Fit the linear model to the data and test for the presence of outliers.

(c) Fit the robust linear model to the data and compare the parameters estimates.

9. **Pulp fiber.** In an experiment measurements were collected on aspects pulp fibers and the paper produced from them. Four properties of each are measured in sixty-two samples are arithmetic fiber length (X1), long fiber fraction (X2), fine fiber fraction (X3), zero span tensile (X4) and one of the outcomes was breaking length (Y1). The pulpfiber data are available from the R library `robustbase`.

   (a) Fit the linear model to the data and test for the presence of outliers.

   (b) Fit the robust linear model to the data and compare the parameters estimates.

10. **Ash.** The dependent variable Softening Temperature (SOT) of ash is modeled by the elemental composition of the ash data by the mass concentrations of the elements P2O5, SiO2, Fe2O3, Al2O3, CaO, MgO, Na2O. The data consist of 99 ash samples, originating from different biomass, with the comprise of an experimental SOT (630-1410 degrees centigrade) (Varmuza and Filzmoser, 2016). The ash data are available from the R library `chemometrics` and from `ash.txt`. Note that it is not possible to add the element K2O to the model due to near singularity.

    (a) Fit the linear model to the data and test for the presence of outliers.

    (b) Fit the robust linear model to the data and evaluate it.

Part V

Appendix

A

Basics of R Computing Environment

The freeware programming language R for analyzing chemical data is very flexible from the level of the user. In a beginning stage one might want to use mainly predefined functions to solve chemical problems in just a few steps. Such holds for statistical testing, (non)-linear modeling, model diagnosis, or even bootstrapping. At an intermediate level, one might want to program a simulation study for saving time and money by efficiently designing a chemical experiment. At an even higher level of proficiency, one may want to combine predefined functions with new functions to develop innovative statistical methods for analyzing chemical data. Once such new developments are successful one might consider to develop an R library on the basis of C++ code making applications faster.

In this chapter you learn

- In Section A.1 how to use basic properties, such as installing packages and reading data.
- In Section A.2 how to use predefined functions and how to program new ones yourself.
- In Section A.3 how to bridge the gap between mathematical model formulation and model definition within the programming language R.
- In Section A.4 how and where to find help for solving your problems.

A.1 R Basics

All computing procedures in this book are performed in the high level computer language R. It is part of freeware, very flexible, and contains much help to the beginner. R is useful for plain calculations, numerical matrix type of procedures e.g., for optimization, up to date data plotting, and it is best in the statistical analysis of chemical data.

For the beginner, browsing R its website can be useful as well as employing its seek function for finding examples of pre-programmed functions or for an overview of functionalities from a reference card. For chemometric type of work

DOI: 10.1201/9781003178194-A

R provides many built-in-functions to perform high level statistical analysis of chemical data.

The system is flexible in allowing users to build their own functions. Larger pieces of code can be developed in a basic text editor such as NotePad, Emacs, or R-Studio and transferred into R by copy pasting or running a source file. The wide scope of using R becomes clear once its basic ideas are properly understood. The active user will soon discover that by combining a few compound type of functions a lot can be computed in just a few lines.

R can be freely downloaded and installed from the CRAN website on any operating system. For e.g., Windows its operating system simply choose "Download R for Windows," "base," and "Download R x.x.x for Windows." After starting R, its command prompt symbol `>` occurs and the language is simply waiting for you to give instructions in the form of command lines.

R is a so-called object oriented language interpreting, scalars, vectors, data frames, lists, plots as objects on which you actually can program. Output produced from applying functions to objects will be printed to the screen, unless this is saved as an object by the assignment `<-` or equality (=) sign. For example, the code

```
x <- rnorm(10,0,1)
```

saves ten pseudo random numbers (realizations) from the standard normal distribution into the vector `x`. The latter is now an object that you can print to the screen by simply typing `x`.

A.1.1 Installing packages

R has many libraries (packages) available that will allow the usage of more complex functions on chemical data. Type `install.packages ("packagename")` to install a single library. To install a complete set of libraries from the CRAN Task View, we may use

```
chooseCRANmirror()
install.packages(c("ctv"),repo="http://cran.r-project.org",dep=TRUE)
library("ctv")
install.views("ChemPhys")
```

This will install the `ctv` package, loads it, and next runs the `install.views` function making all libraries available which are summarized at CRAN its task view Chemometrics and Computational Physics.

Some libraries useful for our purposes are

- `deSolve`: solving Ordinary Differential Equations (ODEs)
- `NISTnls`: fitting non-linear functions (models) to data
- `outliers`: statistically test for presence of outliers among measurements

- `car`: testing for outliers and application of delta theorem for standard error of function of estimated parameters
- `investr`: inverse regression for calibration curves
- `pls`: partial least squares
- `robustbase`: to properly fit models to chemical data in the presence of outlying data observations.
- `DoE.base,FrF2,planor`: designing and analysing experiments

A.1.2 Reading data

We often start an analysis by entering or reading data into the R system. This can be done in several ways.

Some ways acces data by and from the R system are by

- Manually typing data using `c()` - for "concatenating" data into a vector. For example, `x <- c(2,1,3,4)` will create a numerical vector x containing the numbers 2, 1, 3, 4.
- Loading build-in data from an R library by e.g.,

```
library(MASS)
chem
```

  makes 24 determinations of copper in wholemeal flour (parts per million) available as a vector, see (Committee, 1989).

- Reading data from a `.csv` Excel type of file

```
dfa <- read.table("Roszman.csv", header = TRUE,
        sep = ",", row.names = 1)
```

  as a `data.frame` or matrix. The same function reads `.txt` data. Note, however, such a function only works properly if the file is actually available at the current working directory, to be found by `getwd()`. Simply type `?read.table` for more ways to read files.

- The libraries `foreign` and `haven` give various functions to read data from filetypes typically used in software applications.

It is handy to set the working directory by, e.g., the command

```
setwd("D:/university/course/ComputationalMethods/data")
```

and to find the directory you actually are working on by `getwd()`.

To read and save data under the name "`dfa`" after browsing along directories on your device, use

```
dfa <- read.table(choose.files(),header = TRUE)
```

A.1.3 Types of objects

In R everything is an object! Examples of objects are a number, a vector, or a matrix but also a plot, or a function. To investigate the type of an object you simply use `str`, `print`, or `class`.

A.2 Useful Functions

Generic functions as `print`, `summary`, or `plot` work on all types of objects.

A.2.1 Functions on scalars

Some useful built-in functions are `abs`, `log10`, `log`, `exp`, `atan`.

A.2.2 Functions on vectors

If `x` is a vector with chemical measurements, then useful functions for

- mathematics are: `length`, `min`, `max`, `sort`, `order`, `sum`, `round`,
- statistics are: `mean`, `var`, `sd`, `median`, `quantile`, among others.

The distinction between a vector containing numbers and a factor indicating group membership is important for purposes of modeling, testing, computing, or plotting.

A.2.3 Functions on matrices of data frames

If `x` is a matrix, then

- its ij-th element is given by `x[i,j]`.
- its i-th row is given by `x[i,]`.
- its j-th column is given by `x[,j]`.
- the functions `colnames`, `rownames`, `head`, `tail` are useful.
- If the matrix `x` contains numerical measurements, then the transpose and the dimensionality are computed by: `t(x)`, `dim`, and statistical information is computed by, e.g., the functions: `cor`, `var`, and `pairs`,
- If `x` is a square matrix, then functions like `det`, `diag`, `solve`, `diag`, `chol`, `eigen`, and `svd` implement concepts from linear algebra.

TABLE A.1
Some distributions and their corresponding R functions. Use the `help()` command to see their exact syntax.

Distribution	CDF	PDF	Inverse CDF	Draw randomly
Uniform	punif()	dunif()	qunif()	runif()
Normal	pnorm()	dnorm()	qnorm()	rnorm()
χ^2	pchisq()	dchisq()	qchisq()	rchisq()
F	pf()	df()	qf()	rf()
t	pt()	dt()	qt()	rt()

Rows or columns of a matrix can be selected by their index, name, or logical TRUE/FALSE statement, to be used whichever deemed the most convenient. The logical evaluation of a statement is TRUE/FALSE, but this is internally also stored as 1/0, making computations on these easy, such as summation or mean.

A.2.4 Some statistical functions in R

Information on a number of chemical measurements becomes available by applying functions such

- `mean(),median(),max(),min(),sd(),var()` – computes the mean, median, max, min, standard deviation, and variance of an object, respectively.
- `nls(), lm(), glm(), lme()` - fit a (non-)linear model to chemical data. That is,

```
> summary(mod <- lm(y ~ x))
```

  fits the simple linear model and stores all its relevant output in an extensive list under the chosen name `mod`. The generic function `summary` extracts the most important information and prints it to the screen in a form according to common habits in science.

- Table A.1 displays a few basic functions for using important statistical distributions. Any such function starts with `p` for computing probability, `d` for density, `q` for quantile, and `r` for random number generation.

See also Chapter 8 of "An introduction to R," which becomes available after starting the browser by `help.start()`.

A.2.5 Writing your own functions and source code

In almost all cases, you want R to execute a series of commands instead of typing and executing commands line by line. To tackle bigger programming jobs, R allows the user to write a series of commands in a separate editor (such as Emacs, TextEdit, WinEdt, or Notepad) and then R will "source" the code and run it line by line. Some basic details for writing your own functions and programs are given below.

A.2.6 Writing a function

You can define your own function in R using the `function()` command. For example, in Michaelis-Menten kinetics the reaction rate v is a function of the concentration of a substrate s defined as

$$v(s) = \frac{V_{max} \cdot s}{K_m + s}.$$

Suppose we take the constants $V_{max} = 17$ and $K_m = 126$. Then to define v, plot it to the screen for domain $s \in (0, 100)$, and to compute its integral from 0 to 100, we use

```
v <- function(s,Km=17,Vm=126){ Vm*s/(Km + s)}
plot(v,xlim=c(0,100),col='blue')
integrate(v,0,100)
```

A.2.7 For and while loops

For simulation studies on chemical data, we need to be able to generate pseudo measurements from a normal distribution. To generate a sample of 1000 normally distributed measurements with mean 3.5 and variance 0.5^2 and to observe how its sample mean converges to 3.5, we use the "for" loop

```
mu <- 3.5; sigma <- 0.5; n <- 1000; means <- NULL
y <- rnorm(n,mu,sigma)
for (i in 1:n) { means[i]<- sum(y[1:i])/i }
plot(means)
```

Within the loop the sample size i to compute the mean of the measurements is taken from 1 to 1000 to compute the 1000 sample means, which next are plotted in a separate screen. Typical of a for loop is that the number of steps is defined on before hand.

If we seek a zero of a function by Newton's method, then the number of steps is typically unknown, so that we use a "while" loop. We define the function f, its first-order derivative f', and the updating function g to illustrate the generation of a convergent sequence of numbers that zeros the function f, as follows

```
f  <- function(x){x^3+2*x^2+10*x-20}
df <- function(x){3*x^2+4*x+10}
g  <- function(x){x-f(x)/df(x)}
```

```
n  <- 1; x  <- c(10,g(10))
while (n<100 & abs(x[n+1]-x[n])>.0001) {
  n<-n+1
  x[n+1]<-g(x[n])
}
```

The loop stops its iterative steps when the conditional statement between parenthesis becomes FALSE.

A.2.8 Logical arguments

Logic is quite useful for chemical programming. It is a remarkable fact that the very basics of logic are built upon the elementary operations "AND" &, "OR" |, and negation !, assigning the attributes `TRUE` or `FALSE` to a (composed) statement. These can conveniently be applied to vectors or rows of a matrix to check for properties of objects, see also the manuals by `help(Control)` and `help(Syntax)`.

A.2.9 Functions for plotting

By R state-of-the-art scientific figures can be plotted and saved in various file types for report writing purposes. A basic principle making this possible is that a plot is interpreted by R as an object on which you can actually program to make it camera ready.

Important basic functions for plotting are

- `plot(x,y)` – plots the points in x and y on a scatter plot.
- `title("string")` – adds the "string" as the title of the figure
- `xlim,ylim,xlab,ylab` – give domain, range, and labels to axis x and y
- `curve` – adds a curve to a plot of data points
- `legend()` – adds a legend to the current graphic.
- `hist`, `boxplot` produce a histogram or box-and-whiskers plot

Since much scientific work is brought to life by plotting, it is well worth to spend some time exploring the possibilities. High level libraries such as `ggplot2` and `lattice` provide various functions to program scientific figures of data e.g., incorporating a graph with the predicted values from a (non)-linear model.

To produce a pdf type of file of a basic scatter plot of data, we may use

```
pdf("./myfigure.pdf")
plot(x,y)
dev.off()
```

The `dev.off()` function tells R that the code to produce the plot is complete and stores the pdf file on the current working directory for further processing in LaTeXor Word.

To produce a two-by-two panel of diagnostic plots for a linear model to be used into a Word type writer, one may use

```
mod <- lm(y ~ x)
jpeg(filename = "DiagnosticPlotLinearModel.jpg", quality = 100)
par(mfrow = c(2,2)); plot(mod); par(mfrow = c(1,1))
dev.off()
```

A.3 Model Notation

R its model notation is the crucial link between mathematical model formulation and fitting of models to data for chemical hypothesis testing. Part of model notation is the outcome y, continuous explanatory variables x_1 and x_2, and factors that indicate the level of an observation or run. In a chemical experiment the factor `a` could for instance represent temperature at the level 100 ^{o}C and 200 ^{o}C, and a factor `b` could represent whether in a run a catalyst is added or not. For each run of the experiment the levels of the factors specify exactly under which experimental conditions the outcome is measured. Model notation can also be used for various other purposes such as computing or plotting. A brief summary is

```
y           : outcome;
x, x1, x2 : predictive variables
a,b         : factors indicating membership of certain experimental level
mod <- lm(y ~ x1 + x2)          # linear regression of y on x1 and x2
mod <- lm(y ~ x+I(x^2)+I(x^3)) # polynomial regression
mod <- lm(y ~ a)                # one-way analysis of variance (anova)
mod <- lm(y ~ a*b)              # two-way anova with interaction
mod <- lm(y ~ a+b)              # two-way anova without interaction
nls(y ~ b1 - b2*x - atan(b3/(x-b4))/pi, data = dfa)
                                # non-linear least squares
boxplot(y ~ a+b)                # box-and-whiskers plot
aggregate(y~ a+b, mean))        # compute of means of y for each level
 of factor
```

The illustrates using model notation for plotting and computation of, e.g., means with respect to combinations of levels factors `a` and `b`.

A.4 Finding Help

A direct way to find answers for your questions comes from the help button on top of your R screen. Another way is to browse the HTML page after by typing

```
> help.start()
```

and pressing the manual "An Introduction to R."

If you have forgotten about the specific syntax of an R function, but you do know its name, then simply type `help(function)` or `?function` for R to browse to its HTML documentation from the library. As an example, typing `?rnorm` will return documentation about generating normally distributed "pseudo measurements" which are useful for simulations of chemical data or models. Many libraries explain the input and output of functions and give references to the literature for further study. If you know the concept to implement, but not the corresponding function name, then the search button can be quite useful.

Advanced information about Using R for chemical purposes is available from the Task View ChemPhys.

A.5 Exercises

Here are some exercises for students interested in learning more on statistical programming.

1. **Chem data**. The data set `chem` from the `MASS` library gives 24 determinations of copper in wholemeal flour Committee (1989), in parts per million.
   (a) Compute the minimum, maximum, median, and mean of the measurements. Hint: It is handy to save the data as `y`.
   (b) Plot the data to the screen. Sort the data in increasing order and plot the data.
   (c) The standard deviation s is defined as

$$s = \sqrt{\frac{1}{n-1}\sum_{i=1}^{n}(y_i - \bar{y})^2}, \ \bar{y} = \frac{1}{n}\sum_{i=1}^{n} y_i$$

   Program its value for the copper in wholemeal data and compare with the R function `sd`.

   (d) Remove the largest measurement from the data and compute the mean and the standard deviation. What do you conclude?

2. **Molecular descriptors**. GC-retention indices of polycyclic aromatic compounds (y) were modeled by 467 molecular descriptors (X). The data are named `PAC` and are available from the `chemometrics` library. After installing the library, the descriptors can be extracted, e.g., by using PAC$X.

   (a) Construct a box-and-wiskers plot for the HATS descriptors (variables) in column 366 to 371 (columns in PAC$X).
   Hint: Use the function `boxplot`.
   (b) Construct a histogram for HATS0v descriptor in column 366.
   Hint: Use the function `hist`.
   (c) Construct a quantile-quantile plot where the theoretical quantile come from the normal distribution for columns 366.
   Hint: Use the function `qqnorm` and `qqline`.
   (d) Construct a quantile-quantile plot where the theoretical quantile come from the normal distribution for columns 366 to 371.
   Hint: Use the function `qqnorm` and `qqline`.
   (e) Use the `pairs.panels` frunction from the `psych` library to plot the descriptors in column 366 to 371 and briefly explain your observations?
   (f) Compute the eigenvalues of the correlation matrix from the descriptors in column 366 to 371.
   Hint: Use the functions `cor` and `eigen` and extract its values to print these to the screen.
   Hint: Its computation takes a single line.
   (g) Compute the correlation of `y` with each of the descriptors and store the in a vector and plot a histogram of these. How many are larger in absolute value than 0.80?

3. **Pseudoephedrine concentrations.** Fourteen concentrations of pseudoephedrine are measured together with their peak areas from chromatography. The data can be found under the name pseudoephedrine from the library `quantchem` or read the data from `ibuprofen.txt`.

   (a) Plot the data to the screen.
   (b) Estimate the simple linear model using the `lm` function and add the estimated line to the plot.

4. **Michaelis-Menten.** Plot the Michaelis-Menten reaction rate functions

$$v(s) = \frac{V_{max} \cdot s}{K_m + s}$$

for K_m parameter equal to $5, 10, 15, \cdots, 45, 50$ in a single figure taking domain $s \in (0, 100)$.

5. **Concentration curve**. Consider the above concentration function with parameter values $a = 0.221363$, $b = 1.229524$, $K = 0.095948$.

   (a) Plot the function $y(t) = at^b e^{-Kt}$, add a title, add a vertical line at $t = b/K$ to the plot and save its as a jpeg file.

   (b) Use R its `integrate` function to compute the integral

   $$\int_0^{100} at^b e^{-Kt} dt.$$

   Remark: Integration by parts does not work here.

6. **Finding Information**. Finding functions for solving problems.

   (a) Find the standard mathematical functions tan and exponential.

   (b) Find R functions for solving calibration problems.

   (c) Find R functions for optimization and root finding.

   (d) Find the library on:

      i. Fractional Factorial Designs with 2-Level Factors,
      ii. Ordinary differential equations,
      iii. Linear mixed models.

   (e) Find papers in the journal of statistical software explaining functionalities on:

      i. Fractional Factorial Designs with 2-Level Factors,
      ii. Ordinary differential equations,
      iii. Linear mixed models.

Bibliography

Abramowitz, M., Stegun, I. A., and Romer, R. H. (1988). Handbook of mathematical functions with formulas, graphs, and mathematical tables.

Aller, A. (1998). Speciation of selenomethionine and selenourea using living bacterial cells. *Analyst*, 123(5):919–927.

Apostol, T. M. (2007a). *Calculus*, volume 1. New York: John Wiley & Sons.

Apostol, T. M. (2007b). *Calculus*, volume 2. New York: John Wiley & Sons.

Bain, L. and Engelhardt, M. (1992). *Introduction To Probability and Mathematical Statistics.* Duxbury, Pacific Grove, second edition.

Belsley, D. A., Kuh, E., and Welsch, R. E. (2005). *Regression Diagnostics: Identifying Influential Data and Sources of Collinearity*, volume 571. John Wiley & Sons.

Billingsley, P. (2012). *Probability and Measure.* John Wiley and Sons, second edition.

Bliss, C. I. (1935). The calculation of the dosage-mortality curve. *Annals of Applied Biology*, 22(1):134–167.

Box, G. E., Hunter, W. H., Hunter, S., et al. (1978). *Statistics for Experimenters*, volume 664. John Wiley and Sons, New York.

Box, G. E. P., Hunter, W. G., and Hunter, J. S. (2004). *Statistics for Experimenters.* Wiley, New York, NY.

Byers, C. H. and Williams, D. F. (1987). Viscosities of binary and ternary mixtures of polyaromatic hydrocarbons. *Journal of Chemical and Engineering Data*, 32(3):349–354.

Cantoni, E. and Ronchetti, E. (2001). Robust inference for generalized linear models. *Journal of the American Statistical Association*, 96(455):1022–1030.

Casella, G. and Berger, R. L. (2021). *Statistical Inference.* Cengage Learning.

Castillo, J., Lanaja, J., Martinez, M. C., and Aznarez, J. (1982). Flame atomic-absorption spectroscopic determination of antimony in atmospheric particulates by using direct atomisation of the covalent hydride. *Analyst*, 107(1281):1488–1492.

Chambers, J. and Hastie, T. (1992). Linear models. Chapter 4 of statistical models in s. Wadsworth & Brooks/Cole.

Chatterjee, S. and Hadi, A. (1988). *Sensitivity Analysis in Linear Regression.* Wiley Series in Probability and Mathematical Statistics: Applied Probability and Statistics. Wiley.

Chwirut, D. J. (1979). Recent improvements to the astm-type ultrasonic reference block system. Technical report, National Bureau of Standards Gaithersburg MD.

Claeskens, G. and Hjort, N. L. (2008). *Model Selection and Model Averaging.* Cambridge Series in Statistical and Probabilistic Mathematics. Cambridge University Press.

Collett, D. (2003). Modeling binary data. Technical report, Florida, US: CRC Press.

Committee, A. M. (1989). Robust statistics–how not to reject outliers. Part 1. Basic concepts. *Analyst*, 114:1693–1697.

Committee, A. M. et al. (1987). Recommendations for the conduct and interpretation of co-operative trials. *Analyst*, 112(5):679–686.

Cook, R. D. and Forzani, L. (2019). Partial least squares prediction in high-dimensional regression. *The Annals of Statistics*, 47(2):884–908.

Cook, R. D. and Weisberg, S. (1982). *Residuals and Influence in Regression.* Chapman and Hall, London.

Corradini, F., Franchini, G., Marchetti, A., Tagliazucchi, M., and Tassi, L. (1997). Viscosimetric studies on 2-methoxyethanol+ 1, 2-dimethoxyethane binary mixtures from- 10 to 80° c. *The Canadian Journal of Chemical Engineering*, 75(3):494–501.

Cronauer, D. C., Shah, Y. T., and Ruberto, R. G. (1978). Kinetics of thermal liquefaction of belle ayr subbituminous coal. *Industrial & Engineering Chemistry Process Design and Development*, 17(3):281–288.

Daudin, J., Duby, C., and Trecourt, P. (1988). Stability of principal component analysis studied by the bootstrap method. *Statistics: A Journal of Theoretical and Applied Statistics*, 19(2):241–258.

Davison, A. C. and Hinkley, D. V. (1997). *Bootstrap Methods and Their Application.* Cambridge University Press, Cambridge.

Demidenko, E. (2013). *Mixed Models: Theory and Applications with R.* John Wiley & Sons, New York.

Dobson, A. and Barnett, A. (2018). *An Introduction to Generalized Linear Models.* Chapman & Hall/CRC Texts in Statistical Science, CRC Press.

Douglas, C. L., DeMarco, G. J., Baghdoyan, H. A., and Lydic, R. (2004). Pontine and basal forebrain cholinergic interaction: implications for sleep and breathing. *Respiratory Physiology & Neurobiology*, 143(2-3):251–262.

Draper, N. R. and Smith, H. (1998). *Applied Regression Analysis*, volume 326. John Wiley & Sons, New York.

Efron, B. and Tibshirani, R. J. (1993). *An Introduction to the Bootstrap*. Number 57 in Monographs on Statistics and Applied Probability. Chapman & Hall/CRC, Boca Raton, Florida, USA.

Fan, J. and Li, R. (2001). Variable selection via nonconcave penalized likelihood and its oracle properties. *Journal of the American Statistical Association*, 96(456):1348–1360.

Ferguson, T. (1996). *A Course in Large Sample Theory*. Chapman & Hall Texts in Statistical Science Series, Taylor & Francis, London.

Fisher, R. (1925). *Statistical Methods for Research Workers*. Oliver and Boyd, Edinburgh, UK.

Foote, J. W. and Delves, H. T. (1983). Distribution of zinc amongst human serum proteins determined by affinity chromatography and atomic-absorption spectrophotometry. *Analyst*, 108(1285):492–504.

Fox, J. (1997). *Applied Regression, Linear Models, and Related Methods*. Sage.

van de Geer, S. A. (2000). *Empirical Processes in M-estimation*, volume 6. Cambridge University Press, Cambridge.

Gillespie, D. T. (1977). Exact stochastic simulation of coupled chemical reactions. *The Journal of Physical Chemistry*, 81(25):2340–2361.

Giri, S. K., Shields, C. K., Littlejohn, D., and Ottaway, J. M. (1983). Determination of lead in whole blood by electrothermal atomic-absorption spectrometry using graphite probe atomisation. *Analyst*, 108(1283):244–253.

Golub, G. H. and Van Loan, C. F. (1996). *Matrix Computations*. The Johns Hopkins University Press, third edition, Baltimore.

Graham, R. and Stevenson, F. (1972). Kinetics of chlorination of niobium oxychloride by phosgene in a tube-flow reactor. application of sequential experimental design. *Industrial & Engineering Chemistry Process Design and Development*, 11(2):160–164.

Greenwell, B. M. and Schubert Kabban, C. M. (2014). investr: an r package for inverse estimation. *R Journal*.

Grubbs, F. E. (1969). Procedures for detecting outlying observations in samples. *Technometrics*, 11(1):1–21.

Györfi, L., Kohler, M., Krzyzak, A., and Walk, H. (2002). *A Distribution-Free Theory of Nonparametric Regression.* Springer Series in Statistics, Springer.

Hairer, E. and Wanner, G. (1992). *Solving Ordinary Differential Equations I: Nonstiff Problems.* Springer Berlin Heidelberg.

Hairer, E. and Wanner, G. (2010). *Solving Ordinary Differential Equations II: Stiff and Differential-Algebraic Problems.* Springer Berlin Heidelberg.

Hall, P. (2013). *The Bootstrap and Edgeworth Expansion.* Springer Science & Business Media.

Hampel, F. R., Ronchetti, E. M., Rousseeuw, P. J., and Stahel, W. A. (2011). *Robust Statistics: The Approach Based on Influence Functions*, volume 196. John Wiley & Sons.

Hastie, T., Tibshirani, R., and Friedman, J. (2001). *The Elements of Statistical Learning.* Springer Series in Statistics, Springer New York Inc., New York, NY, USA.

Horn, R. A. and Johnson, C. R. (2012). *Matrix Analysis.* Cambridge University Press, Cambridge.

Huber, P. (1981). *Robust Statistics.* Wiley Series in Probability and Statistics. Wiley, New York.

Huber, P.J., R. E. (2009). *Robust Statistics.* Wiley, New York.

Indahl, U. G., Sahni, N. S., Kirkhus, B., and Næs, T. (1999). Multivariate strategies for classification based on nir-spectra—with application to mayonnaise. *Chemometrics and Intelligent Laboratory Systems*, 49(1):19–31.

Janssens, K. H., Deraedt, I., Schalm, O., and Veeckman, J. (1998). Composition of 15–17th century archaeological glass vessels excavated in antwerp, belgium. In *Modern Developments and Applications in Microbeam Analysis*, pages 253–267. Springer.

Jennrich, R. I. (1969). Asymptotic properties of non-linear least squares estimators. *The Annals of Mathematical Statistics*, 40(2):633–643.

Jolliffe, I. (2002). *Principal Component Analysis.* Springer Verlag, New York.

Jureckova, J. and Picek, J. (2005). *Robust Statistical Methods with R.* Taylor & Francis, London.

K.E. Brenan, S.L. Campbell, L. P. Petzold (1996). *Numerical Solution of Initial-Value Problems in Differential-Algebraic Equations.* North-Holland, New York.

Knuth, D. E. (1997). Seminumerical algorithms. *The Art of Computer Programming*, 2.

Konishi, S. and Kitagawa, G. (2008). *Information Criteria and Statistical Modeling.* Springer Series in Statistics, Springer.

Kreyszig, E., Kreyszig, H., and Norminton, E. J. (2011). *Advanced Engineering Mathematics.* Wiley, Hoboken, NJ, tenth edition.

Lancaster, P. and Tismenetsky, M. (1985). *The Theory of Matrices: With Applications.* Elsevier.

Lawson, J. (2014). *Design and Analysis of Experiments with R*, volume 115. CRC Press.

Lehmann, E. (2014). *Elements of Large-Sample Theory.* Springer.

Lehmann, E. L. and Romano, J. P. (2005). *Testing Statistical Hypotheses*, volume 3. Springer.

Lenth, R. (2010). Response-surface methods in r, using rsm. *Journal of Statistical Software*, 32(32):1–17.

Li, X., Sudarsanam, N., and Frey, D. D. (2006). Regularities in data from factorial experiments. *Complexity*, 11(5):32–45.

Lock, R. H. (1993). 1993 new car data. *Journal of Statistics Education*, 1(1).

Luenberger, D. G. (1997). *Optimization by Vector Space Methods.* John Wiley & Sons.

Maclaren, N. (1989). The generation of multiple independent sequences of pseudorandom numbers. *Journal of the Royal Statistical Society: Series C (Applied Statistics)*, 38(2):351–359.

Magnus, J. R. and Neudecker, H. (2019). *Matrix Differential Calculus with Applications in Statistics and Econometrics.* John Wiley & Sons, New York.

Maronna, R., Martin, D., and Yohai, V. (2019). *Robust Statistics: Theory and Methods.* Wiley Series in Probability and Statistics. Wiley.

Maronna, R. A. and Zamar, R. H. (2002). Robust estimates of location and dispersion for high-dimensional datasets. *Technometrics*, 44(4):307–317.

Mazumdar, S. and Hoa, S. (1995). Application of taguchi method for process enhancement of on-line consolidation technique. *Composites*, 26(9):669–673.

McCullagh, P. and Nelder, J. (1989). Generalized linear models ii.

Milliken, G. A. and Johnson, D. E. (2017). *Analysis of Messy Data, Volume II: Nonreplicated Experiments.* Chapman and Hall/CRC.

Montgomery, D. C. (2013). *Design and Analysis of Experiments.* Wiley, New York, NY, eighth edition.

Montgomery, D. C. and Runger, G. C. (2018). *Applied Statistics and Probability for Engineers*. Wiley, Hoboken, NJ.

Morris, R. A. and Watson, E. F. (1998). A comparison of the techniques used to evaluate the measurement process. *Quality Engineering*, 11(2):213–219.

Mukerjee, R. and Wu, C. J. (2007). *A Modern Theory of Factorial Design*. Springer Science & Business Media.

Myers, R., Montgomery, D., and Anderson-Cook, C. (2009). *Response Surface Methodology*. Wiley, New York.

Myers, R., Montgomery, D., Vining, G., and Robinson, T. (2010). *Generalized Linear Models: With Applications in Engineering and the Sciences*. Wiley Series in Probability and Statistics, Wiley.

Nelder, J. A. and Wedderburn, R. W. (1972). Generalized linear models. *Journal of the Royal Statistical Society: Series A (General)*, 135(3):370–384.

Peng, R. D. and Welty, L. J. (2004). The nmmapsdata package. *R News*, 4(2):10–14.

Pinheiro, J. and Bates, D. (2006). *Mixed-effects Models in S and S-PLUS*. Springer Science & Business Media, New York.

Potvin, C., Lechowicz, M. J., and Tardif, S. (1991). The statistical analysis of ecophysiological response curves obtained from experiments involving repeated measures. *Ecology*, 72(1):22–22.

Quarteroni, A., Sacco, R., and Saleri, F. (2010). *Numerical Mathematics*, volume 37. Springer Science & Business Media, Berlin.

Racine-Poon, A. (1988). A bayesian approach to nonlinear calibration problems. *Journal of the American Statistical Association*, 83(403):650–656.

Rao, C. (2009). *Linear Statistical Inference and its Applications*. Wiley Series in Probability and Statistics, Wiley, New York.

Reckman, G. A., Navis, G. J., Krijnen, W. P., Van der Schans, C. P., Vonk, R. J., and Jager-Wittenaar, H. (2019). Whole body protein oxidation unaffected after a protein restricted diet in healthy young males. *Nutrients*, 11(1):115.

Riemann, B. (1867). *Abhandlungen der Koeniglichen Gesellschaft der Wissenschaften zu Goettingen*, volume 13, chapter Ueber die Darstellbarkeit einer Function durch eine trigonometrische Reihe, pages 87–132. Dieterich.

Rocke, D. M. and Lorenzato, S. (1995). A two-component model for measurement error in analytical chemistry. *Technometrics*, 37(2):176–184.

Rosen, M. I. (1995). Niels hendrik abel and equations of the fifth degree. *The American Mathematical Monthly*, 102(6):495–505.

Roszman, L. (1979). Quantum defects for sulfur i atom. National Institute of Standards and Technology (NIST), US Department of Commerce, USA.

Salibian-Barrera, M. and Yohai, V. J. (2006). A fast algorithm for s-regression estimates. *Journal of Computational and Graphical Statistics*, 15(2):414–427.

Seber, G. and Wild, C. (1989). *Nonlinear Regression.* Wiley Series in Probability and Statistics, Wiley, New York.

Sheldon, F. (1960). Statistical techniques applied to production situations. *Industrial & Engineering Chemistry*, 52(6):507–509.

Soetaert, K. E., Petzoldt, T., and Setzer, R. W. (2010). Solving differential equations in r: package desolve. *Journal of Statistical Software*, 33.

Stewart, J. (2007). *Calculus.* Available 2010 Titles Enhanced Web Assign Series. Cengage Learning.

Thisted, R. A. (2017). *Elements of statistical computing: Numerical computation.* Routledge.

Treloar, M. A. (1974). *Effects of Puromycin on Galactosyltansferase of Golgi Membranes.* PhD thesis, University of Toronto.

Tukey, J. (1949). Comparing individual means in the analysis of variance. *Biometrics*, 5(2):99–114.

Varmuza, K. and Filzmoser, P. (2016). *Introduction to Multivariate Statistical Analysis in Chemometrics.* CRC press.

Vasilev, N., Schmitz, C., Grömping, U., Fischer, R., and Schillberg, S. (2014). Assessment of cultivation factors that affect biomass and geraniol production in transgenic tobacco cell suspension cultures. *PLoS One*, 9(8):e104620.

Venables, W. and Ripley, B. (2000). *S Programming.* Springer, New York.

Venables, W. and Ripley, B. (2002). *Modern Applied Statistics with S.* Springer, New York, fourth edition.

Wahba, G. (1990). *Spline Models for Observational Data.* SIAM.

Wang, Z., Liu, H., and Zhang, T. (2014). Optimal computational and statistical rates of convergence for sparse nonconvex learning problems. *Annals of Statistics*, 42(6):2164.

Welch, B. L. (1947). The generalization of 'student's'problem when several different population varlances are involved. *Biometrika*, 34(1-2):28–35.

West, B. T., Welch, K. B., and Galecki, A. T. (2006). *Linear Mixed Models: A Practical Guide using Statistical Software*. Chapman and Hall/CRC, London.

Press, W. H., Flannery, B. P., Teukolsky, S. A., and Vettering, W. T. (2007). *Numerical Recipes: The Art of Scientific Computing*. Cambridge University Press, New York, third edition.

Wilhelm, T. (2009). The smallest chemical reaction system with bistability. *BMC Systems Biology*, 3(1):90.

Williams, D. A. (1987). Generalized linear model diagnostics using the deviance and single case deletions. *Applied Statistics*, pages 181–191.

Williams, E. (1959). Regression analysis. John Wiley and Sons, Inc., New York. 214pp.

Wit, E., Heuvel, E. v. d., and Romeijn, J.-W. (2012). 'all models are wrong...': an introduction to model uncertainty. *Statistica Neerlandica*, 66(3):217–236.

Wood, S. N. (2017). *Generalized Additive Models: An Introduction with R*. CRC Press.

Woods, H., Steinour, H. H., and Starke, H. R. (1932). Effect of composition of portland cement on heat evolved during hardening. *Industrial & Engineering Chemistry*, 24(11):1207–1214.

Ziegel, E. R. (2000). Handbook of *Chemometrics and Qualimetrics, Part b*.

Index

9781032013244